A SCIENTIFIC VIEW OF REALITY

Second edition

R. Lindsey

QUODLIBET ROCK

First published October 2017
This edition published October 2020 by Quodlibet Rock

e-mail: quodrock@gmail.com

ISBN 978-1-8380314-0-4

Dewey 501

CONTENTS

Preface

In the course of human history there have been many different views of reality, the most successful being found until recently in the teachings of the world's main religions. Nowadays, religion-based viewpoints are increasingly being superseded by what is known as science. This is particularly so since the 1950s as dramatic discoveries in biochemistry, cosmology and particle physics, and less dramatic but equally important discoveries in neurobiology have placed science in a position to put forward a convincing account of the nature of the universe and of human existence.

That account is the subject of this book. The topics that will be covered amount to a description of the reality which scientific exploration seems to be uncovering. Some of the topics will be concerned with secure scientific knowledge; others will be speculations on what lies at, or even beyond, the frontiers. The book as a whole will, I hope, be able to convey a feeling of how the world looks when seen through scientifically guided eyes — to suggest a philosophical approach, an attitude to the universe we live in.

In writing a book of this kind it is necessary for me to have a 'typical reader' in mind. You are someone with an interest in science; someone with an enquiring nature and an openness to new ideas. No specialist knowledge is required, though I have assumed you will already be familiar to some degree with science's way of expressing things, and so will be comfortable with elementary algebra, exponent notation, metric units and their abbreviations, and simple chemistry.

I have done my best to ensure there are no errors — especially factual ones — in the text but cannot claim to be infallible. I trust you will forgive me for any shortcomings in this respect. If you wish, in that connection, to check on what I have written or would like to enquire further into any of the subjects covered, you will find a selected bibliography at the back.

A number of preliminary general points about the treatment of material can be made here.

Firstly, it should be kept in mind that none of the subjects addressed in this book is in any way intended to be treated comprehensively. The topics dealt with under each heading are those which have the greatest bearing on our understanding of the universe or our place in it, and the remainder, constituting the bulk of each subject, have unavoidably had to be omitted. Furthermore, those parts of each subject that have been covered have generally been considerably simplified, with attention being paid to the most pertinent principles and concepts rather than to the precise details.

Secondly, although I talked above about a 'philosophical approach', this is not a book about philosophy. The practice of philosophers in general is to draw conclusions about some chosen aspect of reality by creating clear definitions and then using logic and close reasoning; in other words, by using pure thought. The practice of scientists, in contrast, is to concentrate on describing what they observe. It is a fundamentally different approach and it is the scientists' approach I will be concerned with here. Philosophy has its place in this book, but when it is called upon it will not be a matter of logic and close reasoning. Instead it will always be directed towards aspects of scientifically observable human behaviour (sections 2.2 to 2.6, sections 8.2 to 8.6 and chapter 9) or else with attempting to explain subjective experience and beliefs in ways compatible with scientific knowledge (section 2.1, chapter 7 and sections 10.4 and 10.5).

Thirdly, it is not my intention to be authoritative in those parts of the book where I depart from observation and observation-based theory and enter the realm of scientifically guided philosophical speculation. Rather my aim is to be thought-provoking, to show you that these issues *can* be approached scientifically, and to encourage you thereby to reach your own conclusions. In those places where the opinions expressed are my own, I state this clearly to be the case.

Lastly, it will be noted that the book is entitled '*A* Scientific View of Reality'. I have called it *a* view rather than *the* view because I do not wish to claim that it describes the only scientific way of looking at things. While the scientific community is generally agreed on the facts in each particular area of knowledge, the conclusions to be drawn from those facts are open to differences of opinion. Fortunately these differences tend to be fairly subtle (though this is not always so — one occasionally comes across a scientist who dissents on quite fundamental matters), but the fact

PREFACE

remains that, since all human beings are individuals with a unique collection of experiences, each person is bound to have a viewpoint which differs to some extent from every other person's. So *a* scientific view it is.

R Lindsey
Norfolk, UK
June 2020

1

DYNAMICS

1.0 OBJECTIVE: To discover by way of what will mostly be simple observations some of the key concepts and laws of physics.

Taking the overwhelming complexity of the terrestrial world as the unpromising starting point for our observations, a few fairly self-evident laws will be proposed that are apparently obeyed universally. We will find that these laws lead to concepts of force, mass and energy — concepts that will in turn enable the laws to be expressed more rigorously in the form of what are known as conservation laws.

The next step will be to review some of the evidence for the existence of atoms; and then to address the question of whether the conservation laws and the concepts of force, mass and energy work as well on an atomic scale of distance as they do in the familiar terrestrial-scale world. This will enable us to see how the theory of the nature of heat has arisen. It will also make it possible to describe what causes solids, liquids and gases to differ from one another. The electromagnetic force and electric charge will be introduced to account for the various interactions that occur between atoms, including the class of interactions that come within the field of chemistry. Finally, the concept of waves as a form of energy will be mentioned, and it will be shown how electrically charged objects can make these waves.

In the course of this chapter we shall see how the world's apparent complexity can be understood in terms of a few simple ideas. The very fact that this is possible is a vindication of scientific method, and also epitomizes the aim of science: to explain the complex in terms of underlying simplicity. This chapter should serve to justify our trust in scientific method before we proceed to murkier waters. It also provides essential information for the ensuing parts of the book.

DYNAMICS

1.1 ACCELERATION, FORCE AND MASS

INITIAL OBSERVATIONS AND LAWS

The terrestrial world appears utterly chaotic, consisting of countless objects, made of a near infinity of different materials, many moving in random ways, swirling in the wind, falling to pieces, bumping into one another. It is a world of extreme complexity endlessly in motion.

At first sight there seem to be no reasons whatever to think that this turmoil is in any way obeying laws. Yet if the world was truly disordered, predictions of the 'what will happen if....' kind that are so important for people's survival in the environment would be impossible to make. The fact that predictions can normally be made quite reliably indicates that beneath all the chaos there are indeed laws being followed.

Some of these laws are so obvious they usually pass unnoticed. Here is an important example. At any given place and time there can only be one object — two or more objects never occupy the same place at the same time.

Another fairly obvious law is that an object which has been stationary for a period of time will remain stationary unless some other object in its vicinity moves in such a way as to cause the stationary object to move. Stationary objects never begin moving spontaneously. Whenever a stationary object begins moving, the motion is always caused by something else; it never just happens.

Moving objects also obey a law, though one which is less easily discerned. The law states that moving objects which are not being affected by anything in their surroundings always travel in straight lines with a constant speed. Neither speed nor direction changes spontaneously. There is always some other object to blame whenever a moving object changes speed and/or direction.

The two laws concerning stationary and moving objects can easily be combined, for a stationary object is merely a moving one whose speed just happens to be zero. The combined law can be expressed as: no object ever changes its speed and/or direction of movement unless something causes it to do so.

Except in very special circumstances, objects moving about in different directions are bound to collide with each other. It is indeed easy to

7

observe collisions happening on the Earth all the time. Think for example of wind rattling a window, or waves crashing into a cliff, or Isaac Newton's famous apple striking the ground.

The consequence of all this activity is that almost no object is ever stationary or travelling with a constant speed and direction for long. And it is the unending collisions which are the cause of these changes of speed and/or direction.

Of crucial importance here is that when objects collide they don't simply pass through one another and go on their way. Because they can't both be in the same place at the same time, they must necessarily affect each other.

The precise nature of the effect can of course vary. A collision may cause objects to bounce off each other — that's the simplest case — but it may also cause objects to become deformed or break up. These more complicated collisions involve changes of speed and/or direction in which different parts of an object are caused to move at different speeds and/or in different directions. Whatever the details, it remains the case that collisions cause changes of speed and/or direction. And that's all they do.

At this point, a leap of the imagination will help. Putting aside until later sections the many yes-but-what-about questions that will undoubtedly spring to mind, suppose provisionally that the *only* way distinct objects are able to affect one another is by means of collisions which change one another's speed and/or direction of movement. This is admittedly an over-simplification, but it is worth pursuing to see where it leads. It implies that everything that happens in the universe — every event, in other words — will be resolvable into changes of speed and/or direction brought about by collisions.

The first revealing consequence of this hypothesis concerns timing. For an event to happen at this moment rather than at some earlier time requires that the objects taking part in the event were set on their pathway to collision at some moment in the recent past. What set them on that path? Earlier collisions, of course. And these earlier collisions were themselves caused by still earlier collisions. Thus in principle every event — that is, every change of speed and/or direction, every collision — can be traced backward in time in an unbroken sequence right back to the beginning of the universe.

The conclusion can be expressed in terms of another law: all events are the product of past events, and in turn produce future events.

As will hopefully become apparent as the topics in this book unfold, this law lies at the heart of the scientific view of reality. If it was ever to be proved false, almost everything else that follows in these pages would be seriously compromised. Aside from some sub-atomic processes which will be discussed in section 3.3, no violation of the law has yet been discovered.

(It is interesting to note, regarding the way science works, that if an event was observed which was spontaneous and without cause it would be assumed that the event had been caused but that the cause had somehow escaped detection. The evidence would have to be overwhelming before this particular scientific law would be abandoned.)

VELOCITY AND ACCELERATION

It would be useful here to introduce two technical terms. The first is *velocity*. Velocity combines speed and direction in a single word. Two objects will have different velocities if they are travelling at different speeds. They will also have different velocities if they are travelling at the same speed but in different directions. An object's velocity is defined by both speed and direction.

In most collisions the objects involved are in contact only briefly. In contrast, the velocity each object acquires as a result of a collision lasts much longer. This implies that the collision is only responsible for the *change* in velocity, not for the velocity itself. This leads to the second technical term. A change in velocity is called an *acceleration*. In its technical sense, it should be noted that acceleration means *any* change in velocity, to slower as well as to faster speeds, or to the same speed in a different direction. It is the technical definition of acceleration that will be used in this book.

FORCE

It is time to start addressing the issue of over-simplification. Not all acceleration (and thereby, movement) is produced by collisions. For example, a stretched spring can produce an acceleration in anything attached to it; and friction between two objects moving against one another can make them accelerate. And so on. The one feature all these things have in common is that the accelerating objects are in contact during the period of acceleration. The contact may be a matter of pulling, rubbing, pushing or colliding. Thus it is physical contact, rather than solely collisions, which can be identified as the producer of acceleration.

Since no object ever accelerates without being forced to by physical contact with a second object, anything which produces an acceleration is called a *force*. There are thus a number of forces. The momentary pressure of a collision, sustained pressure, the pull of a stretched spring, friction, are all examples of force. Force is what passes between two objects when they cause each other to accelerate. So wherever an acceleration is observed, there also shall a force be found.

The relationship between force and acceleration can be made more rigorous by drawing on an everyday experience. To get a heavy object to accelerate is more difficult than to get a light one to. It is easier to set a pram in motion by pushing it than it is to set a car in motion by the same means. Both can be made to accelerate, but if each is pushed equally hard the car will accelerate much more slowly than the pram. Conversely, if the car and the pram were to be caused to accelerate at the same rate, much greater force would have to be used on the car. The pram would only need to be pushed comparatively gently.

To express the connection between force and acceleration in a precise way requires that a law be formulated. In order to do this it is necessary to observe both force and acceleration in as much detail and as accurately as possible. Furthermore, a means must be found of measuring force and acceleration independently of one another. (If the measurements of force and acceleration were not independent of each other, then evaluations of the observations would require the law connecting force and acceleration to be known in advance.)

Measurements of acceleration are quite straightforward, requiring only a fixed distance to accelerate an object over and a clock to register how long it takes for the object to cover the distance.

Measurement of force is a little more complicated, but at this stage what is required is not the absolute values of forces but only their relative strengths. This can be determined, for example, by taking several identical springs and producing an acceleration by pulling using only one of them, and then two together, then three together, and so on, the relative strengths of the force in each case being proportional to the number of springs being employed.

Once a collection of data about amounts of force and corresponding accelerations has been obtained and analysed, it becomes possible to see a pattern in the figures. It is found that the acceleration imparted to any one object is proportional to the force employed. This is the law being sought. It may be stated symbolically that **F** is proportional to a for any one object, where **F** is the force and a is the acceleration.

(Note that heavy type is used to indicate that the quantity symbolized is a *vector*; i.e. it is a quantity which has an orientation in space as well as a size. A force doesn't just cause an object to accelerate; it causes it to accelerate in a definite direction. Hence force and acceleration have an orientation in space; they are vectors. And since velocity, as explained above, is speed in a particular direction, it too is a vector. Compare this with mass and energy, below, which are 'pure' quantities possessing only a size.)

MASS

Having established a precise relationship between force and acceleration, the data that has been collected can be used in a further capacity. The question can be posed: how are force and acceleration related when different objects are involved (as for instance the car and the pram)?

A good hypothesis would be this: every object possesses a quantity which is equivalent to the amount of force needed to produce some fixed acceleration. This quantity is called *mass*. If two objects are taken and made to accelerate at an equal rate, and it is found that twice as much force, say, must be used on the first object as on the second, then this means the first has twice as much mass as the second. Stated symbolically as above, **F** is proportional to the mass (m) for any one value of a.

The first encouraging thing to discover is that the two statements of proportionality, **F** proportional to a and **F** proportional to m, are compatible. **F** is always proportional to a for any one mass regardless of the size of the mass, and **F** is always proportional to m for any one amount of acceleration regardless of the size of the acceleration. As a result, the two statements can be combined, giving

$$\mathbf{F} = \mathrm{m}\textit{a}.$$

Even more encouraging is that the mass of an object turns out to be precisely proportional to its weight. If there is any departure from exact proportionality it is too tiny to be measured with today's technology, (which is accurate to one part in 10^{13}). This strongly suggests that mass is more than just an abstract property of objects which relates the force used on them to their resultant acceleration. An object's mass can be evaluated quite independently of acceleration simply by weighing it.

This is bringing order out of chaos indeed! The concepts of acceleration, force and mass, formulated in a mathematical equation, provide a way of describing the ceaseless chaotic motion with which this chapter began. The description is both simple and decidedly unchaotic. Given that two objects cannot be in the same place at the same time, all the activity in the terrestrial world can be seen as an endless series of physical contacts between objects.

The situation can be summarized as follows.

- Whenever an event occurs it is always as a result of an earlier event.
- When objects affect each other (i.e. when an event occurs), they do so through the agency of force, producing acceleration.
- All objects possess a characteristic called mass.
- An object's mass is the measure of how much force is needed to produce a given acceleration. $\mathbf{F} = \mathrm{m}\textit{a}$.

1.2 ENERGY

GRAVITY

So far, a number of forces have been encountered (pressure, stretched spring, friction), all of which require contact between the force exchanging objects. But not all forces are of this kind. One of the observations most blatantly ignored in the preceding section is of objects initially in mid-air which begin to accelerate earthward without the apparent involvement of another object causing them to do so. If the ideas developed so far are to be made more general and applied in this new context, then they imply that something must be forcing the falling objects to accelerate. The question is to decide what.

As always in science, the answer can only be found by means of systematic observation. It is found that almost all objects, when released in mid-air, accelerate towards the Earth perpendicularly; that is, towards the Earth's centre. The most obvious hypothesis which accords with this is that the force arises from the Earth itself. It can therefore be proposed that the Earth exerts a force on all objects which causes them to accelerate towards it by the shortest possible path. The force is called *gravity*.

Gravity is unlike the other forces that have been mentioned so far because it doesn't require the force exchanging objects to be physically touching one another. The force can be felt across empty space. As a result, gravity is impossible to avoid. Wherever an object is, whether on the ground or up in the sky, the object is always subject to the force of the Earth's gravity.

One consequence of gravity's pervasive nature is that objects are impelled to accelerate earthwards even when they are not in fact able to do so. An object suspended from a rope would fall to the ground if the rope wasn't holding it in place. So would an object resting on a weighing machine except for the weighing machine being in the way. Which leads back to the concept of mass.

As was pointed out at the end of the previous section, the mass of an object (mass from $F = ma$) is exactly related to its weight. Now, the weight of an object — such as is measured by a weighing machine — must be due to the gravitational force (F_g) exerted on it by the Earth. F_g can be defined as being proportional to this weight. Expressed in a logical form, these facts and the conclusion they lead to can be summarized in the following way.

F_g is proportional to weight (by definition)
AND
Weight is proportional to mass (by observation);
THEREFORE
F_g must be proportional to mass as well as to weight.

Recall now that any force F is always proportional to m for any *one amount* of acceleration. It follows that if F_g is proportional to mass, there must be 'one amount' of acceleration which is always the same regardless of the sizes of F_g and m; i.e.

$$F_g = m \times \text{fixed rate of acceleration.}$$

The prediction that gravity produces a fixed rate of acceleration can easily be tested and is indeed found to be correct. Providing allowance is made for the effects of air getting in the way, all objects, regardless of their mass, fall earthward with the same acceleration. At the surface of the Earth the fixed rate of acceleration has the value $9 \cdot 8$ m s^{-2}. The symbol used for this particular value of acceleration is g. $F_g = mg$.

A generalization is possible here. The force of gravity, like all other forces, is an interaction between two objects, of which the Earth is one. But what is so special about the Earth? Can it not be expected that any two objects will exert a gravitational force on each other? Certainly it can. It is possible to demonstrate gravitational attraction between objects in a laboratory, given suitable equipment.

Analysis of observational data leads to the conclusion that the gravitational force between two objects is proportional to both their masses and is also affected by how far apart they are:

$$F_g = \frac{Gm_1 m_2 (-\hat{r})}{r^2}$$

In this equation m_1 and m_2 are the masses of the two objects, and r^2 is their distance apart multiplied by itself. The quantity which connects the two sides of the equation, called a constant of proportionality, is the gravitational constant, G, which has the value $6 \cdot 673 \times 10^{-11}$ m^3 kg^{-1} s^{-2}.

(That leaves the symbol $\hat{r}$ to explain. For simplicity it is often omitted, but technically it needs to be there for a mathematical reason. F_g is a vector, so the expression on the right must also be a vector or the equals sign will be invalid. Unfortunately m_1 and m_2 and G are not vectors. Neither is r^2, for although r is a vector, the product of two vectors — as in r times $r = r^2$ — is not itself a vector. So for F_g on the left of the equation to equal the expression on the right, the quantity on the right must include a vector from somewhere. $\hat{r}$ fulfils this role. It has the same orientation in space as F_g does, while the 'hat' on top of the r indicates it is a unit vector

— that is, it has a value of 1. The minus sign, '$-\hat{\mathbf{r}}$', reflects the convention that repulsive forces, those which cause **r** to increase as time passes, are in the positive direction, thus making forces in the opposite direction — attractive forces like gravity — negative.)

There is one characteristic that this equation must have if it is to be acceptable. Suppose one of the masses is much larger than the other, as is the situation when the force is between the Earth and comparatively small objects near its surface. The equation must predict that the small objects, regardless of their mass, will all have the same acceleration towards the Earth. Only if it makes this prediction will the equation be in accord with observation. Look again at the expression on the right hand side. So long as the value of **r** changes over only a comparatively tiny range, the one thing on the right hand side of the equation which can vary is m_1 (taking m_2 to be the mass of the Earth). So $\mathbf{F}_g$ is proportional to m_1, yielding a fixed rate of acceleration ($\boldsymbol{g}$), as shown earlier.

KINETIC ENERGY

Because gravity does not require the gravitating objects to be in contact, it is able to operate continuously. While the force that arises in, for example, a collision only lasts as long as the collision does, the force of gravity between objects lasts forever. Not surprisingly, continuous force leads to continuous acceleration. Objects coming together due to the force of gravity move faster and faster as they approach. How fast they are moving when they finally meet depends on how far apart they were when they started. It is, furthermore, intuitively obvious that the more massive they are — note that 'massive' in this book will be used to mean 'possessing mass', without the usual implication of great size — the bigger will the crash be when they do meet. Putting these things together yields another concept to add to force, mass and acceleration: the concept of *energy*.

An example will help to give an idea of what energy is. Think of a stone falling to the Earth. The energy acquired by the stone by the time it hits the ground is a measure of how much damage it will be capable of doing. The more energy the stone has, the greater will be its ability to affect anything it collides with. The falling stone achieves its damaging effects by means of a force arising the instant it hits the ground. This force produces acceleration in the bits of ground being struck. Hence assigning a value to energy is the same as quantifying the amount of acceleration an object is able to impart to things it collides with: the greater the energy, the greater the acceleration. The energy of the stone is a measure of the quantity of events the stone can bring about.

To turn energy into a scientific concept, it is necessary to measure it. That in turn requires it to be mathematically analysed first. Although I avoid giving mathematical derivations in this book, I will make an exception here. That's because the derivation is an easy one and also because it provides an informative illustration of the interplay between mathematics and physics which underlies all the science in this chapter and chapters 3 and 4.

Consider the falling stone. Assume it falls to the ground from a height $\mathbf{H}$. To make the mathematics easy, take $\mathbf{H}$ to be small enough, and ground level to be far enough from the centre of the Earth, that $\mathbf{r}$ (in the equation $\mathbf{F_g} = Gm_1m_2(-\hat{\mathbf{r}})/r^2$) hardly changes as the stone falls. As long as that is the case, the force exerted on the stone by the Earth is approximately constant. The energy acquired by the stone can then be defined as the force $\mathbf{F_g}$ on the stone, multiplied by the distance $\mathbf{H}$ travelled by the stone while feeling the force; i.e.

$$E = \mathbf{F_g H}.$$

(Just as the product of $\mathbf{r}$ times $\mathbf{r}$ mentioned earlier is not a vector, neither is the product of $\mathbf{F_g}$ times $\mathbf{H}$. Energy too is not a vector. It has size but no orientation in space.)

It should be noted that as $\mathbf{F_g}$ is proportional to m_1 the equation meets the two intuitive requirements of the concept of energy listed above, that an object's energy depends on how far it falls for ($\mathbf{H}$) and on how massive it is (m_1).

$E = \mathbf{F_g H}$ can be applied to other forces than gravity; i.e. $E = \mathbf{FH}$ generally. For most forces, $\mathbf{FH}$ is of no use when it comes to determining an accelerating object's energy. This is because the value of $\mathbf{F}$ can sometimes be hard to measure directly, and it may also not be known how much distance $\mathbf{H}$ the force has been in operation over. What are known and easy to measure are the mass and velocity of the object. Fortunately it is possible to find out what $\mathbf{FH}$ equals in terms of these two quantities, using gravity as a model force. There are four steps in the process.

16

DYNAMICS

1. The distance travelled by, say, a falling stone, is fairly obviously the product of its average velocity and how long it was falling for; that is,

$$H = \tfrac{1}{2}(v + u)t.$$

Here v is the stone's final velocity, u is its initial velocity (so that $\tfrac{1}{2}(v + u)$ is its average velocity) and t is how long it was falling for. If the stone was initially stationary (i.e. $u = 0$) then

$$H = \tfrac{1}{2}vt.$$

2. Recalling that $F_g = mg$, $E = F_gH$ becomes

$$E = mg(\tfrac{1}{2}vt) = \tfrac{1}{2}mvgt.$$

3. It should also be fairly obvious that the final velocity of the stone is equal to how much it was accelerating by (g) multiplied by how long it was accelerating for (t):

$$v = gt.$$

4. This gives finally

$$E = \tfrac{1}{2}mv^2.$$

With this expression, the energy of objects can be observed and evaluated. The concept of energy can thus be put on a firm scientific footing. It can be remarked, incidentally, that the expression for energy derived above is not intuitively obvious. It is to be hoped, however, that all the expressions used to obtain it ($F_g = mg$, $E = F_gH$, $H = \tfrac{1}{2}(v + u)t$, and $v = gt$) are obvious. Such a situation is commonplace in mathematical science.

So energy can be investigated systematically by way of mass and velocity. Such investigation soon makes it clear that energy exists in more than one form. The energy of movement, which is what has been discussed so far, is called *kinetic energy* to distinguish it from the others.

Kinetic energy is energy due to motion. Any moving object possesses kinetic energy.

CONSERVATION OF ENERGY

One thing to emerge from observations of kinetic energy is that it can be made out of nothing. A falling stone has no kinetic energy before it starts falling, but may have a lot by the time it reaches the ground. Furthermore, if the stone was to strike springy ground, its kinetic energy, instead of being absorbed by the ground, would in effect be reflected. The stone would bounce, rising into the air and slowing down until coming to a halt, at which point it would again possess no kinetic energy.

In a sense though, the bouncing stone does still have some energy when it has come (momentarily) to a halt in the air. It has, so to speak, *potential energy*. Because of its position above the ground it has the potential to acquire kinetic energy by falling. Potential energy can be defined as being equivalent to the amount of kinetic energy an object would acquire if it was to fall the full distance available to it. With potential energy defined in this way, it can be worked out how much potential energy an object experiencing a force has. It is equal to the product of the force ($\mathbf{F}$) and the distance over which the force acts ($\mathbf{H}$), so long as the object does not actually start to cover this distance. In this connection the object may be momentarily at rest or it may be fixed in position by some means.

Once again

$$E = \mathbf{FH},$$

where $\mathbf{F}$ is, as usual, any force, including gravity, and $\mathbf{H}$ is not the distance travelled while being accelerated by the force, but the distance that could *potentially* be travelled during the acceleration.

In this situation, which is static for at least an instant, measuring $\mathbf{F}$ and $\mathbf{H}$ is much easier than is the case for moving objects possessing kinetic energy. Hence the equation need not be converted into some other form this time. For potential energy, $E = \mathbf{FH}$ will do nicely.

Once an object begins to accelerate due to a force, it acquires more and more kinetic energy. At the same time, the distance remaining for the force still to act over decreases. Because the value assigned to both types of energy originates from the same equation (E = **FH**), each unit of kinetic energy acquired by the object is exactly offset by the loss of a unit of potential energy. Whenever the kinetic energy goes up (or down), the potential energy goes down (or up) by a matching amount. This means that if the potential and kinetic energy are added together, the resulting total energy (total energy = potential energy + kinetic energy) always has the same value.

If the total energy possessed by two interacting objects never changes, then it follows that the combined energy of lots of mutually interacting objects will also never change. This lends support to the proposal that, in any situation, the total energy of all the participating objects taken together is a fixed quantity, regardless of the details of what happens. If this is so, it means energy can be converted from one form to another and redistributed without limit, but the total amount of it never decreases or increases. Technically it is said to be *conserved*; that is, it cannot be created or destroyed. This fact, which can be verified experimentally, is called 'the law of conservation of energy'.

It has to be admitted that this conservation law only works if potential energy is accepted as being real. But is it? It looks awfully like a book-keeping exercise, a fiddle designed to make energy appear conserved when it actually isn't at all. But there is one crucial point about potential energy that provides a justification for regarding it as more than a convenient fiction. It comes from the following observation. If an object moves towards or away from a source of force, the kinetic energy gained or lost by the object as a result of the movement is exactly cancelled if it is returned, by no matter what route, to its starting place. If an object rising into the air from height x to height y loses all its kinetic energy, then when it returns to height x that kinetic energy will be exactly regained. It is not unreasonable to regard it as stored in a different form, i.e. as potential energy, at height y.

The law of conservation of energy is a restatement of one of the laws proposed at the start of this chapter. It was said there that all events are the product of previous events and produce further events. Now, it has been seen that all events involve acceleration, which implies the presence of force, which in turn implies the existence of potential energy due to the force, and thereby kinetic energy. Hence events and energy are inseparable. Indeed an event can be regarded as simply a redistribution, by means of acceleration, of energy amongst the objects that are taking part in the event. To say that the energy needed to produce an event must be obtained from somewhere is to say that the event is produced by some previous event (which supplies the energy). Hence the law that events are, and can only be, produced by other previous events is essentially the same as the law of conservation of energy.

It is also interesting to note that this law receives support from an entirely different quarter. It can readily be observed that the laws of physics do not vary from one day to the next. This observational fact implies that energy *must* be conserved. Such a conclusion is a consequence of a piece of mathematical reasoning called Noether's Theorem, which reveals that there is a deep connection between energy and time. This connection will be encountered again in section 3.3.

CONSERVATION OF MASS

The law of conservation of energy contains within it a hint of a second conservation law, one to do with mass.

For energy to be conserved it is necessary that no source of force is ever created or destroyed. If such a creation or destruction was to happen, it would alter the amount of force and thereby the amount of potential energy in the universe without there being a compensatory alteration in the amount of kinetic energy.

Now, all forces emanate from objects. The characteristic of any object that is most inextricably associated with the forces which have been encountered so far is the object's mass. It may be suggested therefore that mass is, directly or indirectly, the source of all force; that is, that all forces arise from massive objects. (It will be seen in chapter 3 that this suggestion is indeed broadly correct.) If all forces do arise from massive objects, then to say that no source of force is ever created or destroyed may be to say that no mass is ever created or destroyed. It can be proposed that mass, like energy, is conserved.

This line of reasoning is not rigorous. It is suggestive rather than inevitable and needs thorough observational verification. Observation does indeed verify it. A law of conservation of mass can be added to that of energy.

The two conservation laws, of mass and of energy, are not, let it be stressed, laws in the sense of laws made by human societies, which can always be broken. Physical laws have never been known to be broken. Even the most ingenious criminal has always obeyed them. Everything obeys them. They show that beneath the confusion and complexity of the terrestrial world lies a basic simplicity. What could be simpler than the statements: overall change in mass = 0; overall change in energy = 0? The simple interplay of mass and energy, adhering to the two conservation laws, and carried out by means of force and acceleration, produces the Earth and everything that happens on it.

1.3 ATOMS

The explanation of the behaviour of objects given above is certainly beautifully simple. But in one particular way it is unsatisfactory: it places no lower limit on the size of the objects concerned. Just how small can they be? The objects were also taken to be solid. What about liquids and gases? The answers to these sorts of questions depend on the small-scale structure of objects. What exactly are objects made of?

To start with, a basic observation: all objects consist of one or more substances. A pencil, for example, is made of three substances: a tube of wood, a covering of paint, and a central core of graphite or some similar material. Paint, wood and graphite are three substances. Clearly, when an object is made of more than one substance it can be broken up along the dividing lines between the various substances in its makeup. But what about an object made of only one substance? Given such an object, a sheet of glass or a lump of chalk, say, experience reveals that it can be broken into bits. The key question from the point of view of small-scale structure is how small these bits can be.

A question like this can only be resolved scientifically by experiment. It is necessary to take a sample of the substance, whatever it is, and systematically break it into smaller and smaller pieces. The results, however, are disappointing. Even using the most powerful optical microscopes available, there seems to be no limit to the breakability of substances. Any substance can be broken up until the pieces are so small they become invisible.

Direct experiment as a means of determining small-scale structure is thus inconclusive. All it really shows is that that structure, whatever its nature, is very small indeed; too small to be seen under a microscope. Some other way to investigate this question must be found.

One possible alternative line of enquiry rests on the assumption that the small-scale structure of substances has an effect on their large-scale properties. If this is so — and it seems reasonable — then what needs to be done to pursue this possibility is to classify the enormous number of substances found on the Earth and see if any patterns can be discovered in the data.

First attempts to classify substances in the terrestrial environment are more or less innate and based on function. Substances are classed as safe or dangerous, hot or cold, edible or inedible, and so on. This functional classification, while essential for survival, has its drawbacks. What, for instance, is to be made of the thirst-quenching liquid called water when one cold day it turns into a hard glassy kind of stone? Or the wood that goes black and crumbly if put in a fire? From the point of view of function, such transformations are disturbing. If substances can change their form, and hence their function, any function-based system of classification is compromised. Such a system is of no use in investigating the structure of substances. What is required is a classification based on the permanent features of substances rather than on ephemera.

(There is at this stage, it should be noted, no reason other than intuition for thinking that an alternative classification actually exists.)

The search for a new classification proceeds with the compilation of data documenting under what circumstances any given substance changes into something else. In this way it is to be hoped that the transient features of substances can be separated from their constant ones. From studies of this sort of data, it soon becomes apparent that a set of laws governs the transformation of different substances into each other. Any one substance can be turned into only a small number of other substances.

There are two means whereby a substance can be transformed. One is to alter its temperature by heating or cooling it. The other, possibly combined with the first, is to mix it with selected other substances. This latter approach forms the subject of chemistry, and it yields the more interesting quantitative results. Of particular importance is the observation that if two substances, A and B, are mixed together such that they turn into a new substance, C, then the relative amounts of A and B that are needed never vary. If too much A is mixed in, some of A will be left over after the formation of C is complete. Similarly, if too little A is mixed in, some of B will be left over.

To explain these observations simply, a hypothesis is proposed that all substances are made of tiny grains. When the transformation of a substance occurs, such as of A and B into C, the idea is that the grains of the involved substances unite together in some way to make super-grains. The number of grains that go to make up one super-grain of any particular substance is fixed. Hence the relative amounts of A and B needed to make C do not vary.

Obviously, some substances are made of individual grains while others must be made of super-grains. The grains are called *atoms*. The super-grains are called *molecules*. A molecule consists of two or more atoms united together. The transformation of a substance by means of this hypothesized atomic union is called a *chemical reaction*.

In terms of substances A, B and C, what happens can be written:

$$A + B \rightarrow C.$$

A and B mixed together turn into C. A molecule of C contains exactly one atom of A and one atom of B. That is why, if too much A is put in the mixture, some of A will be left over after the reaction is complete; some of the A atoms will remain unpartnered. If too little A is put in, it will be some of the B atoms that are unpartnered. The hypothesis that all substances are made of atoms, singly or combined, can thus account for the quantitative aspects of chemical reactions straightforwardly.

However, it may be quite feasible to construct a non-atomic explanation of the nature of substances which accords just as well as the atomic hypothesis with what is observed. The atomic hypothesis is to be preferred on account of its simplicity, but caution is required at this stage. After all, no one has ever seen an atom directly. So more evidence would be helpful. What other observations can be made which might support the atomic hypothesis? Three in particular are worth mentioning here.

The first concerns substances which undergo chemical reactions to form more than one possible resultant substance. Suppose the substances A and B react together to form either substance C or substance D. Which of the product substances is actually formed depends, amongst other things, on the relative amounts of A and B used. It might be found that, given a fixed amount of B, twice as much A is needed to produce D as is needed to produce C. In many chemical reactions with more than one possible product substance, the amounts needed to make one product compared to the amounts needed for the other differ in this simple way. The ratio between the amounts will be twice as much, or three times as much, or two thirds, or some such fraction, but not a complicated ratio like 79/113. The atomic hypothesis accounts for this observation quite easily. In the case of A and B above, the chemical reaction between them to produce D can be symbolized:

$$2A + B \rightarrow D.$$

A molecule of D contains 2 atoms of A and 1 atom of B. The quantity of A making D is thus exactly twice as much as the quantity needed to produce C. By proposing that molecules generally consist of only a few atoms, the absence of complex ratios is accounted for, since ratios like 79/113 would necessitate huge molecules. (Such molecules will in fact be encountered in chapter 5.)

Another observation that reflects favourably on the atomic hypothesis concerns the number of substances which are made of individual atoms like A and B, and the number which are molecules like C and D. The conclusion to be drawn from an enormous amount of chemical experimentation is that almost all substances are made of molecules. Only about 90 different types of atom have been found on the Earth (the exact number depending on how 'found' is defined). There are also a number of man-made atoms. The atomic hypothesis thus reduces the near infinity of substances to combinations of roughly 100 atoms.

Finally, the most convincing directly observable evidence for the existence of atoms is a phenomenon called Brownian motion. Fine dust particles in water, viewed under a microscope, are observed to move about erratically. A mathematical analysis of the motion shows it is caused by the dust particles being hit randomly by discrete bits of water; in other words, by water molecules.

In conclusion, although the small-scale structure of substances is invisible to the human eye, the evidence strongly supports the idea that all substances are made of atoms. These atoms can exist either singly or in combination, with all substances being classified according to which atoms they contain and in what combination. Systematic observation has thus made it possible to build up a consistent picture of the nature of substances and their structure, to the extent that it can be considered certain that everything is made of atoms or molecular groups of atoms. The guess that began as the atomic hypothesis can now properly be called the atomic theory.

This elaboration of the nature of substances has been carried out largely in isolation from considerations of force, mass and energy. The next task must be to apply the theory of force, mass and energy to the theory of atoms. If the one can be applied to the other without producing any contradictions or inconsistencies, then both can be considered much more secure. This is so because if either was fundamentally in error, it is unlikely they would be compatible with each other.

1.4 THE PHYSICS OF MOLECULES

APPLYING KINETIC ENERGY TO MOLECULES

Imagine a tank full of water, across the middle of which is a partition. The water on one side of the partition is coloured by dye. On the other side it is clear. The partition is removed very slowly so as not to disturb the water. What happens? Gradually the dye spreads into the clear water until all the water contains dye molecules.

It can be concluded from this experiment that water molecules are not rigidly fixed in place. They are free to move about, thereby permitting the dye molecules in the water also to move about in the tank.

A similar observation can be made of gases. If a coloured gas is introduced into a boxful of air, the coloured gas rapidly spreads throughout the box.

The molecules of the water or air — or of any other liquid or gas — can thus be envisaged as moving about and bouncing off each other in ceaseless but invisible motion.

Now, every object, as well as being made of molecules (in the interests of brevity I shall henceforward refer to molecules, rather than to molecules and/or atoms; the 'and/or atoms' should be understood), possesses a certain amount of mass. It follows that molecules must also have an amount of mass commensurate with their size. Any molecule which is free to move is therefore likely to have associated with it some kinetic energy, subject only to whether the molecule's speed is zero or not at any given moment. (Recall that kinetic energy = $\frac{1}{2}mv^2$.) Applying the theory of energy to atom-sized objects thus leads to the conclusion that any substance whose molecules are not fixed in place — any substance which isn't a solid, in other words — will possess some internal kinetic energy associated with the motion of its individual molecules. This energy is additional to any kinetic energy the substance possesses on account of its overall motion.

The kinetic energy of a molecule depends on its mass and velocity. Thanks to the law of conservation of mass, a molecule's mass can be considered unalterable. But there is no reason to think its velocity is similarly fixed. This means that the internal kinetic energy of a liquid or gas due to molecular motion is variable. In that case, it should be possible to take the kinetic energy of some solid object and transfer it to a liquid or gas, but to do it in such a way that the liquid or gas as a whole remains stationary. The idea is that the energy would instead be acquired by individual molecules so that they move around inside the liquid or gas faster; i.e. more energetically. Energy would be conserved in such a process. What kind of experiment could be carried out to test this possibility?

Imagine the tank of water again. This time, instead of dye, add a propeller. In this new experiment the propeller is immersed in the water. A force is exerted on the propeller, causing it to accelerate and acquire kinetic energy. Next the force is removed. The propeller comes gradually to rest, transferring its kinetic energy to the water as it does so. After local eddies have died down the water becomes still. The energy the water has acquired from the propeller must now be entirely internal, distributed amongst its randomly moving molecules. The experiment has succeeded: the kinetic energy of the propeller has been converted into an increase in the internal kinetic energy of the water molecules.

There is, however, a problem. The trouble is that the experiment lacks measurement. The kinetic energy lost by the propeller can be measured easily enough, but what about the internal kinetic energy of the water?

Unless that can be measured too it is impossible to prove that the propeller's energy has gone where atomic theory says it has.

Unfortunately, atomic theory thus far has provided no clue as to what characteristic of the water reveals its internal kinetic energy. With no idea what to measure, the best course is to guess; that is, to invent a few hypotheses. These hypotheses will indicate what measurements should be made. Doing this it becomes clear that the most consistent change to take place as a result of the propeller experiment is an increase in the water's temperature. It can therefore be proposed that temperature is the measurable quantity being sought. When energy is input into a liquid, the liquid's temperature rises. Ditto for gases.

Strong support for this connection between temperature and energy comes from more elaborate propeller experiments. These show that the increase in temperature of a liquid or gas is directly proportional to the energy that is transferred to it. The conclusion: temperature is indeed a measure of internal kinetic energy.

Armed with this knowledge, the transfer of kinetic energy from a moving object to the molecules of a stationary liquid or gas can be quantified. Gratifyingly, systematic measurements indicate that no energy is lost or created in the transfer process. The law of conservation of energy is fully compatible with the atomic theory.

So far at least, then, the atomic theory has combined well with the theory of force, mass and energy. And as a bonus it has been discovered that the physics of heat, instead of being an independent science, has turned out to be a consequence of the interplay of force, mass and energy on an atomic scale. Heat, whose quantity in any particular substance is indicated by the substance's temperature, is just the kinetic energy of collections of molecules.

APPLYING FORCE TO MOLECULES

As is so often the case in science, the discovery that heat is not a distinct property of substances, but is just kinetic energy writ small, great simplification of physics though it is, brings with it several new problems. Where do solids come into the picture? What makes a gas different from a liquid? Can these questions too be answered in a way that is consistent with both the atomic theory and the theory of force, mass and energy?

Once again take water as a typical substance.

The temperature of water cannot go up without limit. Eventually, as heat is put into a tank of water, thereby increasing the average kinetic energy of the water molecules, the temperature stops rising and the water begins to turn into steam. Steam is a gas. Similarly, if the temperature of water is reduced as heat is taken out of it, reducing the average molecular kinetic energy, the water eventually turns into ice. Ice is a solid. The first conclusion to be reached is thus that the transitions from solid to liquid to gas and vice versa are related to the quantity of heat possessed by a substance. Molecules in gases have more kinetic energy than molecules in liquids, which in turn have more kinetic energy than molecules in solids.

It can be noted in support of this idea that transitions amongst the three types of matter are normally reversible simply by cooling or heating a substance as appropriate.

The difference between a gas, a liquid and a solid though is clearly a matter of more than the kinetic energy of molecules. There is a sharp qualitative difference between them as well.

Consider gases first. If the lid is removed from a tank full of gas, the gas rushes out in all directions. The molecules of gas are moving about randomly inside the tank, so that when the lid is removed those that would have bounced off it suddenly find it is no longer impeding their upward movement. They escape from the tank. Without the lid to act as a restraint, every time a molecule near the top of the tank finds itself travelling upwards as a result of random collisions with its neighbours, it escapes.

Next look at liquids. If the lid is removed from a tank full of liquid, nothing happens. In fact, close observation reveals a kind of skin across the surface of the liquid which holds in the molecules below it, acting as a form of self-made lid. Below this skin the molecules are free to move randomly, as was seen at the beginning of this section.

Lastly solids. Here the molecules are locked firmly in place. Since solids can be at many different temperatures they must contain a variable amount of heat. This means the kinetic energy of the molecules in the solid is not fixed. To accommodate this fact it can be envisaged that the molecules are able to move slightly, but not so much as to significantly change their positions with respect to one another. The motion of a molecule in a solid can best be likened to a vibration, the size of the vibration increasing as the temperature does.

As it stands, the atomic theory is quite incapable of coping with these differences between solids, liquids and gases. Where does the skin on liquids come from? How are molecules in solids locked into place? Without a surface skin all liquids would be gases. Without their molecules locked into place all solids would be gases too. These questions thus boil down to a simple one: why aren't all substances gaseous?

What is needed is a new hypothesis which can be added to the atomic theory. The new hypothesis is this: that all molecules are sticky. Or to put it more scientifically, there is an attraction between them that they only feel when they are almost touching.

In a gas, the molecules have much kinetic energy and so are fast moving. There is so much energy involved in collisions amongst them that the attraction is hardly noticed. By way of analogy, think of two slightly sticky marbles striking each other at great speed. In a liquid by contrast, the molecules are comparatively slow moving. The attraction doesn't stop them completely but it does make them sluggish. Inside the body of the liquid the attraction is felt by each molecule to be coming from all directions, from all its neighbours. But at the surface it is only felt to be coming from one direction — inward into the liquid. Surface molecules are in effect partly stuck to those beneath them, and it is this which causes them to form the surface skin and prevents them escaping like a gas. In a solid, finally, the molecules are completely stuck together.

The idea of a short-range attraction — stickiness — between molecules is certainly satisfactory as a way of explaining the differences between solid, liquids and gases. But how exactly does a short-range attraction work? How does it do what it does?

Think of the way the vibrating molecules in a solid must regard each other. As they vibrate, the distance between them increases. Then the attraction pulls them back together again, causing them to collide and bounce apart. This happens over and over repeatedly. The situation is very similar to that which a very elastic ball is in as it bounces up and down off the ground. The ball is attracted to the Earth by the force of gravity. In a way, the force of gravity makes objects stick to each other. The attraction between molecules can similarly be attributed to a force. In other words, the short-range molecular attraction is due to a force they exert on each other.

How does this attractive force between molecules look when expressed in the language of force and energy? As a molecule moves away from its neighbour following a collision, the force between them pulls on it, causing it to slow down. Its kinetic energy is converted into potential energy due to the inter-molecular force. If its kinetic energy is small enough to start with, the molecule will soon come to rest. It will then fall back towards the attractive neighbour. This is what happens to vibrating molecules in a solid. On the other hand, in liquids, and even more so in gases, the kinetic energy of the molecule as it recedes from its neighbour is so large that the attractive force is not strong enough to bring it to a halt. It becomes unstuck.

The development of the hypothesis of inter-molecular force is potentially very fruitful. All the forces of section 1.1, the collision, the stretched spring, friction, must eventually be resolved into forces between molecules if the atomic world is to be successfully understood. The hypothesis that there is a specific force between molecules brings this aim much closer to being realized.

The objective in this section has been to take the concepts discovered in observations of the terrestrial world — force, mass, acceleration and energy — and see if they can be combined with the atomic theory without giving rise to any contradictions. Do atoms violate the conservation laws, for example? So far, no problems have been encountered. The rules that govern how things happen at a terrestrial scale seem to work just as well at the atomic scale. And there have been two unexpected bonuses: the nature of heat has been explained; and a new hypothesis, that of an inter-molecular force, has accounted for the difference between solids, liquids and gases.

This is a very satisfactory state of affairs, spoilt only by the fact that details about the inter-molecular force are far too sketchy. The atomic theory will only be truly secure when this force is properly understood. That then is clearly the next task.

1.5 THE PHYSICS OF ELECTRONS

ELECTROMAGNETIC FORCE AND ELECTRIC CHARGE

What can be said at the outset about the inter-molecular force?

Perhaps the first point to make is that it is nothing to do with gravity. True, the force of gravity is in effect a kind of stickiness between objects, but the size of that force on an atomic scale is extremely small. Quite simple calculations show that molecules' kinetic energy is far greater than the gravitational potential energy between them. Inter-molecular gravity is much too weak to be able to stick molecules together into solids. So the force is not gravity.

Another point arises from observations of what happens if two large objects are placed side by side. The two objects do not immediately stick together. This means that the force between neighbouring molecules is very short in range. While the molecules inside each object are stuck to each other, the force responsible is unable to cross the tiny gap between the surface molecules of one object and another. The force is thus much stronger than gravity but is only felt when molecules are almost touching.

In investigating gravity, three facts proved crucial: the strength of the force, which is given by the constant G; the range of the force, which is infinite; and what its source is, namely, the masses of the attracting objects. In investigating the inter-molecular force, its strength and range have just been commented on, but nothing has been said yet about what its source is. Is it similar to gravity, arising from mass? Or does it come from something else?

The clue to the source of the inter-molecular force is found from studying a mysterious kind of attraction between terrestrial-scale objects. When some solid substances (rubber membranes, for instance) are rubbed vigorously with a cloth, they become sticky to other objects brought near them. It is almost as if the range of the inter-molecular force is somehow increased by the act of rubbing. Suppose that this is indeed what is happening. This hypothesis (which is really no more than a guess) indicates that it may be worth examining these instances of mysterious stickiness for clues about the nature of the inter-molecular force.

(I confess this hypothesis about inter-molecular force is definitely not remotely obvious in advance. I have to plead hindsight.)

The study of this type of stickiness between objects, which is known as electrostatics, led to the concept of *electric charge*. Electric charge is a characteristic of certain objects which gives rise to a force between them. There are two kinds of electric charge, called positive and negative. (They could just as validly have been called left and right, or male and female!)

Positive charges repel each other, and negative charges repel each other. Conversely, a positive and a negative charge attract one another. In normal circumstances, a substance will contain one positive charge for each negative charge, and the sum total of all the repellings and attractings between the electric charges it contains will add up to zero. It will not appear to be sticky. However, in unusual circumstances, such as when a rubber balloon is rubbed by a cloth, the positive and negative charges become unbalanced. An attractive force results, making the balloon sticky.

The study of electric charges led to the elaboration of the mathematical relationship describing the force between the charges. This relationship can be written:

$$\mathbf{F_e} = \frac{Ke_1e_2(\pm\hat{\mathbf{r}})}{r^2}$$

In this equation, $\mathbf{F_e}$ is the force between the electric charges e_1 and e_2, and $\mathbf{r}$ is the distance between them. If e_1 and e_2 are both positive or both negative the force is repulsive: $(+\hat{\mathbf{r}})$. If one is positive and the other negative the force is attractive: $(-\hat{\mathbf{r}})$. Lastly, there is a constant of proportionality, which I have written here as 'K' for simplicity, that plays the same role in this equation as G plays in the force-of-gravity equation in section 1.2.

The force between electric charges needs to be given a name. It is called *electromagnetism*. This name is used because electric charges, when they are in motion, are responsible for producing magnetic phenomena. It will not be practical to go into how they bring these about here.

The mathematical expression for the force between two electric charges is of the same form as that for the force $\mathbf{F_g}$ between two masses. This indicates that the two forces have much in common. In particular, it can be stated that mass is to gravity as electric charge is to electromagnetism. Mass can legitimately be called gravitational charge. The terms 'mass' and 'gravitational charge' will be used interchangeably from here onwards. Gravitational charge and electric charge are both fundamental characteristics of an object.

It can be seen that neither for gravity nor for electromagnetism is there any limit on the size of r^2 in the denominator. Obviously the further apart two attracting (or repelling) objects are, the weaker will be the force between them; but the force never reaches zero, no matter how large r^2 gets. Thus both gravitational and electromagnetic forces have an infinite range.

If the properties of the gravitational and electromagnetic forces are so similar, what distinguishes them? Firstly there is strength. The smallest type of atom is hydrogen. If two hydrogen atoms were each to be given one unit (see below) of electric charge, the electromagnetic force between them would be 10^{36} times greater than the gravitational force between them. Secondly there is the number of charges. Gravitational charges only come in one form, which is always attractive. The effect of an accumulation of massive objects is to produce a strong force, despite gravity's comparative weakness. Electric charges come in two forms, as already remarked, and the two forms — positive and negative — cancel each other out when brought together by their mutual attraction. The effect of an accumulation of positive charges *or* negative charges is to produce a colossal force; the effect of an accumulation of positive charges *and* negative charges is to produce a weak force or no force at all, despite electromagnetism's comparative strength.

It is this second difference between gravity and electromagnetism which is crucial. The idea behind investigating electric charges was to explain the short-range force between molecules. Yet so far all that has been found is another force of infinite range. The existence of two types of mutually cancelling charges is what makes a short-range form of the electromagnetic force theoretically viable.

Suppose each molecule contains an equal amount of positive and negative charge. The attraction of the positive and negative charges will keep them nicely together inside the molecule. The repulsive effect each negative charge has on the negative charges of neighbouring molecules will be cancelled out by the attraction each positive charge has for them. The net result is that there will be no overall electromagnetic force between the molecules.

Next suppose that the positive and negative charges inside a molecule have to obey the law that no two things can be in the same place at the same time; that is, the charges have to be in slightly different locations within the molecule. If this is the case, then when one molecule gets very close to another, and only then, it will begin to feel the other's electric charges. In effect, it will find the attraction and repulsion of the other molecule's electric charges to be coming from slightly different directions and have slightly different strengths. As a result, the repulsive and attractive forces it experiences will no longer exactly balance. This is how short-range inter-molecular forces can be hypothesized to arise.

The notion of electric charge and electromagnetic force is a versatile one. Together with the rule that two things cannot be in the same place at the same time, it can be used to account for all the contact forces so far encountered. All contact forces are essentially just short-range forces between the molecules on the surface of the objects in contact. Whether the contact force is friction, suction, collision, or whatever, it can be described in terms of the short-range forces being exchanged by opposing surface molecules. Consequently, it may be proposed that there are only two fundamental forces in nature: gravity and electromagnetism.

ELECTRONS

The discovery of electromagnetic force and electric charge, tremendous advance though it is, is not yet satisfactory. A detailed explanation still needs to be put forward showing how the short-range by-product forces can be both the cause of an attraction between molecules (e.g. in liquids) and a repulsion between them (as when molecules collide and bounce apart). The formulation of a theory which can account for both these opposite effects hinges on describing exactly how electric charges are distributed inside molecules.

Study of the removal of electric charges from molecules, such as takes place when some kinds of objects are rubbed together, leads to the conclusion that only negative electric charges are ever dislodged by such means. This suggests that negative charges are in some way loose within the molecule. Furthermore, it was discovered that, unlike gravitational charges, negative electric charges come in lumps of a fixed size. Each negative electric charge has the same amount of electric charge as every other negative electric charge. And when the mass of each of these negative electric charges is measured it is also always the same. Each negative electric charge is a tiny particle of mass and electricity of fixed size. These particles, all of them exactly identical, are called *electrons*.

The picture that has emerged of the place of electrons in molecules is that they are found on the outside. The counterbalancing positive charges are inside. The electrons, being loose, are not fixed in place but have the freedom to move around on the surface of the molecule. This is the electric charge distribution which must be used to explain what happens when two molecules attract one another; and to explain what happens when they collide and bounce off each other.

THE SHORT-RANGE ATTRACTION

Picture a typical molecule. It contains a number of atoms, each one consisting of a distinct quantity of positive charge. Over the surface of these positive charges move the electrons. Now, averaged over a period of time the electrons are spread evenly over the outside of the molecule. But at any one instant the electrons may by chance be found to be more on one side of the molecule than the other. For this brief instant the molecule will look electrically charged to its neighbours. Those that are on the side where the electrons have clustered will find it looks like a negatively charged molecule. Those on the opposite side will find it looks like a positively charged one.

This situation leads to an attractive force in the following way. Consider three neighbouring molecules, B, C and D. Let C be the one whose electrons have briefly clustered, and let the side they have clustered on be the side adjacent to D. D, seeing C's apparently negative charge, will find its own electrons are repelled to the far side of it by C's electrons. This partially exposes D's internal positive charges which are then attracted to C's negative charge. D thus finds itself pulled towards C. Molecule B, on the other side of C, sees C as being apparently positively charged. B's electrons are attracted to C's positive charge. Thus B too finds itself pulled towards C. So C is attractive to both its neighbours.

The tendency of electrons to cluster on one side of a molecule is usually transitory; that is, any clustering which develops rapidly evens itself out again. However, while it lasts it produces the stickiness which gives rise to liquids, and also some solids. (In the case of solids there are in fact several different ways by which molecules are locked into place. Electrons are involved in all of them. See, for example, ionic bonds in section 5.1.)

COLLISIONS

So much for the short-range attraction. When one molecule experiences an attractive force from a neighbour it begins to accelerate towards it. Eventually the two are likely to collide. In the collision the two molecules bounce off one another. The act of bouncing amounts to a change of direction, which must involve acceleration away from the object being bounced off; i.e. a repulsion. That in turn requires a force to operate, producing the acceleration. The source of the force must once again be the electrons on the outside of the molecule.

Although electrons can cluster on one side of a molecule, there is a limit to how far this process can go. A molecule's internal positive charges are never fully exposed. Consequently, as two molecules approach each other as a result of the short-range attraction, there comes a time when their respective sets of electrons begin to come into contact. They then run up against the law that no two things can be in the same place at the same time. This results in a repulsive force between the electrons of one molecule and those of the other. As the collision proceeds, the repulsion rapidly overwhelms the earlier attraction. The molecules then accelerate away from one another. They have, in effect, bounced apart.

This description of inter-molecular collisions carries with it the consequence that molecules never touch. The outside of a molecule is not solid but is made of the electrons moving loosely around the interior positive charges. Collectively these electrons more closely resemble a cloud than a solid barrier. When two molecules meet, they are prevented from merging or passing through one another not by a hard impact but by the inability of their respective clouds of electrons to be in the same place at the same time.

(A note for enthusiasts: the repulsion arising from different electrons' inability to be in the same place at the same time is *not* of electromagnetic origin. Electrically neutral particles such as neutrons (see section 3.1) also experience the same repulsion arising from the same inability. In essence, for two particles to be in the same place at the same time requires one of them to increase its potential energy. This means it must reduce its kinetic energy; i.e. slow down. Which is tantamount to its being repelled.)

CHEMICAL REACTIONS

It was seen, when investigating the concept of energy as it applies to molecules, that success was accompanied by a bonus; namely, an explanation of what heat and temperature are. Investigating the concept of electromagnetic force also brings a bonus with it; namely, an explanation of the entire subject known as chemistry.

Perhaps the most fundamental observation made by chemists is that molecules can be combined together in what are called chemical reactions. Use has already been made of this in section 1.3. How would a process of this kind appear when viewed in terms of electromagnetic force and molecules made of electrons plus interior positive charges?

Consider again the chemical reaction:

$$A + B \rightarrow C.$$

At a molecular level there are the two molecules, A and B, initially leading separate existences. (A and B could equally well be atoms. For simplicity, and because the vast majority of substances are molecules, I will continue to refer to 'molecules' rather than the more accurate 'atoms or molecules'.) The two molecules will begin to feel the usual short-range attraction when they approach each other to within 10^{-8} m — that being the experimentally determined maximum range of the inter-molecular stickiness. As a result, they will get closer still until their clouds of electrons begin to overlap. Repulsion then sets in.

Now, although the repulsion between colliding molecules is normally about a thousand times stronger than the attractive force which produces inter-molecular stickiness, it does not increase without limit as the molecules approach. Eventually, if they manage to get close enough to each other, their two sets of electrons will begin to merge. When that happens, the electrons avoid their being in the same place at the same time by redistributing themselves. As the electron redistribution round both molecules occurs, it once again produces a net attractive force. This renewed attraction is what binds A and B together in the molecule of C. Thus *every chemical reaction is at root a rearrangement of electrons.*

To make the position clearer, the sequence of events can be represented diagrammatically. Figure 1.1 is a graph of the sum of the kinetic energies of A and B as they approach each other. Note that as the energy involved in chemical reactions is so much greater than that producing the inter-molecular attraction, the latter is too feeble to show on the graph.

Consider the numbered points on the graph in turn.

1. The two molecules are rushing towards each other. Their velocities are high. They have much kinetic energy.
2. The repulsion between their respective sets of electrons slows the two molecules down. Their kinetic energy falls, being converted into potential energy. If it falls to zero before point 3 is reached, no chemical reaction occurs. The stored potential energy is converted back into kinetic energy as the two molecules accelerate away from each other. The molecules have collided and bounced.
3. If the kinetic energy does not fall to zero before point 3 is reached, then the chemical reaction begins to take place. The molecules continue to approach each other, their sets of electrons starting to rearrange themselves as they pass point 3.
4. The chemical reaction has taken place. The kinetic energy of the two molecules has increased as a result of the rearrangement of their two sets of electrons. Depending on the type of molecules that A and B are, they may now have more kinetic energy than they started with (4a), or less (4b). In either case, the two molecules continue to approach each other.
5. Eventually repulsion sets in. This arises because the interior parts of the two molecules are much more resistant to being rearranged than are their clouds of electrons. Consequently they repel each other. The kinetic energy is converted into potential energy and the molecules come to a halt.

The graph of kinetic energy and distance can also be drawn from the point of view of potential energy. See figure 1.2. As long as the existence of only two kinds of energy, kinetic and potential, is recognized, the law of conservation of energy requires that the sum of the potential energy and the kinetic energy never changes. Figure 1.2 is therefore a mirror-image of figure 1.1. Whenever the kinetic energy goes down, the potential energy goes up by exactly the same amount, and vice versa.

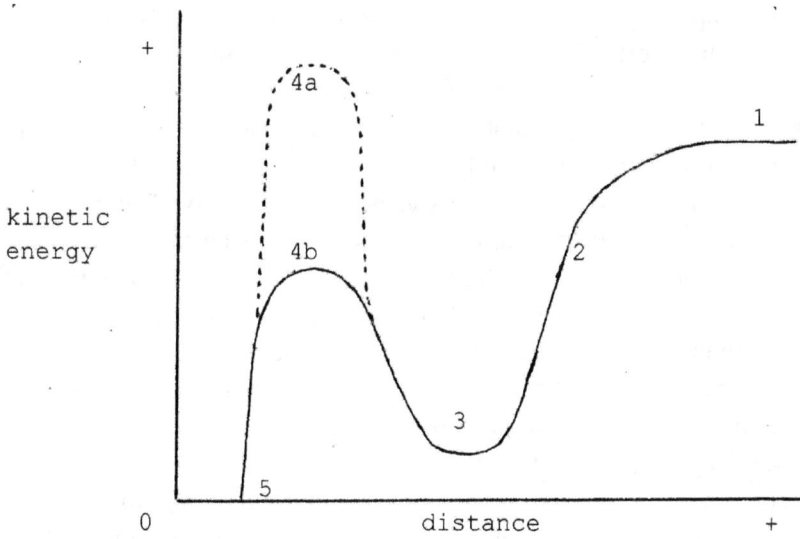

Figure 1.1 The kinetic energy of two molecules as they approach each other and undergo a chemical reaction.

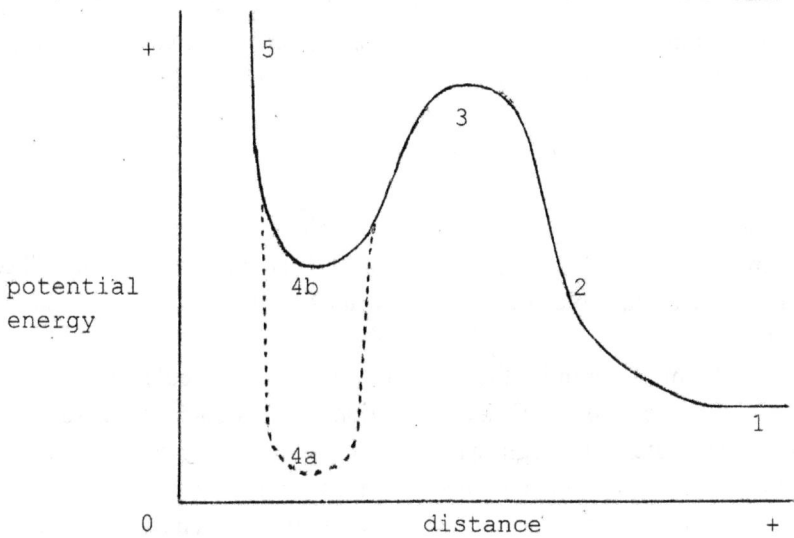

Figure 1.2 The potential energy of two molecules as they approach each other and undergo a chemical reaction.

As colliding molecules go from point 1 to point 3 on the potential energy graph they slow down, much as a ball does as it rolls up a hill, and for the same sort of reason. Because of this, point 3 on the potential energy graph can be thought of as being at the summit of a potential (energy) hill. Similarly, points 4a and 4b lie at the bottom of a potential (energy) well. For a molecule lying at the bottom of the well to break up — that is, for its constituent parts to become free of one another — the molecule must have enough kinetic energy to roll up the hill to point 3 without stopping and falling back down to the bottom. A stable molecule is one with all its constituent atoms trapped inside the potential energy well with insufficient energy to get from point 4a or 4b to point 3.

The graph of potential energy against distance represents one of the most important chemical properties of molecules. Just as all objects attracted by gravity try to turn their gravitational potential energy into kinetic energy by falling, so too do molecules try to turn their electromagnetic potential energy into kinetic energy. In graphic terms, they try to get as low as possible on the potential energy graph. The shape of the graph is such as to prevent molecules at point 4b from breaking up into their constituents at point 1, and to prevent molecules at point 1 from combining into larger molecules at point 4a. The potential energy well is what makes molecules stable. As we observers are ourselves made almost entirely of stable molecules, we owe our very existence to it.

EXOTHERMIC REACTIONS

This neat theory of how molecules combine in chemical reactions contains, unfortunately, a catastrophic flaw. This is most easily described with respect to 4a molecules; that is, to those with less potential energy than their constituents.

Consider more carefully the reaction of the two molecules A + B → C. Take it to be the case that C in this reaction has less potential energy than A and B separately. To begin with, A and B will happily co-exist without reacting so long as collisions between them aren't energetic enough to take them from point 1 to point 3. External energy needs to be supplied for this to happen. If they do come to be given enough energy to reach point 3, they combine, and in so doing fall to the bottom of the potential well at point 4a. The conversion of potential energy into kinetic energy this entails means that A and B are now travelling towards each other at quite a speed. They expend their kinetic energy by climbing up the far side of the potential well to point 5 where they come to a halt.

So far so good. And now for the catastrophic flaw. As has been seen, objects with high potential energy — as at point 5 — always try to turn their potential energy into kinetic energy. Consequently the newly formed larger molecule at point 5 falls back down into the well to point 4a. When it reaches this point its constituent parts A and B are travelling exactly as fast as they were before (because energy is conserved), but this time they are heading away from each other, towards point 3. Since they had enough kinetic energy to get over the hill at point 3 on the way in, they must have enough to get over it on the way out (because energy is conserved). So they pass point 3, then point 2, and arrive back at point 1, completely separate molecules again. What has happened can be written

$$A + B \rightarrow C \rightarrow A + B.$$

Far from the chemical reaction producing a stable molecule, it would seem to have led to nothing more than a very brief liaison. As it stands, the theory of chemical reactions doesn't work. It is not in accord with observation.

What is needed to save the situation is a means whereby the two reacting molecules arriving at point 4a can convert their kinetic energy into some other form — a form which can be expelled from the newly created molecule C without taking the two molecules, A and B, with it. The energy expelled in this way would enable energy to be conserved while at the same time permitting chemical reactions to produce stable molecules rather than transitory ones.

What form could the expelled energy take? The obvious hypothesis is that the newly combined molecule passes its energy to nearby molecules in the form of molecular kinetic energy; i.e. heat. The nearby molecules would then function as an accessory to the reaction between A and B making C, playing a mainly passive part in the molecule-forming collision and emerging from it travelling much faster than previously.

As usual, observation decides the issue. It turns out to be readily observable that when molecules at point 4a are formed, their surroundings get hot. Chemical reactions which heat their surroundings are called *exothermic reactions*. Observation of this heat shows that the large amount of kinetic energy produced in arriving at point 4a is expelled from the newly-formed molecule C and transformed into an increase in the speed of its neighbours. Thus the flaw in the theory of chemical reactions is removed. Chemistry can safely be seen as a part of the larger subject of molecular physics. It can confidently be stated that the chemistry of any given substance is the physics of the electrons on the outside of its molecules.

1.6 WAVES: A NEW FORM OF ENERGY

Providing a theoretical basis for the science of chemistry is not the only bonus the concept of electromagnetic force brings with it. It also provides one other scientific benefit which it would be useful to end this chapter by describing. It is a third form of energy to add to kinetic and potential energy.

Imagine a particle of some kind producing a force. It doesn't matter whether this is a particle with mass producing gravitational force, or one with electric charge producing electromagnetic force. Near this particle a second charged particle is placed. The second particle feels a force due to the first one. The direction of the force lies along a straight line connecting the two of them. This is always the case no matter where one particle is in respect to the other. So any particle which is a source of force can be pictured as having an infinite number of straight lines coming out of it in all directions. These imaginary lines, which represent the direction of the force, are called *lines of force*.

All the time the source particle is stationary or moving with a constant velocity, the lines of force are also stationary or moving with the same constant velocity. But what happens to these lines when the particle concerned changes its velocity? Whatever causes the particle to change its velocity accelerates the particle but not its lines of force. The particle will initially leave its lines of force behind. They will become bent and stretched. To help visualize the process, see figure 1.3. Remember that lines of force produced by a particle are imaginary lines along which other particles feeling the force are urged to accelerate.

What has this to do with energy? Sometimes charged particles are forced to undergo rather drastic changes in velocity. This can happen, for example, when fast-moving molecules collide. Although somewhat over-simplifying things, the charged particles can be thought of as being shaken; that is, accelerated first in one direction, then in the opposite direction. This produces ripples, or waves, in the lines of force that

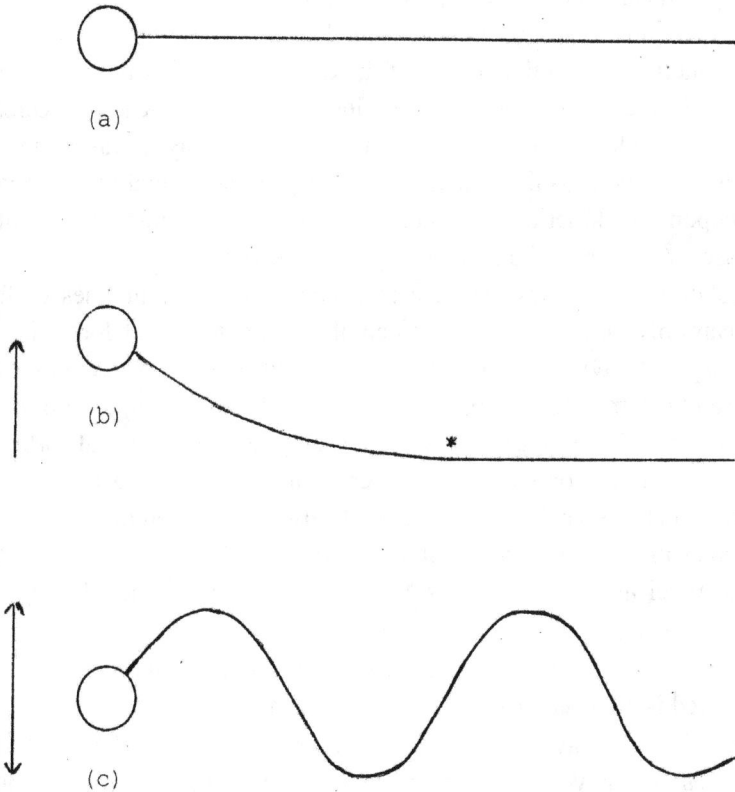

Figure 1.3 Lines of force associated with a source of charge moving in various ways.
(a) A line of force arising from an unaccelerating charge.
(b) A line of force arising from an accelerating charge. Points to the right of *
have yet to reflect the acceleration.
(c) A line of force arising from a vibrating charge.

emanate from the particles. These waves are able to carry away some of the kinetic and/or potential energy of the particles as they accelerate.

It may not be immediately obvious that waves are a form of energy. But think of making waves on the surface of water. It is clear that energy must be put into the water to make the waves. And anyone who has ever been buffeted by sea waves knows that they can cause acceleration and must therefore possess energy. Waves in water thus take energy in the making and can yield that energy up in appropriate circumstances.

So it is with waves in lines of force. They are made through reductions in the kinetic or potential energy of the object generating the wave. When a charged object makes a wave, its kinetic or potential energy decreases. Conversely, when a charged object is encountered by a wave, the wave may be absorbed — the exactly opposite process to making a wave — whereupon the kinetic and potential energy of the intercepting object increases. As always, the total energy is conserved.

It should be emphasized here that since waves are in lines of force, they can only be produced by objects that are a source of force; i.e. that are charged. Gravitational waves are produced only by objects possessing mass, and electromagnetic waves are produced only by objects possessing electric charge. Similarly, waves can only affect or be absorbed by likewise charged objects. Again, gravitational waves affect massive objects; electromagnetic waves affect electrically charged ones.

Waves in lines of force travel at a well-defined speed. Electromagnetic waves travel at $2 \cdot 998 \times 10^8$ m s^{-1}. The speed of gravitational waves has not been measured — indeed gravitational waves are so feeble it is only recently they may have been detected — but for strong theoretical reasons their speed is also believed to be $2 \cdot 998 \times 10^8$ m s^{-1}.

The idea that waves can be formed in imaginary lines of force without the waves themselves being imaginary is an outlandish one which requires careful experimental verification. The wave energy given out by an accelerating or vibrating charged object must be detected. To detect such a wave requires a means of converting it back into kinetic energy. This is because observers, and all measuring devices, are made of massive objects. The only way wave energy can affect massive objects is to increase their kinetic energy, either directly or via their potential energy as an intermediate step. So to detect the energy of a wave, it is necessary to measure the kinetic energy it produces in objects that are affected by it.

Detection of wave energy thus involves the reverse of making the wave in the first place. It relies on charged objects being able, as already remarked, to intercept the wave and turn it into kinetic energy.

Detection is quite straightforward. If electrons, for example, make the waves, so electrons in molecules can detect them. When an electromagnetic wave encounters an electron, it may give it so much kinetic energy that it knocks it right out of the molecule. Such an event can be detected with suitable equipment. Alternatively it may simply jolt the electrons in a collection of molecules, indirectly causing inter-molecular collisions to become more energetic and thereby increasing the temperature of the molecules. Again this is an observable effect.

Closer study of electromagnetic waves reveals that, like waves in water, they can possess a variable amount of energy. They can be large or small — a property known as *amplitude*. They can also be close together or far apart — the property in this case being called *wavelength*. Electromagnetic waves with a larger amplitude and/or a shorter wavelength carry more energy than those with a smaller amplitude and/or a longer wavelength.

Electromagnetic waves of many different lengths have been investigated. In one particular wavelength range — around 500×10^{-9} m — they have just the right amount of energy to affect electrons in certain chemicals located at the back of the human eye. This effect is detected by the brain and experienced as sight (see section 6.4). In other words, *light* is one kind of electromagnetic wave. The twin concepts of electromagnetic force and electromagnetic waves provide a theoretical basis for the science of light.

1.7 OBJECTIVE REALITY AT THE END OF CHAPTER 1

It will be helpful to end this chapter by taking stock. What are the components of objective reality as they are now conceived?

Four fundamental components have been identified. They are mass, force, energy and conservation laws.

- <u>Mass</u> is something possessed by every object, from the tiny electron up to stones and stars.

- Aside from the electron, every object is made of one or more atoms, usually grouped together in molecules.
- Atoms are the basic units of mass. There are about 100 different types of atom so there are about 100 different types of mass, plus the electron.
- Massive objects can move with respect to one another. When they do so they possess a velocity. A change in the velocity of a massive object is an acceleration.

- Force is the means by which one object interacts with another. One object can affect another only by means of force.
- There are two fundamental forces, gravity and electromagnetism. To be a source of gravitational force an object must have a gravitational charge. This is identical to its mass. To be a source of electromagnetic force an object must have an electric charge. Similarly, to respond to another object's force, the responding object must have the same sort of charge. An object with no electric charge, for instance, is completely unaffected by electromagnetic force.
- The more charge an object has, the more force it produces and the more strongly it responds to other objects' force.
- There is only one sort of gravitational charge, which produces an attractive force.
- There are two sorts of electric charge, negative (possessed by electrons) and positive (found inside atoms). The force between like electric charges is repulsive; that between opposite electric charges is attractive.

- Energy is passed between charged objects when they feel the force due to each other's charges. If force is how two objects affect each other, energy is what they affect each other with.
- There are three types of energy. Potential energy is possessed by an object which feels another's force but has not fully responded to it. Kinetic energy is possessed by an object because it is moving. Wave energy is a kind of pure energy, separate from objects but associated with the force they produce. As there are two fundamental forces, there are two corresponding forms of wave energy, gravitational wave

energy and electromagnetic wave energy. Light is a form of the latter.

- The three kinds of energy may be changed one into another.

- <u>Conservation laws</u> govern the ways mass, force and energy can interact.
- There are two conservation laws: one for mass and one for energy. These state that the total amount of mass in any given situation never changes; neither does the total amount of energy.

There are, then, around 100 types of mass, possessing one sort of gravitational charge and two sorts of electric charge. These masses interact by means of two forces, gravity and electromagnetism, and exchange energy in three forms, kinetic, potential and wave, with all this activity being governed by the two laws of conservation of mass and energy.

Given the confusion and complexity of the terrestrial world, it is astonishing that it can be reduced to the interplay of a mere 100 masses, 2 forces, 3 types of energy and 2 laws of conservation. The original — and scientifically fundamental — hypothesis that there is some orderliness behind the world's apparent chaos has been borne out far beyond initial expectations. A whole range of subjects which were quite distinct historically have all proved to be different aspects of this law-abiding interplay of mass, force and energy. Heat, light, chemistry, electricity and magnetism have all become part of a single theoretical structure, which this chapter has called dynamics.

One of the most significant things about the bringing of the various different subjects into the one theory of dynamics is the very fact that it can be done at all. There is no obvious reason why such disparate disciplines should prove to be related. That the knowledge gleaned in each subject area is compatible with all the others, and can be shown to spring from an embracing theory of dynamics, is very encouraging. At the least it gives no reason to think that science is on the wrong track.

This chapter set out to discern the simplicity and orderliness underlying the turbulence of the terrestrial world. That task has been accomplished. Scientists' belief in its being possible has been well and truly shown to be justified.

But why stop with the terrestrial-scale world? The challenge is to go on and apply scientific method to all other aspects of reality: to the inside of atoms; to the universe at large; and to ourselves. These topics will be tackled starting in chapter 3.

In the next chapter, however: a digression. Having used the tools of the scientific trade to such impressive effect, and having hopefully thereby convinced you that scientific method deserves to be taken seriously, it would be worth giving the tools of science a closer look. What *is* scientific method? What are its basic assumptions? What are its limitations? What exactly are hypotheses and laws and theories? *Why* does scientific method work? What makes it superior to other ways of disclosing the truth about existence? To a degree, the value of scientific knowledge rests on answers to these questions. It would accordingly be as well to address them sooner rather than later.

2

SCIENTIFIC METHOD

2.0 OBJECTIVE: To examine the means by which we are endeavouring to discover the nature of reality.

We begin by considering two kinds of reality, subjective and objective. Appreciating the differences between them — and the similarities — will make it possible not only to state the most basic assumption of science, but also to advance an essentially pragmatic justification of it. Next, an account will be given of how we deal innately with the objective world. It will be found that this innate approach to reality contains within it the seeds of scientific method, which will then be described using a simple example. Having established what the rules of scientific method are, we shall go on to look at various aspects of the method, both in terms of its principles and its practice.

2.1 SUBJECTIVE REALITY

Let me begin at the very beginning. I exist. I know it beyond any possibility of doubt. Indeed for me to make any statement about anything requires as a prerequisite that I must exist. It is absolutely certain to me that I do. In the same way, every thought I think also exists, as do my impressions of the world around me, and my desires in respect of that world. Then there are internal experiences, things such as pain, happiness, hunger, love and other emotions, and my dreams. All these various things are components of my existence. They are real by virtue of my experiencing them.

This kind of reality is called *subjective reality*. It is entirely personal to me and is not available to anyone else. In reciprocal fashion, other people's experiences are entirely personal to them. I cannot know what it feels like to be someone else; I may imagine it as part of my own experience, but I cannot know it directly. I cannot experience someone else's existence and they cannot experience mine. Such is subjective reality. It is something I have sole possession of and sole access to.

Subjective reality is a fitful kind of reality. When I go to sleep I cease to be aware of the existence either of myself or of the things I experience while awake. I, and they, cease to exist. On the other hand, while asleep I experience dreams. Dreams are subjectively real. The extent to which dreams diverge from wakefulness serves to emphasize subjective reality's fitful nature.

If subjective reality was the only kind of reality then nothing would be real except my experiences. To talk of things existing in any other form would be meaningless. I would have to conclude for instance that other people only exist while I am experiencing them as part of my own existence; that is, they exist in my mind and nowhere else. In writing this book I would only be talking to myself. I may even be right in thinking this.

But subjective reality is not the only possible kind of reality. It is reasonable to propose that the sensations I am aware of subjectively — vision, hearing, taste, smell, touch, etc. — are representative of objects which exist independently of my experiencing them. If that is the case, then the world exists whether I happen to be experiencing it or not. This kind of reality is called *objective reality*.

There is the potential here for these two kinds of reality to be mutually exclusive. Subjective reality consists purely of my experiences. Therefore the world which I experience exists within me. Objective reality, conversely, consist purely of objects and events. Therefore I exist as an object taking part in events within the world. Clearly the universe cannot be both inside me (subjective reality) and bigger than me (objective reality). There is a choice to be made. Is the universe one part of me or am I one (tiny) part of the universe?

The key to resolving this most fundamental of questions lies in one crucial feature of subjective reality, namely the extent to which it is constrained. The nature of the constraints can best be appreciated in a situation in which they are absent, during dreaming.

The events that take place in dreams are entirely capricious. When compared to events occurring during wakefulness, their lack of constraints is obvious. Events in the wakeful world are rational and consistent while events in dreams are neither of these things. Now if both the wakeful world and the dream world exist purely subjectively it is not clear why one should differ from the other. That they plainly are different suggests the wakeful world is not exclusively subjective but is being constrained by something external to subjective reality — something which forces wakeful subjective reality to be rational and consistent. It is objective reality which provides that external constraint.

It must be made clear that the existence of objective reality is not proved by this line of reasoning. Since all knowledge is acquired through subjective experiences, it is impossible for me to establish the objective existence of anything, and that includes the existence of any supposed external constraint.

This fact — that it is impossible to prove the world is objectively real — can be turned on its head. It is also impossible to prove it is not objectively real. Certainly the constraints imposed on wakeful subjective reality force me to behave at all times as if the world exists independently of me. Say I am hungry. In that case, I must put food in my mouth in order to cease being hungry. I cannot bring the food into existence just by thinking about it. Instead I must go through a set of subjective procedures — going to the cupboard and opening a tin, for example. These procedures are indistinguishable from those I would have to go through if the world is objectively real. Such are the constraints imposed on wakeful subjective reality. They are both rigorous and (in my experience anyway) inviolable.

Since I must, in fact, treat the world as if it is objectively real regardless of whether it actually is or isn't, nothing is gained by denying objective reality. I shall therefore not deny it. The apparent contradiction between a subjective reality made of experiences and an objective reality made of objects and events will be resolved by taking the relationship between them to be as follows.

- Subjective reality exists within objective reality; that is, I exist within the universe.
- At the same time, an *image* of objective reality exists within subjective reality; that is, I experience an image of the universe within me.

All this may seem trivial. Nobody seriously denies the existence of objective reality. Yet the assertion that subjective reality has an objective counterpart is arbitrary and unprovable and it is important to be clear on this point since scientific method is wholly concerned with objectivity. Acceptance of the postulate of objective reality is an essential precondition for the pursuit of science. In this book it will be accepted unreservedly from here onwards. The universe is objectively real.

Before proceeding to look at the basis of scientific method it would be worth emphasizing one very important aspect of the relationship between subjective and objective realities. It concerns the assertion that the subjective world contains an *image* of the objective one. Images represent objects without being equal to them. As a result it is clearly possible that an image may be inaccurate or distorted or just plain false. Consider, for example, the effect on mental images of experiencing such things as optical illusions, hallucinations and mirages. The subjective experience of these things is absolutely certain. When I experience a perception I do not doubt its subjective reality. However, there is an element of doubt about the accuracy with which images in my mind represent what is happening in the objective world. Consequently I cannot be certain that any given perception exactly corresponds to something objective. There is always the possibility that my subjective image of an object or event is partly or wholly false.

To put this more rigorously, an observer may describe the situation at some place (*x*) and time (*t*) by a statement. The statement of what is actually, objectively the case at *x* and *t* can be symbolized as $O(x,t)$. The statement '$O(x,t)$' is the truth. But observers can only know of what is at *x* and *t* through the impact it makes on their senses. The statement of what is experienced to be the case at *x* and *t* can be symbolized as $S(x,t)$. $S(x,t)$ is an observer's image of $O(x,t)$. Given the unreliability of images, it can clearly not be asserted with certainty that $S(x,t)$ exactly corresponds to $O(x,t)$. The subjective image formed of any particular aspect of objective reality does not necessarily match the objective truth. Whenever an observer makes a statement about the objective world, an equality is assumed between $S(x,t)$ and $O(x,t)$ which may be unwarranted. It should always be kept in mind that objective reality and an observer's image of it may not correspond. We may sometimes be mistaken.

2.2 THE BEHAVIOURAL BASIS

Given the objective reality of the world, people — and to varying degrees most other living things — instinctively seek to acquire knowledge about it. The reason for this is not hard to see. The world is a dangerous place in which all manner of challenges to our physical integrity may arise. Failure to meet these challenges successfully is often subjectively painful. Pain is a punishment to be avoided. Even worse, failure can be fatal, and death is also to be avoided, especially if it appears imminent. Our chances of avoiding pain and imminent death are greatly enhanced if we know about the kinds of challenges we can expect to encounter. We are then less likely to be taken by surprise or off our guard and more likely to respond correctly to a dangerous situation. It is thus in our interests to acquire knowledge about the world we live in.

How do we set about this? The process can be divided into five steps.

1. Observation. The first step is taken as soon as our sensory organs are able to provide us with information, albeit possibly distorted information, about our surroundings. By means of our sensory organs, especially our eyes, we observe the world about us.
2. Recognition. Images of the objects we observe are remembered so

that the objects can be recognized. If, for instance, a particular object is found to be harmful, then providing it can be recognized when it is next encountered, steps can be taken to deal with it before it can do further harm.

3. Classification. Different objects are classified into groups according to their similarities. This is beneficial in the following way. Suppose a particular object is found to be useful. Clearly all other identical objects will be recognized as being equally useful. But what about objects which are quite similar but not identical? (Goats and sheep, for example.) Without classification, each similar but not identical kind of object would have to be investigated from scratch. But by noting the similarities and classifying objects into groups by what they have in common, it is possible for us to know about objects never seen before. All that is required is that they resemble something already familiar. If we know dogs are carnivores, then the first time we encounter a wolf or a fox we can be pretty sure they are carnivores too.

4. Prediction. It may not be enough to identify something as dangerous. Suppose for instance we happen to be in the unfortunate position of being in the direct line of sight of a crocodile. We may immediately recognize the crocodile and classify it as dangerous, but then what? Unless we can predict how the situation is going to develop, we run the risk of hastening calamity rather than averting it. If the crocodile is heading away from us it is probably better to keep still than to flee. Conversely, if it is approaching, keeping still is unlikely to be the best course of action. On the other hand, if there is an electrified fence between us and the crocodile we can probably ignore it whatever it is doing. The ability to predict enables us to know what course of action is likely to be most appropriate.

5. Explanation. In the course of the above four activities we ask questions about things. Through asking questions — by being curious — we come to take things to pieces, to experiment with them, to see what they can do. And by formulating answers to our questions, we take the fifth step in acquiring knowledge about the world: we construct explanations. This last step is beneficial because it can feed back into classifications and predictions, making them more reliable and accurate.

These then are the steps we take in order to deal with our environment. Firstly we observe what takes place around us. This involves all our senses but especially vision. Secondly we learn to recognize objects and thirdly to classify them by what they have in common. Fourthly we learn to predict how objects will affect each other, how they change as time passes, and so on. Fifthly we try to establish why particular objects behave in the way they do, why they possess this or that characteristic, and how their properties arise.

The sequence of these five steps is not fixed. All of them interact with each other. Acquisition of knowledge about the world is achieved by observing, recognizing, classifying, predicting and explaining in such a way that each of the five processes draws on and gives support to the other four.

Observations, recognitions, classifications, predictions, and explanations can all be expressed in the form of statements. These statements are subjective; that is, the statements are arrived at as a result of subjective experience. To use the formalization of section 2.1, they are $S(x,t)$ statements. When we hold them to be true we are asserting that our image of objective reality, $S(x,t)$, corresponds to the real thing, $O(x,t)$.

It can be argued that our innate abilities of observing, recognizing, classifying, predicting and explaining will indeed lead most of the time to $S(x,t)$ statements which correspond closely to objective reality. This is because discrepancies between reality and our subjective image of that reality would produce inappropriate behaviour on our part. If what we think is happening is not what is actually happening, any action we undertake is likely to be mistaken. And mistakes in a dangerous environment are potentially fatal. The fact that the great majority of us do manage to cope satisfactorily with the environment over long periods of time indicates that our subjective image of objective reality is at least close to the truth.

The argument breaks down when statements about things of little or no relevance to personal survival are considered. Despite survival not depending on our making observations or constructing explanations in such cases, our inborn abilities in these respects lead us to do so anyway. Statements about how the universe began, about the core of the Earth, about bats being birds or mammals, about gods, are examples in this category. Where personal survival is not at stake as a result of the failure of S(*x,t*) to correspond to O(*x,t*), the discipline imposed by our interacting with the world disappears. It becomes possible for different people to create widely varying subjective images of some particular O(*x,t*) without there being any untoward consequences. And precisely because in these circumstances the existence of S(*x,t*) statements which are untrue does not impair the survival prospects of people who subscribe to them, such statements are able to persist and spread indefinitely. A situation can then arise in which there may be widespread disagreement about which S(*x,t*) statement out of the many purporting to correspond to a single O(*x,t*) statement is the true one. The problem is how best to resolve these disagreements. This is where scientific method comes into play.

2.3 QUESTIONS AND ANSWERS

TWO REASONS TO USE SCIENTIFIC METHOD

There are a number of methods available to determine which of two or more rival subjective statements best corresponds to objective reality. In the absence of science the most widely used tends to be duress. An individual or group of people find themselves being bludgeoned, metaphorically or literally, into agreeing that their subjective viewpoint is less realistic than the viewpoint of those disagreeing with them. The most extreme form of this method of determining truth is trial by combat.

This book rejects such options in favour of scientific method. Can this decision be justified? From a position of hindsight the answer is a clear yes. The truths determined by science have enabled humans to transform vast areas of the Earth for their short-term benefit. No other approach has ever had anything like as much effect. Put at its simplest, scientific method works.

Scientific method can also be justified in that it bears a striking similarity to the innate processes encountered in the previous section (as will be seen shortly). This is not surprising, for science succeeds in large part by enhancing our prospects of survival and our personal comfort, which is exactly what the innate processes do. The innate processes *must* be effective in matching subjective image to objective reality if we are to avoid the pain and death that the objective world can inflict on us. They are in a sense fashioned by the very objectivity that they seek to comprehend. So any method which bears a close resemblance to them can also be expected to be largely successful in matching image to reality. The choice of scientific method as the means of determining truth is therefore rational as well as effective.

AN EXAMPLE OF SCIENTIFIC METHOD

When someone draws up a statement about some aspect of objective reality, what are they doing? How do they produce the statement?

Such statements are always answers to questions. The questions are things such as: what is it? what is it made of? how does it work? why does it do that? what is it going to do next? The answers, except in the simplest cases, consist of mentally 'dismantling' the target of the question and describing connections amongst its dismantled parts which then come together to produce whatever it is that gave rise to the question. Since mentally taking things and events apart and rebuilding them is easy, all manner of statements can be drawn up in answer to any one question. Choosing the correct answer from amongst a range of different statements is equally easy where immediate practical issues are involved. But for matters of a more academic nature the choice is harder to make.

To illustrate how science goes about resolving disputes over primarily academic matters, here is a straightforward question and various possible answers to it. The ensuing discussion will provide a basic illustration of what the method of science is.

The question is this. The sky is nicely consistent, blue with sun during the day, black with stars at night, the moon following its own regular day-and-night pattern, clouds coming and going randomly. And then one day there's a rainbow. This is not a common sight and is not random, being a geometrically pure arc of a circle. It demands to be explained. What is it?

Now, in isolation the phenomenon contains nothing that will assist in creating the desired answers. It is necessary to broaden the observations to include the context in which the rainbow was seen. The rainbow and its context are the target of the question: the thing that must be mentally dismantled and reassembled. The main problem that arises immediately is that there is an awful lot of context. Which parts of the context are connected to the rainbow and which parts are nothing to do with it? The need here is for observations of rainbows on different occasions, searching for common factors in the various contexts encountered. It might, for instance, be noticed that the appearance of some objects in the sky is related to the weather. It might also be noticed that the mood people are in is affected by how the sky looks. So when observing rainbows particular attention is paid to these factors, and to any others that might be relevant. Say for the sake of argument that after seeing rainbows on several different occasions only two things are observed to be common to each case: that the weather was rainy, and that a lot of people were in a bad temper.

So far the process has been no more than guesswork. Any answers now formulated to the question of what causes rainbows will be guesswork likewise. The use of the term 'guesswork', incidentally, is not intended to be disparaging. It merely serves to emphasize that at this stage any explanation, while constrained by relationships noted in the observations, is nothing more than conjecture. The scientific name for a guess of this kind is *hypothesis*.

So here are a number of hypotheses which might provide a rudimentary answer to the question of what causes rainbows.

Hypothesis 1: a rainbow is produced by rain; when it's raining a rainbow can always be seen somewhere.

Hypothesis 2: a rainbow is produced by a supernatural being — God (a word to be defined and discussed in section 10.4) — to indicate that the rain isn't about to be in such quantity that the world is drowned by a deluge.

Hypothesis 3: a rainbow is an unreal hallucination seen by people when the weather makes them fed up, whether they know they're fed up or not.

Hypothesis 4: when enough people are in a bad mood a rainbow is generated in the atmosphere by human brain emanations.

Hypothesis 5: a rainbow is produced by sunlight when it shines on falling water droplets.

It will be noticed that none of the five hypotheses explains *how* the proposed cause leads to the creation of a rainbow. The hypotheses serve only to suggest what to look for next time a rainbow appears, and indeed when to look for a rainbow.

Taking the rational standpoint that only one of the hypotheses can be correct, a way must be found to choose which one. The key to the choice as far as scientific method is concerned is further observations. Observational tests need to be carried out. Does observation match hypothesis?

In the case of hypothesis 1 it plainly does not. It can be observed that rainbows do not appear whenever it is raining, but only occasionally. No matter what other compelling reasons there may be for liking this hypothesis, it is observably false and must be rejected.

Hypothesis 2 on the other hand would seem to be in accord with observation; rainbows appear whenever God thinks people need to be reassured they're not all about to drown in a flood. However, hypothesis 2 is not in fact remotely satisfactory from a scientific point of view. That's because it actually reveals nothing more about rainbows than we already know. True, it explains why rainbows appear when it rains, but the statement 'God is reassuring people' is exactly interchangeable with 'a rainbow appears'. It neither gives any clue as to how God makes rainbows, nor provides any observational way of testing whether rainbows are indeed caused directly by God. It ultimately comes down to the following circular dialogue.

"A rainbow has appeared in the sky because God is reassuring people."

"How do you know God is reassuring people?"

"Well, if God wasn't reassuring people, the rainbow wouldn't have appeared."

Since hypothesis 2 is a dead-end, leading to no further observations, it can be declared unscientific and rejected accordingly.

Hypothesis 3 is more interesting. Observation does indeed support the view that people tend to be fed up when the weather is bad. Here are two people discussing this hypothesis.

"I know the rainbow isn't really there," says Person 1. "What I'm seeing is a trick my brain is playing on me because I'm fed up with the weather."

"Hold on though," says Person 2. "I can see the rainbow and I'm not fed up. That disproves the hypothesis."

"No, it doesn't," rejoins Person 1. "All it means is that you're suppressing your feelings. If you think you're seeing a rainbow, you're fed up even if you've managed to convince yourself you aren't."

Couched in these terms, hypothesis 3 is unassailable. It can claim to be correct if you admit you're fed up; and it can claim to be correct if you deny you're fed up. And that is where it runs into difficulty. What test can be carried out that hypothesis 3 could possibly fail? A hypothesis which is constructed so that it will always pass every observational test regardless of outcome is not really being tested at all. Scientific tests must be able to give either of two verdicts: pass or fail. One-verdict tests simply won't do. Hypothesis 3 is thus rejected not because it is necessarily wrong but because it lacks a crucial characteristic it must have if it is to be considered scientific: it must be disprovable.

Requiring that an explanation is neither observationally falsified, nor barren, nor constructed to fit the observations no matter what happens, has eliminated the first three hypotheses. But there are still two remaining in play. The scientific method of choosing between them is to have recourse to even more observations. Observation *always* decides the issue. The new observations should be made as systematically as possible. The position of the sun and the accompanying rainbow may be carefully noted whenever they appear. The crime rate and the incidence of public disorder may be monitored as a measure of bad temper in order to see whether there is any correlation between peaks of bad temper and the appearance of rainbows. In short, detailed measurements need to be made of anything that the hypotheses suggest should be measured in order to confirm them.

With a mass of quantitative and qualitative data having been accumulated it may be possible to discard one of the two surviving hypotheses. It might turn out, for instance, that there is no relationship between indicators of public bad moods and the appearance of rainbows. However, suppose there is a significant correlation: that when the sun is visible and people are generally bad-tempered, rainbows appear, provided also that it is raining at the same time. This would seem to keep hypothesis 4 in play so long as it is modified to say that bad temper can only produce a rainbow in the presence of sunlight and rain.

At this point a scientific principle going under the somewhat strange name of *Occam's razor* can be invoked. This says that when two rival explanations both fit the observed data, the simpler is to be preferred. The principle is somewhat dubious. It implies that the universe, at its most basic level, is as simple as it can be. It is not an unreasonable assumption. Nevertheless it *is* an assumption. It will be mentioned again in the next section when commenting on wholes and parts.

For now, use of Occam's razor means that hypothesis 4 can be provisionally rejected because it requires three things to be connected (sun, rain, bad temper) while hypothesis 5 has only two (sun, rain). It must be emphasized though that hypothesis 4 has not been found to be observably false, but only that it is considered less likely to be true than its rival. On that understanding, hypothesis 5 is adopted as the best one to pursue further.

In the course of this scientific investigation of rainbows lots of measurements which fit hypothesis 5 satisfactorily have been amassed. What is to be done with them next? In order to turn the hypothesis (which is a guess, remember) into something more formal, the measurements must be analysed. A search must be made of the data to see if any kind of pattern can be found. It will be noticed that the rainbow is part of a circle whose centre is exactly opposite the sun in the sky. This is the case in all observations, plus or minus small values arising from inaccuracies in making the measurements. This regularity or pattern is called a *law*. Using the law it can be predicted where to look for the sun if a rainbow is in sight, and vice versa.

It is important to realize that laws do not *explain* anything. They are valuable because they enable predictions to be made, but they do not explain why those predictions prove correct. A law is a statement of 'this is what has always happened' without any indication of why. It is descriptive, not explanatory.

The drawing up of a law provides something concrete on which to build a formal explanation. Once again the scientist has resort to guessing. A possible explanation of the law is proposed. The scientist then tries, often with the use of mathematics, to see if the proposal will generate the same data that was actually observed and that the law predicts. Such an explanation is called a *theory*. The theory of the cause of rainbows would have much to say about the properties of light and of water droplets. It would call on geometric constructions to show how the paths of light waves of different wavelengths are bent by different amounts when encountering dense but transparent materials.

The formulation of the theory has some interesting consequences. Because it refers not specifically to sunlight and rain but to white light and water droplets, it contains within it novel predictions. One is that it should be possible to make a rainbow experimentally by shining light onto a fine spray and positioning oneself appropriately. It should be possible to go even further and substitute glass for water as the transparent material and attempt to make a kind of rainbow that way. Predictions such as these, which go beyond the scope of the original question, are the hallmark of a good theory.

The predictions a theory makes lead to the final step in the use of scientific method. Having devised a theory which fits the observations and explains the law, the theory must be tested. It must be established whether what actually happens is what the theory predicts should happen. And every bit as important, whether what the theory predicts can't happen, doesn't. If the theory's predictions are not borne out by observation, then the theory is erroneous. If for example it was to prove impossible to make a rainbow by shining white light onto a spray, then the theory of rainbow formation outlined above would not be tenable. Nor would it be tenable if a rainbow was observed in the absence of clouds.

With this last step the six components of scientific method have been covered. Scientific method always starts from an observation about which a question has been asked. Science proceeds to answer the question as follows.

1. Observations are made of whatever is the subject of the question, looking particularly for anything that might provide a clue to its nature; for instance, the circumstances in which it is observed.
2. One or more tentative, observationally testable answers to the question — hypotheses — are put forward.
3. Using the hypotheses as a guide, the subject of the question is studied by observation, and also by experiment where possible.
4. A law is formulated which describes the patterns found in the observations and/or experiments.
5. The original question is answered by devising a theory — a detailed answer which encompasses the law and makes predictions about observations that have not yet been made.
6. The predicted observations, both of what should happen and what should not, are made, and the theory is thereby tested.

These six steps constitute, in an idealized form, the method of science. If they are compared with the five innate processes described in section 2.2, of observation, recognition, classification, prediction, and explanation, it can be seen that there is a lot of similarity between them. However, scientific method is not merely a formalization of the innate processes. It is far more methodical and more elaborate, as will become clear when considering some aspects of scientific method in greater detail in the next section.

2.4 SCIENCE IN PRINCIPLE

THE STATUS OF SCIENTIFIC LAWS

Laws, as has been seen, are statements of what is observed without any indication of why. They vary according to the kind of observations they describe. Some laws are fundamental; some are not.

In the latter category are laws relating to biology and to complex structures in general. They describe what happens to be, rather than what has to be. For example, a law may be propounded that all biological catalysts are made of proteins (see sections 5.2 and 5.3). It is not inconceivable that had chance dictated otherwise, some catalysts might have been made of quite different chemicals. Another law of this kind is that all reptiles and mammals have four limbs. Again there is no known reason why this must be so; the number could just as well have been six or even five (think of animals which use their tails as an extra limb). And there can be exceptions (e.g. snakes).

Laws describing what has to be — fundamental laws — are those of physics. For example, the law that, all other things being equal, the volume of a quantity of rarefied gas halves when the pressure on it is doubled is fundamental. It has never been observed to be violated and chance has no part to play in the situation. And so it is with most laws of physics, especially those for which a theoretical explanation exists. It is fundamental laws that have the greatest bearing on how science views reality, and these laws, therefore, are the ones this book is primarily concerned with. The word 'law' henceforward will always refer to laws which are fundamental.

The significance of (fundamental) laws is sometimes misunderstood, so it is worth being clear on this issue. Laws are not statements of what *must* happen. They are statements of what has always been observed to happen in the past. Inasmuch as the universe is consistent from one day to the next, it can reasonably be expected that laws derived in the past will also apply in the future. But expectation is all there is. Laws do not tell the universe what to do; they tell observers what to expect it to do. No more than that.

In spite of this, laws of physics, as has been made clear, are inviolable. What this means is that if in future some law was observed to be violated against expectations, then that law would cease to be valid. A new law would have to be devised to replace the discredited one and provide new and non-violated expectations. And as always in science, observation will be the final arbiter on the survival of the new law as it was of the old.

In a nutshell, the status of scientific laws is that they exist only until they are violated (if ever). It must never be forgotten that observation determines the laws; the laws do not determine observation.

APPARENT EXCEPTIONS TO LAWS

When an attempt is made to discover some law or other operating in the physical world, it is often necessary to ignore inconvenient observations. You have already encountered this in chapter 1 where proposals for laws may well have invited the riposte: 'Yes, but what about so-and-so?' Surely a refusal to countenance things which seem from the outset to invalidate a proposed law is very unscientific.

The defence against this charge is in two parts. In the first place, the world is so very complex that any attempt at explaining it must begin with a limited number of observations. Laws usually start life not as universal laws but as laws valid for a range of situations, where the range is defined by those observations which have not been ignored.

In the second place, 'ignore' is really too strong a word. Awkward observations are not discarded but rather are put to one side; as further observations add to the growing fund of knowledge, so it eventually becomes possible to modify the law and to incorporate the awkward observations. If some observations persist in being contrary to a law even then, the validity of the law concerned must indeed be severely questioned.

There is, in a way, an art to science. It lies in the choosing of the limited range of observations within which a law is to be sought. If the choice is wrongly made, any law that is apparently discovered is likely to be incapable of being extended to cover a wider range of observations. In this book, scientific knowledge is being looked at with the benefit of hindsight. All the incorrect choices of the past, and the failed laws derived from them, will go unmentioned. What is left are the laws which have proved to be generalizable; that is, those which can be modified to cover a steadily widening range of observations, including the inconvenient observations that were initially put to one side. The observations that were apparently ignored in chapter 1 were not ignored arbitrarily but are those which can, with the benefit of hindsight, be temporarily left out. All the yes-but-what-about queries that may have occurred to you will hopefully be dealt with in the following chapters, as the laws applicable to limited circumstances are gradually modified and generalized.

THE STATUS OF SCIENTIFIC THEORIES

Every explanation starts life as nothing more than a guess of what lies behind a particular set of related observations. From this uncertain beginning it is elaborated into a detailed proposal or theory whose aim is to provide an explanation of why such and such a thing or event is the way it is, or happens the way it does.

When a guess is made about something, it would clearly be foolish to jump straight away to the conclusion that the guess must be true. Any guess obviously has a fair chance of being wrong. Now, since a theory is at heart a detailed guess, it too must be assumed to be possibly erroneous. However, as time passes, the theory will be subject to observational tests. If it fails these tests then it is incorrect; the original guess was wrong. But suppose it passes the tests; that is, the test observations are successfully predicted by the theory. It is still possible that the theory is wrong and that its predictions coincide with the observations by pure chance. But each time the theory passes some new test by predicting another observation, it becomes increasingly improbable that it is doing so by accident.

How many tests must a theory pass before it can be considered certain? The answer depends on what is meant by certain: certain to all intents and purposes; or absolutely certain?

In the former case, the answer is largely subjective. Different people have different degrees of credulity and skepticism (I am skeptical about the U.K.'s rationally indefensible spelling of sceptical!) and cease to doubt a given theory when they personally are convinced that further testing would be pointless. As such, the stage at which a theory becomes effectively certain is arbitrary. But arbitrary only within limits. Anybody who was especially credulous or especially skeptical would be less likely to survive the challenges of the environment than their contemporaries. Hence people's degree of credulity and skepticism is constrained by the very objective reality that scientific theories seek to explain.

But what of absolute certainty? Can a theory ever pass so many tests that it is impossible for it to be in error in any respect and in any (combination of) circumstances? The answer, unfortunately, is no. It's true that as a theory passes more and more tests it becomes increasingly unlikely that it is doing so by some sort of compound coincidence. But it is also true that while the probability of coincidence may become tiny it can never reach zero. In addition, there always remains the possibility that one day some new observation will be made which the theory cannot explain. A theory can only be certain in an absolute sense when every test has been performed. There are an infinite number of such tests, every test needing to be carried out in every conceivable differing situation. In short, no theory can be regarded as being true in every respect and in all circumstances beyond any possibility of doubt. *Scientific method cannot discover absolute certainty.*

The best way to regard scientific theories is to believe they are true, subject to the proviso that future observations, however improbably, may falsify any given theory or require its range of applicability to be altered. The edifice of scientific knowledge is not chiselled in rock. Nor, going to the opposite extreme, is it spelled out by smoke signals in the sky. It is securely written with pencil on paper in the firm but fallible conviction that the writing will never need to be crossed out.

(The question of human fallibility is something of a side issue, but it's worth remarking on briefly. Lack of absolute certainty is not a drawback of science alone. To claim possession of absolute knowledge about the world, by whatever means, is to claim to be incapable of making mistakes; i.e. to be infallible. No method of deriving explanations which acknowledges human fallibility can rationally claim to be discovering absolute objective knowledge. And it is, of course, an observational fact (a law) that humans are fallible. A theory to explain this is easy to construct. Of course, there's always the possibility that an infallible human will be found one day. However, until it happens, the law and the theory and the consequences that follow from them can safely stand.)

There is one other factor which has a bearing on the degree of certainty that can be attached to any given theory. Historically, scientific theories began life addressing very limited sets of observations. Each subject area was a separate field of knowledge covered by a separate theory. Thus there was a science of mechanics, a science of heat, a science of light, a science of chemistry, and so on. But as science has progressed and the quantity of observations has grown, it has been found possible to combine subject areas which were previously distinct into a single theoretical structure. To some degree, much of chapters 1, 3 and 4 describes this gradual unification. So successful has this approach proved that it has become one of the aims of scientific method. The idea is to extend the scope of scientific theories such that the number of theories decreases while those remaining cover a widening range of observations. This inevitably means surviving theories incorporate more and more (previously distinct) topics. In other words, the aim of science is to create theories which are as universal as possible. To have one theory which explains an extensive collection of disparate observations is better than to have many theories each of which explains only a few observations.

This reflects on the issue of the certainty of scientific theories because the more universal a theory is the more testable it is. A theory which explains a small number of observations can only be tested within the limits of those observations or perhaps a bit beyond those limits. Contrastingly, a theory which explains a large and diverse collection of observations can be tested over a much wider spread of future observations. The more universally applicable a theory is, the more extensively it can be tested and — providing it passes its tests — the more likely it is to be true.

(Consider here also remarks made above about extending laws to cover an ever increasing range of observations. Making laws more comprehensive goes hand in hand with making theories more universal.)

THE RELATIONSHIP BETWEEN WHOLES AND PARTS

There are two other characteristics of scientific theories which are particularly noteworthy. These are the characteristics of simplicity and completeness.

Take simplicity first. The scientific preference for simplicity — an assumption that objective reality is fundamentally simple — has already found expression in Occam's razor, as mentioned previously. But the belief about simplicity runs deeper, or goes further, than that. Every explanation connects different events or objects together. A given event can only be explained in a non-circular way by referring to other facets of the universe which combine to produce it. Similarly, when explaining the nature of an object, reference will be made to its components and how they are connected together to make the whole. It is implicitly assumed in this procedure that the things being connected in the explanation are in some sense simpler, or easier to understand, or more fundamental, than that which they are an explanation of. Whenever a scientist seeks an explanation it is always in terms of breaking a complex phenomenon down into simpler components.

The second characteristic, completeness, has as its basis an assumption that wholes are made entirely of parts. It is regarded as essential that a complete explanation of some object or event is one which fully accounts for the observation in question. If the combining of the parts of something into a whole was found to produce some mysterious new property not derivable from the parts, that new property would itself become an object to be explained. The explanation would again be in terms of simpler parts.

Simplicity and completeness underlie the way wholes and parts are connected: wholes are made completely of simpler parts. The twin assumptions of completeness and simplicity have so far served science well, enabling very successful theories to be devised. Nonetheless it would be advisable to remember that assumptions, even these two successful ones, can, like theories, never be considered absolutely certain.

THE PRIMACY OF OBSERVATION

Scientific method involves an interplay between observation and theory. An observation is made, a theory is constructed to account for the observation, more observations are made to test the validity of the theory. The relationship between observation and theory is not however an equal one. True, theory is a guide to what observations are made and observations are a guide to what theories are put forward. But when theory and observation are in conflict it is always the theory that is discarded in the end. Observation is paramount in science.

The requirement that observation is both initiator and determinant of theoretical explanation imposes limits on what scientific method can be used to investigate. To be open to scientific enquiry a phenomenon must be observable, directly or indirectly, and more, it must be measurable and repeatable.

The need for measurability arises because of the possibility, mentioned in section 2.1, that subjective experience may be distorted. The use of a measuring instrument is a great help in countering this possibility and thereby putting the objective reality of any given phenomenon on a firmer footing. Also, consider the situation which would arise should some phenomenon be unmeasurable by any instrument that imagination and ingenuity can devise. It is hard to see in such circumstances on what basis a detailed theory could be constructed and then tested. In the absence of measurement it cannot even be made clear what the phenomenon is that needs explaining. (Measurability, which is of crucial importance to science, will be discussed further in sections 9.2 and 10.1.)

The accompanying need for repeatability arises because errors may be made in the course of making the measurements. Repeated measuring reduces the risk of some kinds of error, as the same error is unlikely to occur every time the measurement is carried out. In addition, in order to build up the quantity of systematic data necessary for the formulation of a theory, many measurements must be made. Whatever the phenomenon is, it must be repeatedly observed if the necessary quantity of systematic data is to be obtained.

These various considerations show — and this cannot be repeated often enough — that observation is absolutely central to science. Indeed, the simplest way to determine whether a given topic is a science or not is to discover the extent to which it rests on systematic observation. In most cases the distinction between science and non-science is quite stark.

A field of knowledge can only be called a science if it depends on conformity with observation for its continued existence. Scientific theories do not survive because they are cleverly constructed or because much artistry went into their making. They survive because what they say should happen is what is observed to happen. And because what they say should not happen, doesn't. A field of knowledge which exists independently of observation — or worse, in defiance of observations which have repeatedly contradicted it and thus shown it to be fraudulent — is not a science.

THE USE OF SPECULATION

In sections 4.6 and 5.5 and in various other places in this book, particularly chapter 7 and some parts of chapters 8 and 9, ideas will be advanced which will be described as speculation. What role does speculation have in science?

Inasmuch as they are attempts at explanations, speculations are related to theories and hypotheses. Just as hypotheses can be thought of as rudimentary 'pre'-theories, so speculations can be thought of as rudimentary 'pre'-hypotheses. Unlike hypotheses, speculations are based on inadequate observations, vary widely in the amount of detail they possess and make predictions which have not yet been tested. The lack of testing is often due to their predictions being beyond the reach of available technology. The ability of speculations to guide future observations is likely to be compromised for the same reason.

Despite the weakness of its observational support, a speculation can still with a little charity be considered scientific. This is so long as it doesn't run counter to existing observations or established theories. Since the observational base is unsatisfactory, a scientific speculation is usually an extrapolation of current lines of theoretical thinking. Its role is to plug the gaps in present scientific knowledge and thereby to enable the impatient to see what the overall scientific picture may be like. Speculations should always be regarded as doubtful (in the sense to be defined in section 8.4).

A comment on fancy is also in order here. Fancy is based on few or no observations, is not detailed and makes general and untestable predictions. It is scientifically indefensible and has no place in scientific reality. Fancy will be resorted to only once in this book, in section 4.6, when confronting the very first moment of the universe's existence. The reason for employing it there will hopefully be apparent at the time.

2.5 THE PRACTICE OF SCIENCE

While science itself can be viewed as an ideal, the practice of science cannot be discussed without reference to its practitioners. A practitioner of science is called a *scientist*. It should be noted that a scientist, in the sense to be understood throughout this book, is anyone who accepts and uses the scientific method as the means of determining what is true about the objective world. Scientists, by this definition, need not work in a laboratory or earn their living from science (any more than Christians must work in a church or earn their living from religion). A scientist is someone who approaches objective reality scientifically, no more and no less.

The outline of scientific method given in sections 2.3 and 2.4 describes how the method ought to work rather than how it actually does in practice. Every scientist adheres by definition to a basic notion of scientific method, but the practice of science may sometimes nonetheless diverge markedly from its principles. (See Bibliography: Chalmers for examples and a full discussion of this topic.)

To start with, people like being right. We are uncomfortable when it comes to admitting we've got something wrong. Some of us hate admitting it. We can be jealous, arrogant, opinionated. And scientists are as much prey as anyone else to these human foibles. For working scientists there are also additional pressures: they need the approval of their superiors and mentors, without which they could find themselves out of work; they want a secure career path, something likely to be imperilled if they challenge the prevailing scientific theories too radically; and so on. Human nature, in these and numerous other ways, has a big impact on how scientific method proceeds in practice.

Another big impact is due to the fact that hypotheses, laws and theories are all made of ideas. Ideas can be observed (as you are currently observing the ideas in this book), which makes them a kind of object. All objects have associated with them a number of properties. In the case of idea-objects, these properties are largely independent of an idea's origin, scientific or otherwise. For this reason, the fate of a scientific idea is determined more by the nature of ideas than by the nature of scientific method. Ideas and their properties will be discussed at length in chapter 8.

A good example of how ideal and practice diverge in science occurs when a theory fails some observational test. In principle the theory must of course repeatedly fail the test before being judged incorrect, since any one observation may be faulty; only by repeated testing can it be hoped that observational errors will cancel each other out or otherwise be eliminated. But once a theory has repeatedly failed a test then it must obviously be rejected. There is, however, a caveat: it is not possible to be absolutely sure a theory is wrong any more than it is possible to be absolutely sure a theory is correct.

Because of this loophole it is tempting for scientists who have long advocated a particular theory to resist discarding it. They may argue that when their theory repeatedly fails a test it is the test that is at fault. Perhaps the equipment being used to perform the test doesn't work exactly how it's supposed to. Or perhaps there are hidden — and false — assumptions underlying the interpretation of the test results. Or if the test cannot be faulted, theory loyalists may describe the test results as 'anomalies' and dismiss them. Or they may contrive some ad hoc modification to their theory which leaves it fundamentally intact. It is thereby the case that scientific theories which fail tests are seldom discarded without a struggle by those who subscribe to them.

There are, incidentally, some instances where a failed theory is retained quite legitimately. These arise when there is no other theory available to replace the failed one. It is better to have a theory which fits most of the observations than to have no theory at all.

And in fairness it also needs to be mentioned, to the embarrassment of scientific method devotees, that occasionally a theory is retained despite failing its tests because enough scientists feel (probably on the grounds of its simplicity or explanatory power) that it 'ought' to be correct.... And, lo and behold, eventually it does indeed turn out that the tests were misleading, the anomalies were anomalies, the contrived modification is in accord with observation, and that those who held to the theory when it should have been rejected were right all along.

Another less clear-cut way in which principle and practice diverge in science concerns the questions scientists seek to answer. Here, what is of concern is not so much the resistance of theories to being discarded, but rather the way people's existing ideas determine what ideas they seek to acquire in future. The questions scientists ask (i.e. the observations they are seeking to explain in a theory) are largely determined by theories they already possess. In other words, existing theories greatly influence the formulation of new theories.

Scientists do not ask questions they consider to be stupid. Nor do they ask questions that they think they already know the (unverified) answers to. To give a trivial example, if it is raining and a scientist observes that the gutter on a building is overflowing, the question of why it is overflowing does not immediately suggest itself as a sensible thing to ask. Even if the gutter has never overflowed before, the scientist knows it must be blocked at some point along its length. Given the answer is already clear, no scientific investigation of what is actually happening will be carried out. It is remotely possible that the gutter is not blocked and that some hitherto unknown Nobel-prize-worthy physical process is being called into play. But the scientist will never make the discovery because existing theories give rise to the belief that there is no discovery to make.

On a less trivial level there are two examples worth mentioning. The first concerns the Special Theory of Relativity. In the days before the theory appeared, nobody questioned the nature of time much. It was the sort of subject that was considered to be self-evident. All the answers were already known. It was only when observations were made which could not be accommodated within any existing theory that questions which would previously have been considered stupid began to be asked. The unaccommodating observations, be it noted, were not made in connection with the nature of time, but were concerned with the speed of light. It was consequently quite accidental that scientists discovered their ideas about time were wrong. (This topic will be discussed at length in chapter 10.)

The second example concerns what may loosely be called the supernatural. There is a general disinterest amongst the scientific community in such things. The problem is that supernatural events violate the laws of physics by definition. Yet as scientists confidently expect the laws never to be violated, to ask how the violations occur or to investigate them would almost certainly prove to be a waste of time. Hence the disinterest. Human trickery and the inevitable discrepancies between $S(x,t)$ and $O(x,t)$ statements already offer ready-to-hand and credible ways of accounting for supposed supernatural phenomena.

In any case, scientific investigation of the supernatural would be self-defeating. For scientific method to operate in respect of any phenomenon requires it to occur repeatedly and measurably, as already remarked. Repeated measurements of a supernatural phenomenon could be used as the basis for an explanation. Indeed the whole point of a scientific investigation would be to explain the phenomenon. Once explained it would no longer be supernatural. Even if it was discovered that some major law of physics was being broken, all that this would signify is that the law was incorrect. A new law would be formulated to encompass the once supernatural, and now perfectly natural, event. Science and the supernatural are mutually exclusive. The former must always seek to explain away the latter.

2.6 THE END OF SCIENCE

The ultimate goal of scientists is to construct a single explanation which completely accounts for all that can be observed. There seems to be no reason in principle why this goal should not eventually be achieved. And yet at least three question marks attend the enterprise.

Firstly there is the thought that every explanation, even an all-embracing one, gives rise to a new question: why this explanation and not some other? Could scientists ever be content with a final answer of 'that's just the way it is'?

Secondly there are problems looming about testability. Examples found in this book concern particle physics, the Big Bang, and the origin of life. In particle physics, as the distances and particles under investigation get smaller, ever higher energies are needed to test the theories pertaining to these almost infinitesimal sizes. What will happen when the energies needed are higher than can be achieved in today's universe? How can rival theories be tested and decided amongst in those circumstances? Similar considerations arise in constructing theories about the extremely high energies believed to be present in the first moments of the Big Bang. And how are theories about the origin of life on the Earth to be tested when they make predictions involving sterile oceans and atmospheres of planetary size and timescales of millions (at least) of years? What it boils down to is this: if theories devised to answer the most fundamental of questions about the universe and life necessarily extend into realms where they can't be tested, how can they be considered properly scientific? (See Bibliography: Baggott for a skeptical discussion of this issue.)

Thirdly, if an ultimate explanation is devised, there are doubts about how comprehensible it will actually be. Already many of the quantum phenomena which dominate particle physics can only be visualized — if at all — in terms of analogy. And it must be said that when physicists attempt to describe the reality behind quantum events, the analogies they most favour are extremely unconvincing. (See Bibliography: Rae (2004) for accounts of these analogies.) The position seems likely to get worse. As science penetrates phenomena further and further removed from everyday experience, what is discovered is, not surprisingly, increasingly unlike the world people know through their largely unaided senses. Yet to understand something, most people (if not all people) need to be able to picture it in terms of the commonplace. It is conceivable that it may eventually become impossible for anyone to do that. The value of an ultimate explanation in such a situation would be questionable.

Science has yet to reach a stage where these problems are insurmountable. For the time being scientists continue to probe the objective world ever further, not knowing how far away the ultimate goal is, or even if such a goal is anything more than a figment of their imagination.

To conclude this chapter, here is a summary of the steps which have thus far enabled science to describe objective reality. The steps are: to pose a question about something that has been observed; to propose an answer (a hypothesis); to make measurements; to work out the law being followed; to explain the law with a theory; to test the theory.

It is a method which unlocks the door to great knowledge. And to a degree of understanding.

3

PARTICLE PHYSICS

3.0 OBJECTIVE: To describe current theories about the nature of particles, forces and energy.

Obviously if a complete scientific description of objective reality is to be formulated, it is important that the most basic components of that reality are discovered. In searching for these components the techniques, and more importantly the lines of thought, employed in chapter 1 will continue to be used. Proceeding in this way can be justified by the fact that there is no dividing line between the directly and near-directly observable components of reality discussed in the opening chapter, and the unobservable objects of this chapter. The former shades gradually into the latter.

We will start by discovering the constituent parts of atoms, and show that atoms' chemical properties depend on electrons taking up a shell-like configuration enclosing an atom's interior. In creating this picture, we will be forced, as we contemplate such tiny sub-atomic distances, to use concepts of considerable abstraction. We will find ourselves confronting particles which behave collectively like waves, forces which depend on energy being fleetingly created out of nothing, and a micro-universe where probability and uncertainty hold the key to everything that happens. Immersion in this strange world will lead us to discoveries of new forces and new particles and new conservation laws.

The way the world of particles differs from what we experience through our unaided senses detracts to a degree from the scientific advances that have been made. It does, however, highlight an important aspect of science: if observation leads us to theories which fly in the face of common sense, or which cannot be explained using terrestrial-scale analogies, or which we simply don't like, then so be it. The value of a theory is determined by how well it describes and accounts for what is observed, not by how easy it is to understand or how pleasant it is to contemplate.

Note that the several tables referred to in various places below are to be found together at the end of the chapter.

3.1 ELECTRONS, PROTONS AND NEUTRONS

<u>THE PERIODIC TABLE</u>

The theory that all substances on the Earth are made of combinations of about 100 types of atoms gives rise to an obvious question: is it possible in turn to reduce the 100 atoms to combinations of a few even more basic constituents? In other words, are atoms made of simpler things?

As usual, the way to answer this question requires detailed observations to be made — in this instance of each type of atom. The resulting data can then be analysed with a view to discovering if there is any sort of pattern present. The task involves performing qualitative and quantitative measurements of very many chemical reactions amongst the different elements — an *element* being any substance made of only one type of atom. On the basis of the properties revealed by the measurements, different atoms then need to be grouped together by what they have in common. In addition, the weight of each type of atom must be determined. This makes it possible to find out if there is any connection between an atom's mass and its chemical properties.

The mass of an atom is not as difficult to obtain as might be thought. There are several methods available to do it. One involves weighing fixed volumes of gases whose molecules consist of only one or two types of atoms. Another requires one or two electrons to be separated from an atom, thereby making the atom electrically charged. Next, the atom is exposed to an electromagnetic force of known size which makes it accelerate. Its mass can then be calculated using $\mathbf{F} = m\mathbf{a}$. By means such as these it is possible to obtain the masses of each of the 100 or so types of atom.

Once the mass of the various atoms has been evaluated, they can be listed in order of ascending mass. The grouping of atoms by what they have in common then needs to be added to the list. The idea is to write down each element in order of ascending atomic mass, but in such a way that those elements with similar physical and chemical properties are found near to each other. Gratifyingly, the result of setting out the elements in this way is a very pronounced pattern. One form of it will be found in table 3.1.

Table 3.1 lists the name of each element, together with its chemical symbol and the average weight of each atom in atomic weight units (1 unit = $1 \cdot 66$ x 10^{-27} kg), increasing from left to right and top to bottom. All the elements in any one column have similar properties. Those in adjacent columns are similar to a lesser degree. Those in widely separated columns are very different from each other. Listing the hundred atoms in order of weight, it can be seen that the same physical and chemical properties recur periodically. For this reason, table 3.1 is called the *periodic table of the elements*.

The discovery of a pattern such as clearly emerges from table 3.1 is not in itself an explanation of anything. It does, however, constitute a major clue to the answer to the question: do atoms have constituent parts? The existence of a pattern suggests that those atoms with similar properties must have some ingredient in common which dissimilar atoms do not have. That suggests in turn that atoms do indeed possess constituent parts, amongst which the ingredient, whatever it is, is to be found.

The constituent of atoms which is already known about from section 1.5 is the electron. And electrons are certainly heavily implicated in producing the pattern in the periodic table. This is because of the kinds of properties that are grouped together. Chemical properties (for example, elements in column 1 react with oxygen to form a molecule with the formula X_2O, where X is the column 1 element) and physical properties (for example, elements in the far right column are all gases) both involve the stickiness of atoms to each other, in which electrons play the key role.

Let it be hypothesized then that the pattern found in the periodic table is produced by the electrons which surround each atom. How might they be able to produce such a pattern? One possibility that can be ruled out straight away is that there are different kinds of electron. All electrons, from whatever atom, are found by experiment to be identical. All possess the same size of negative electric charge, and all have the same mass, a minute $9 \cdot 1$ x 10^{-31} kg. This is 1/1837th the mass of the lightest atom, hydrogen, which weighs $1 \cdot 673$ x 10^{-27} kg.

Another suggestion as to how electrons produce the pattern in the periodic table is that it is the *number* of electrons surrounding any given type of atom which is crucial. This suggestion has proved to be compatible with experimental evidence.

The way in which the number of electrons determines an atom's physical and chemical properties will be described in detail in the next section and in section 5.1. For now, the question being asked is: what are atoms made of? What does a variable number of electrons imply about atoms' constituents?

THE NUCLEUS

Except in unusual circumstances, atoms are electrically uncharged. Given that an atom may be surrounded by a number of electrons, it must also contain a number of positive electric charges to exactly counterbalance the negative charges of those electrons. Only then will it appear uncharged. It has already been remarked that the positive charges lie on the inside of the atom. This inside is called the *nucleus*. The nucleus of an atom contains the positive electric charges around which the negatively charged electrons collectively form a kind of outer covering, called a *shell*.

(This shell is not to be thought of as solid. Nor is it to be thought of in picturesque terms as electrons orbiting the nucleus like planets orbit the sun. There is no terrestrial-scale analogy to an electron shell — or indeed to an electron — as will shortly be seen.)

Obviously, the more electrons there are in the shell, the more positive charges there must be in the nucleus. If it is assumed that each positive charge has a mass, then each additional electron in the shell, and thereby each matching additional positive charge in the nucleus, can be expected to increase the weight of the atom containing them. This reasoning leads to a simple theory that the heavier the atom — i.e. the more mass it contains — the more electrons it has in its shell and the more positive charges it has in its nucleus.

The lightest atom is hydrogen. If its electron shell is also the lightest it will contain only one electron. Experiment confirms that only one electron can be taken from a hydrogen atom. It follows that the hydrogen nucleus contains only one positive charge. Further to that, experiment indicates that this positively charged nucleus cannot be broken up into pieces. The conclusion follows that a hydrogen nucleus consists of a single positively charged particle, which is called a *proton*.

The picture of the hydrogen atom that emerges from these investigations is thus of a negatively charged electron forming a kind of shell around a positively charged proton. The proton has exactly as much positive electric charge as the electron has negative electric charge. This contrasts with the marked imbalance between their gravitational charges. The proton has 1836 times as much mass as the electron.

It is a tempting thought that each step up the periodic table simply involves adding a proton to the nucleus and an electron to the shell. Helium, the second lightest atom, would then contain two protons and two electrons. Lithium, the third lightest, would contain three protons and three electrons. And so on. If this is correct it reduces the 100 types of mass to a mere two, the proton and the electron.

Unfortunately the weight of each type of atom does not accord with this idea. Helium weights almost as much as four hydrogen atoms, not two. Lithium weighs nearly as much as seven, not three. With each step up the periodic table, the weight of the atom tends to increase by the approximate equivalent of two or three hydrogens; that is, by the equivalent of two or three protons and electrons.

The explanation of this unwanted mass increase that best matches observation invokes a third particle to add to the proton and the electron. This particle, which has no electric charge and a mass slightly greater than the proton, is called the *neutron*. Neutrons are found in all nuclei except the simple hydrogen nucleus described above.

Neutrons permit the proton-plus-electron idea to be retained. Each element does indeed have one more proton and one more electron than its nearest lighter neighbour. The excess increase in mass is accounted for by adding a neutron or two to the nucleus, along with each extra proton.

One consequence of this picture of atoms is that it can be confidently stated how many protons and electrons a particular type of atom contains. For example, atoms of element number 8, oxygen, contain 8 protons and 8 electrons. In contrast there is no equivalent certainty about the number of neutrons. Within narrow limits the neutrons in the nucleus of atoms of any given element can, and in fact do, vary in number. The fractional part of the atomic weights listed in table 3.1 largely results from each element containing atoms having more than one overall nuclear mass. Some of the nuclei of any one element have more neutrons than others. The atomic weights in the table are thus average weights.

Fortunately the number of neutrons in a nucleus makes no difference to an atom's chemical properties. The periodic table pays no heed to neutrons. It is electrons and protons that matter.

TWO PROBLEMS

It would appear that great progress has been made. The hypothesis that each element is characterized by the number of electrons its atoms have in their electron shells has enabled the number of types of mass to be reduced from about 100 to 3. Instead of 100 atoms there are only the electron, the proton and the neutron to set alongside the two forces and three types of energy. This is surely yet another triumph for scientific method.

But alas, something has gone horribly wrong. The trouble is immediately obvious if this freshly created theory of atomic structure is checked to be fully compatible with the theory of dynamics. In that light, the nucleus is a disaster area. Hydrogen excepted, every nucleus contains more than one proton. Protons are all positively charged and so must repel one another. Protons in a nucleus should immediately fly apart. The existence of atoms other than hydrogen is, quite frankly, against the law (of how electric charges react to one another).

And there is a second disaster to face concerning size. There are various ways to determine the size of atoms. For example, one method involves directing electromagnetic waves at crystals in which the atoms are arranged in layers. When the waves possess a wavelength close to the distance between the crystal layers, partial reflection from adjacent layers in the crystal produces a pattern in the re-emerging waves. Typically, wavelengths needed to see this pattern are around 10^{-10} m, meaning that that is the size of the atoms in the crystal, assuming they are closely packed together.

Compare this size to that found by a quite different experiment. When electrically charged high-kinetic-energy helium atoms are fired at thin sheets of metal foil, the resulting collisions between helium and metal atoms cause some of the helium atoms to change direction quite drastically. This indicates that the impacts involve something much smaller than 10^{-10} m. The theoretical explanation is that when a charged helium atom undergoes a large change in direction it must have gone right through the electron shell of a metal foil atom and hit its nucleus. The size of the nucleus can be calculated to be 10^{-14} m at most.

An atom thus appears to consist of an electron shell 10^{-10} m across, inside which is a 10^{-14} m diameter nucleus. In terms of scale, if the distance from the centre of a nucleus to its surface is one unit, then the distance to the electron shell is ten thousand units further.

This leads to disaster in the following way. Consider an electron in a shell around the nucleus. If the electron is stationary it should fall onto the nucleus. (Opposite charges attract one another.) Alternatively, if it is in motion around the nucleus, the lines of electromagnetic force from the electron ought to be wiggling vigorously, carrying away the electron's kinetic energy in the form of electromagnetic waves, as described in section 1.6. The electron should very rapidly come to a halt and, as before, fall onto the nucleus. Because an atom is so small, this ought to happen in a mere fraction of a second. The atom should collapse down to the size of a nucleus. Yet it doesn't. Left entirely to themselves, electron shells are completely stable. Electrons do not fall onto the nucleus; they keep their distance from it. And they don't make electromagnetic waves.

This is a very serious situation. For the first time there are two theories, both supported by observation, which are incompatible. The theory of dynamics, as it stands, permits neither nuclei with more than one proton, nor electron shells. The theory of atomic structure requires both these things. At least one of the two theories must be wrong. But which?

3.2 QUANTA

'LUMPS' OF ENERGY

Confronted by inconsistencies in theories, the wisest course is to tackle each inconsistency in turn. The problem of having many protons in a nucleus will be left until section 3.4. Here the problem of electron shells will be addressed. The task is to resolve this conflict between the theories of dynamics and atomic structure, and if possible to do it in a way which also explains how electrons give atoms the recurrent properties revealed by the periodic table.

The simplest approach when faced with conflicting theories is to ask whether it might be possible to modify the theories rather than abandon one of them altogether. In what way could a modification be attempted?

When atoms were first discussed in chapter 1, the question was posed whether the theory of dynamics that had been formulated up to that time was applicable to atoms. An answer was given in the affirmative. The theory of atoms did seem compatible with the theory of dynamics. But suppose dynamics does not after all apply to atoms, at least not to the inside of atoms. If the supposition is correct, the theory of dynamics and the theory of atomic structure can co-exist. All that needs to be remembered is that the theory of dynamics only works for objects at least as big as, or bigger than, atoms. Electrons, being much smaller, are exempt.

It would be possible at this point to go ahead and construct an entirely new theory of dynamics just for electrons. But that would be a rather unsatisfactory thing to do. For what would happen if a particle was subsequently discovered with dimensions midway between electrons and atoms in size? Which theory of dynamics should be used in that case? It would be better if the existing theory of dynamics could be modified so that the size of the objects to which it is being applied comes directly into the theory. There would then be a single theory rather than two separate ones. So what kind of change could be made to the theory of dynamics which would enable it to permit stable electron shells to co-exist with wave energy and electromagnetic force?

The theory of dynamics is a 'smooth' theory. Energy can have any value from zero upwards and can change by infinitely small amounts. Suppose this is not exactly correct. After all, mass is not smooth; it comes in proton, neutron and electron 'lumps'. Perhaps the same is true of energy.

Let a definite proposal be made along these lines. The new hypothesis is this: when the kinetic energy of electrons is converted into wave energy, the conversion happens in lumps. If these lumps, which are called *quanta*, are very tiny they will not be noticeable except at the sort of distance scale that electrons are on. On larger scales, the quanta will merge together. The theory of dynamics devised in chapter 1, in which energy is smooth rather than lumpy, remains correct so long as the scale of the measurements is of sufficient size as to make the quanta indistinct. This is how size can be built into the theory. The situation is analogous to a line made of many dots. Close up, the dots are noticeable. From a distance, although the line can still be seen, the individual dots are invisible. They appear to merge smoothly together.

The proposed modification to the theory of dynamics works like this.

In the laboratory, a comparatively large-scale place, moving electric charges convert their kinetic energy into waves in quanta, but these quanta are so tiny they pass unnoticed. As a result, the production of wave energy looks like it is taking place smoothly, exactly as described in section 1.6. The line can be seen but not the individual dots it is made of. Inside an atom, a comparatively minute place, electrons convert their kinetic energy into waves in quanta, but on this scale the quanta are relatively large. The dots can be seen much better than can the line. Electrons do not smoothly slow down and fall into the nucleus. Instead, they approach the nucleus like a ball falling down a flight of steps. And the bottom step is as far as they can fall. The electron shell is normally found on that bottom step. The electrons cannot get any closer to the nucleus, despite being attracted to its positive charge(s), because they have less energy than the minimum sized quantum. Hence the electron shell is stable.

THE ENERGY OF ELECTROMAGNETIC WAVES

The hypothesis that charged particles can only convert their kinetic energy into wave energy in lumps enables the theories of dynamics and atomic electron shells to coexist without contradicting one another. As it allows both theories to be retained it is worth serious investigation. Great effort must be made to convert the hypothesis into a theory. The first step to take in order to do this is to make a lot of systematic measurements of the interaction between electrons and electromagnetic waves.

Picture an electron in a shell around a nucleus. If an electromagnetic wave happens to strike it, the electron may absorb the energy of the wave. (This has already been mentioned in section 1.6.) The electron's total kinetic and potential energy will increase. If a sufficient increase is produced, the electron may acquire enough energy to escape completely from the atom. Experimentally it is found that atoms in the left-most columns of the periodic table are particularly prone to losing electrons by this means. This is exactly the sort of interaction that needs to be measured in order to investigate the quantum hypothesis. What do the measurements reveal?

Take an element from the left of the periodic table and shine a light on it. The amplitude of the light waves (how bright the light is) and the wavelength of the light (its colour) are measured. Also the kinetic energy of the escaping electrons is measured. The same measurements are made over and over again for different wavelengths and amplitudes of light waves. Then the accumulated data is searched for a pattern.

The conclusion that emerges from these experiments is that the kinetic energy of the escaping electrons is determined solely by the wavelength of the light being used. The amplitude of the light waves, that is, the height of the waves as opposed to their length, affects the number of electrons that escape, but not the kinetic energy of each one.

The next thing to do is to quantify how much kinetic energy an electron gains from a wave with a particular length. Again the experimental data is analysed. It shows that the energy gained by an electron is inversely proportional to the wavelength of the light. The longer the wavelength, the less energy the wave gives to any electron it encounters. Mathematically this can be written:

$$E_\lambda = \frac{hc}{\lambda}$$

where E_λ is the energy of the wave and λ is the wavelength. The constant of proportionality which connects the two sides of the equation is the product of two numbers, h and c. h is called Planck's constant after the physicist who first hypothesized its existence. It has a value of $6 \cdot 626 \times 10^{-34}$ J s. c is the speed at which electromagnetic waves always travel when in a vacuum. It is called simply the speed of light and, as already mentioned in section 1.6, has the value $2 \cdot 998 \times 10^8$ m s^{-1}.

ATOMIC ELECTRON SHELLS

The equation for the energy of a wave which has just been formulated is fully compatible with the quantum hypothesis. The next move is to return to the flight-of-steps picture of electrons forming a shell to each atom. How does this tie up with the newly derived connection between energy and wavelength? What observation reveals is unfortunately rather intricate but is worth detailing for the reward it brings in its wake.

Take an atom and shine a light on it. Choose a wavelength for the light which makes it just a bit too long to give the atom's electrons enough energy to escape. According to the quantum hypothesis, any electron which absorbs a wave will increase its kinetic and potential energy by an amount equal to E_λ, moving away from the nucleus as it does so. It will then be rather like a ball thrown to the top of a flight of steps. The next thing the electron will do is to fall back down towards the nucleus, perhaps dropping to the lowest step in one go, or perhaps stopping on one or more intermediate steps. With each distance dropped, a wave will be emitted. The quantum hypothesis says that the steps are of a fixed size. If that is correct then the waves emitted as the electron falls towards the nucleus will also be of a fixed size; that is, the energy of the waves will not vary. This energy can be evaluated by measuring the wavelength and putting it into the $E_\lambda = hc/\lambda$ equation. Does observation support the hypothesis?

Yes, it does. Pick an element and measure the length of the waves emitted by falling electrons; they always possess the same set of lengths. Hence the energy emitted by the electrons has a matching set of values. Each value corresponds to a lump of energy; energy comes in quanta.

In fact, investigating different elements shows that each one has its own set of specific energy values. Each element has, in a manner of speaking, its own unique flight of steps.

So far so good. But in order to develop the quantum hypothesis into a full theory, work needs to be done on the flight-of-steps idea. The notion of steps seems rather contrived. Steps there undoubtedly are, but what actually are they?

The key point to make about the steps is that a falling electron can stop on each one before proceeding to the next. When it stops on a particular step the electron can be regarded as forming briefly an electron shell at the height associated with that step. Each step, in other words, is in effect an electron shell that the electron can occupy on its way down towards the nucleus, albeit an electron shell which is usually empty.

Equating steps to electron shells leads immediately to the question: what exactly is an electron shell? The answer is that an electron shell is a volume of space surrounding a nucleus that an electron can occupy. A shell may have electrons in it or it may be empty. In general the outer electron shells will be empty because electrons try to get as near to the nucleus as possible (subject to a restriction to be mentioned shortly). To elevate an electron to one of the outer shells, energy must be input to the atom. Even then, the elevation will only be temporary. The electron will re-emit the additional energy as electromagnetic waves as soon as it can and return to a lower shell.

What do measurements of E_λ reveal about the energy associated with each electron shell? Once again to observation! Careful examination of the wavelengths emitted by falling electrons indicates that each shell is not as simple in structure as might at first be thought. In any particular shell it turns out the energy an electron can have is not unique — the electron's energy may be one of several distinct discrete values. This affects the size of the quantum of energy the falling electron emits when arriving at or leaving the shell. As a result, the wavelength associated with a fall from one shell to the next can actually be one of several slightly different wavelengths.

One of the causes of this variable electron energy can best be pictured — very inadequately — by considering the analogy of planets orbiting a sun. In its orbit a planet may follow a circular path, or it may follow an oval path. The energy of the planet differs slightly according to which shape of path it follows. Something similar applies to electrons orbiting a nucleus. The path the electron follows may form a spherical shape, or it may form an ovoid shape. (Just as spheres are the three-dimensional equivalent of circles, ovoids are the three-dimensional equivalent of ovals. Ovoids are generally described as having a shape resembling an egg shell.) In an atomic context, the more ovoid the shell shape, the lower the energy of the electrons forming that shape.

There is a limit to how ovoid the shape can be. The innermost shell is too small to accommodate anything other than a spherical shape. The next shell outwards, in contrast, has room for two shapes to co-exist: a modestly ovoid one and a spherical one. The third shell outwards has room for three distinct shapes: seriously ovoid, modestly ovoid, and spherical. And so on. The further from the nucleus a shell is, the more shapes of varying degrees of ovoid-ness are available for electrons to form.

A third factor affecting the energy of electrons occupying a shell is that all electrons are engaged in a permanent spin. There is nothing like electron spin in the world we perceive with our senses. In particular, the spin of an electron can't speed up or slow down; it has a value which is forever

$$\pm \frac{1}{2}\left(\frac{h}{2\pi}\right),$$

where h is, as defined before when discussing the energy of electromagnetic waves, Planck's constant. Usually the $h/2\pi$ in the above expression is omitted, and the spin is written $\pm\frac{1}{2}$ for short. In terms of electron shells, the spin may add to the electron's orbital motion $(+\frac{1}{2})$ or it may subtract from it $(-\frac{1}{2})$. The energy of the electron will differ between the two cases.

Finally, moving electric charges generate magnetism, and this too affects how much energy each electron has.

The energy associated with any given atomic electron depends on the combination of these various factors; that is, shell height, shell shape, magnetic effects, and spin orientation. In each case, the value that can be assigned to a given factor is restricted to a discrete set of numbers, with any particular electron possessing one value from each of the four sets of numbers. Those four values define the state the electron is in, and hence its energy. The values the numbers can have are as follows.

- n: the number of the shell. $n = 1$ is the shell nearest the nucleus; $n = 2$ is the next shell moving outwards, and so on.
- l: the ovoid-ness of the electron orbit within the shell. The bigger the shell number (n) the greater the number of shapes possible. l must be zero or a positive whole number less than n. The lowest l value belongs to the most ovoid shape. For example, for shell $n = 3$, the most ovoid shape has $l = 0$, and the spherical shape has $l = 2$.
- m: the magnetic number associated with the motion of the electron. m can have any integer value between $+l$ and $-l$. For $l = 2$, for example, m can only have the values 2, 1, -0, -1, -2.
- s: the spin of the electron, which can be $+\frac{1}{2}$ or $-\frac{1}{2}$.

With these four numbers, the energy of any electron in a shell can be determined. The quanta of electromagnetic waves emitted by electrons as they jump from one set of values of the four numbers to another set with less energy can be fully accounted for.

THE PERIODIC TABLE REVISITED

The quantum hypothesis is now becoming detailed enough to be regarded as a theory. But to be secure it needs to make some predictions which can be tested. Specifically, can it, as was requested at the start of this section, predict and explain the periodic table? To do that would be a triumph indeed for quantum theory.

To bring a quantum explanation of chemistry within reach it is necessary to add another ingredient to the mixture. The trick is one that has been used before. Recall when solids, liquids and gases were first discussed that it was necessary to add to the theories of atoms and dynamics the hypothesis that molecules are sticky. The same thing needs to be done again here. A new hypothesis is needed that can be added to the theories of quanta and of chemistry in order to bring the two together. The new hypothesis is that no two electrons can be in the same place at the same time.

At first sight this doesn't seem like a new hypothesis at all. It has been used before (see sections 1.1 and 1.5). But what makes it new is the interpretation now to be given to the word 'place'. Most of the space around a nucleus is barred to electrons. So place for an electron moving round a nucleus can't be just about anywhere; electrons have to be in one of the shells. Similarly, inside each shell the places where an electron can be are further restricted by the values of l, m and s that it has. As a consequence of these considerations it can be proposed that place for an electron is *defined* by the numbers n, l, m and s. The place hypothesis can be re-worded to read: no two electrons in a shell around a nucleus can have the same set of values of n, l, m and s at the same time. This law is known as the *Pauli Exclusion Principle*, Pauli being the physicist who first enunciated it.

If this hypothesis is added to quantum theory, what it means is that an electron shell can get full up. The number of electrons which can be in any one shell at any moment is precisely limited. The bottom shell with $n = 1$ has only two possible place values: $n = 1$, $l = 0$, $m = 0$, $s = +\frac{1}{2}$; and $n = 1$, $l = 0$, $m = 0$, $s = -\frac{1}{2}$. There are only two places for electrons in shell number 1 to occupy. The shell can only accommodate two electrons. A third electron cannot enter the $n = 1$ shell unless one of the two electrons already there is ejected to make room for it. The second shell can contain eight electrons. The possible values are: $n = 2$, $l = 0$, $m = 0$, $s = \pm\frac{1}{2}$ (two places); $n = 2$, $l = 1$, $m = 1$, $s = \pm\frac{1}{2}$ (two places); $n = 2$, $l = 1$, $m = 0$, $s = \pm\frac{1}{2}$ (two places); and $n = 2$, $l = 1$, $m = -1$, $s = \pm\frac{1}{2}$ (two places). In the same way, the third shell can contain 18 electrons, the fourth shell can contain 32 electrons, and the fifth shell can contain 50 electrons.

At this point it needs to be emphasized that the statements that have just been made about electron shells are derived from quantum theory; that is, from the study of the waves emitted by atomic electrons. They are *not* derived from studying chemistry. Yet if the 'magic' numbers of electrons per shell — 2, 8, 18, 32 — are noted and then reference is made to the periodic table (table 3.1), it is apparent that the same numbers crop up there. Row one has two elements, rows two and three have 8, rows four and five have 18, row six has 32.

This is clear evidence that the periodic table, and chemistry in general, are somehow determined by atomic shell structure. To get a better idea of how this could be so, think about how an atom will appear to its neighbours. What it looks like to them will determine how they react to it, and will hence determine the chemical properties it is observed to have. The key point is that an atom's neighbours will only be able to see its outermost occupied electron shell, since that shell can be expected to mask any inner ones. Hence the appearance of an atom's outermost occupied shell can reasonably be proposed as the source of that atom's chemical properties.

The most important factor governing the way an atom's outermost shell appears to other nearby atoms is the number of electrons it contains. Consequently it can be further hypothesized that two different elements whose atoms both have, say, four electrons in their outermost shell will have very similar chemical properties. They will both be found in the same column of the periodic table.

To complete the details on this picture three facts are decisive.

- As has been seen repeatedly before, objects with potential energy convert that energy into kinetic energy and/or wave energy whenever possible. Thus each electron preferentially occupies whichever shell and shell shape is not yet full and gives the electron the lowest potential energy.
- In the main, the lower the value of n, the lower the energy of the shell.

There are however some exceptions. For example, the part of shell $n = 3$ that has shape $l = 2$ fills up after the part of shell $n = 4$ with shape $l = 0$. There is nothing mysterious about this. Electrons with $n = 4$, $l = 0$ have less energy than those with $n = 3$, $l = 2$, so electrons occupy the $n = 4$, $l = 0$ places first.

- Because of the restrictions on the values of m and s, for $l = 0$ there are only 2 shell spaces; for $l = 1$ there are 6; for $l = 2$ there are 10; for $l = 3$ there are 14.

Taking these facts into account, it is possible to display the energy of each electron shell and shell shape in tabular format. In table 3.2 the energy associated with each combination of n and l increases from left to right in each row. Also, the left hand side of each row has more energy associated with it than the right hand side of the row preceding it.

Comparing this table with the periodic table, it becomes clear why all the elements in, say, column two of the periodic table (beryllium down to radium) have similar properties. It is because they all have an outer electron shell of shape $l = 0$ with two electrons in it. As another example, elements in the penultimate column on the right (fluorine down to astatine) all have an outer electron shell containing seven electrons, 2 of shell shape $l = 0$, 5 of shape $l = 1$.

The only discrepancy you may notice between tables 3.1 and 3.2 is that helium looks like it ought to be on the left hand side of table 3.1. However, it has a full outer shell (of shape $l = 0$) and this means it is most like the other full outer shell elements — neon, argon etc. — and so goes in their column.

The picture of chemistry to emerge from quantum theory can be summarized in the following statements.

- Each type of atom contains a unique number of protons. If a proton was to be added to a nucleus, the atom would move one step to the right on the periodic table. For example, if a proton was added to a fluorine nucleus the fluorine would turn into neon. If yet another proton was added, the neon would become sodium, and so on.
- The number of protons in a nucleus determines the chemical properties of the atom indirectly by requiring that for each proton there is a matching electron in a shell around the nucleus. The more

protons the nucleus contains, the more electrons will be found in that atom's electron shells.

- Electrons fall to the shell where they have the lowest possible energy, subject to the requirement that no two electrons are ever in the same place at the same time; i.e. that no two electrons ever have identical values of n, l, m and s.
- The number of electrons in the outermost shell is directly responsible for giving an element its particular chemical properties.

The success of quantum theory in explaining the periodic table has been described at some length. This is because the theory of quanta opens the door on a sub-microscopic world where the laws governing what can happen are very different from those derived from observations in everyday life (the sort of laws described in chapter 1). The fully elaborated quantum theory does great violence to ordinary common sense. Such theories are rightly rejected unless there are compelling reasons to accept them. It is to be hoped that the quantum theory's insight into chemistry — one of the most readily comprehensible achievements of the theory — is just such a compelling reason.

3.3 QUANTUM THEORY

WAVE-PARTICLES

The fact that chemistry can be seen as a consequence of quantum theory gives the latter a great deal of credibility. But chemistry is just the tip of the quantum iceberg. To explore quanta fully would go beyond the scope of this book, but certain consequences have a bearing on the many-protons-in-a-nucleus problem. These aspects of quanta will be described in this section.

In section 1.4 it was seen that two distinct subjects, dynamics and heat, were possessed of a common identity. Heat is not a totally separate property of substances from dynamics. Rather, heat is how dynamics at a molecular scale is made manifest in the terrestrial-scale world. Heat and dynamics are experienced differently, but they are nonetheless distinguished only by the scales on which they operate. The achievement of bringing them together in one theory is described as a *unification*.

Quantum theory contains within it the unification of two very disparate objects: waves and particles. As has been seen, the energy transferred to an electron by a wave of length λ is always a fixed quantity. The quantity is independent of the amplitude of the wave (its brightness). As an illustration of this point, imagine a beam of light of one wavelength, say λ = 5 x 10^{-7} m (a greenish-blue), shining onto a metal surface. Electrons emitted from the surface as a result of interacting with the beam of light all have the same kinetic energy. This means that electromagnetic wave energy is being tapped from the beam of light in quanta. For λ = 5 x 10^{-7} m, the quantum E_λ is 2 x 10^{-19} J.

Now, this transfer of energy from electromagnetic waves to electrons in quanta is not a very wave-like property. Waves are generally smooth and continuous, not lumpy. So in this case it is more appropriate to consider the beam of light to be made up, not so much of waves, as of particles. Each particle consists of a quantum of energy. Light made of electromagnetic waves of length λ = 5 x 10^{-7} m can also be thought of as consisting of electromagnetic particles of energy E_λ = 2 x 10^{-19} J. The particles are called *photons*. (Photons, be it noted, are still required to travel, like their wave counterpart, at the speed of light.)

The concept of the photon is one instance of quantum theory violating common sense. It is clear that electromagnetic energy consists of waves. These waves refract, reflect, and interfere with each other just as ripples do in a tank of water. It is true that, unlike ripples in water, electromagnetic waves don't seem to be *in* anything, but that can be overlooked given the weight of evidence in support of the wave theory. Yet it is also true, when dealing with quantities of sub-atomic size, that atomic particles which feel the electromagnetic force always acquire energy from, and lose energy to, electromagnetic waves in quanta. The quanta which are emitted by electrons falling towards a nucleus, and which are absorbed by electrons as they climb away from it, are only two examples of experimental evidence that electromagnetic energy comes in the form of photons.

It is intuitively obvious that waves and particles are mutually exclusive things. Experiment shows that intuition is wrong. Electromagnetic energy consists of either waves or particles depending on the type of experiment being performed. Fortunately, no experiment ever finds electromagnetic energy to consist of waves *and* particles simultaneously. This is some consolation for being forced to accept that the quantum world is becoming rather difficult to visualize.

The wave-particle nature of electromagnetic energy can be generalized to cover gravitational energy as well. Just as photons are the particle equivalent of electromagnetic waves, so there could well be a particle equivalent of gravitational waves: *gravitons*. There is as yet no generally accepted observational evidence that gravitons actually exist. It can only be said that quantum theory indicates they ought to.

And it is possible to take the concept of wave-particles even further. Recognizing that waves sometimes behave as particles invites a reciprocal speculation. Perhaps the sub-atomic particles, electrons, protons and neutrons, sometimes behave like waves. Experiment shows that they do. The wavelength (λ) of an object with mass (m) travelling at a speed (v) is given by

$$\lambda = \frac{h}{mv}$$

where *h*, as usual, is Planck's constant. Note that this equation doesn't apply to particles which travel at the speed of light — photons and gravitons — but only to those whose velocity is less than that ($v < c$).

The discovery that electrons, protons and neutrons sometimes behave like waves is startling enough to warrant further thought. It will be noticed that the equation for λ given above places no restriction on the value of the mass m. This implies that wave properties are not confined to sub-atomic particles but that all objects, even footballs and planets, will in appropriate circumstances have wave-like properties. It is side-stepping the issue to point out that as m in the equation is a divisor the wavelength associated with large masses is extremely tiny and in practice undetectable. What then is to be made of objects-with-a-wavelength?

The first thing to do is to make the rational assertion that a single particle can*not* be a wave. Wave properties, such as the formation of diffraction patterns (dark and light bands produced when waves encounter an object whose size is close to the wavelength) cannot be formed by a single particle. This applies whether the particle is an electron or a photon or a planet. When particles behave as waves they can only do so collectively. An aggregate of particles, whatever they are, acting together simultaneously, or individually over a period of time, can produce diffraction patterns. A single particle acting alone cannot.

When waves produce a diffraction pattern, what is happening is that waves arriving from slightly different directions are combining so that they add together (their peaks coincide) or they cancel out (one's peak and another's trough coincide). From a particle point of view, particles cannot cancel each other out. That would violate the law of conservation of mass. So the particle interpretation of wave diffraction patterns is simply that no particles ever arrive at locations where the equivalent waves cancel each other out. Where the waves cancel, the probability of finding a particle is zero.

Probability, then, provides the link between waves and particles. A particle is always at some point in space. A wave, on the other hand, is spread out and not at any one point. Because particles collectively behave in some circumstances like waves, and waves are not located at precise points, it cannot be known exactly at which point in space any given particle is to be found. It can only be said where the particle probably is. That probability is described mathematically by something called a *wave function*; that is, by a wave expressed as a mathematical formula. This is how it is possible to view waves and particles in a complementary rather than a contradictory way. The dynamics of particle behaviour is 'wavy'.

The discovery that particles can collectively be described mathematically by wave functions enables quantum theory to explain why electrons around a nucleus only exist in specific shells. Each shell has a circumference which is equal to a whole number of electron waves. When an electron moves to the next higher or lower shell, the number of electron-wavelengths per shell increases by one or decreases by one respectively. If an electron was to attempt to occupy a position in between two shells, that is, to occupy a shell-shaped space containing a fractional number of electron-wavelengths, then the peaks and troughs of its wave would cancel each other out. In effect, its probability of being found between shells is practically zero. It is far more likely to be found in a shell than to be found (momentarily) between shells.

Wave-particle unification also gives support to the quantum theory's use of the numbers n, l, m and s to define place for an electron. It is no longer permissible to regard an electron as a particle in the same way as, say, a grain of sand is a particle. An electron in motion cannot be assigned a precise location in space. Instead the electron particle must be described as being at an imprecisely known point somewhere within the volume of space defined by the electron's wave-function. That volume of space encompasses all the possible locations of the electron-particle. Place for an electron is defined by n, l, m and s rather than by the sort of numbers used in the terrestrial-scale world because the electron is not at a precisely known point but somewhere in a volume.

In quantum theory then, waves and particles are different aspects of the same thing. Every object in the universe — perhaps even the universe itself — can be described mathematically by a wave function. The wave describes the probability of where the object may turn out to be, or what the object may do.

MASS-ENERGY

The unification of waves and particles shows that for all the dissimilarities between sub-atomic particles on the one hand and waves in lines of force on the other, their identities are closely connected; they both have a dual wave-particle nature. This shared characteristic can lead to speculation about whether they have any other properties in common. Specifically, the sub-atomic particles have a gravitational charge. Do the photon and the theoretical graviton — does wave energy — also have such a charge?

As ever, observation holds the key. Photons passing close to very massive objects (such as the sun) travel, not in a straight line, but in one bent slightly towards the massive object. Photons are, in other words, accelerated in the direction of the massive object, the acceleration in this case being a change of direction rather than speed. The force producing this acceleration is gravity. Since an object only responds to a force if it has a charge associated with that force, the photon must indeed have a gravitational charge. As the photon is a quantum of electromagnetic wave energy, this means that waves, by their very existence, have a mass associated with them in some way.

Because of their theoretical similarity to photons, it can be assumed that gravitons also have a mass.

The discovery that wave energy has a gravitational charge — that it weighs something, in effect — has a consequence for kinetic energy. It was seen in section 1.6 that electromagnetic wave energy can be made from, and turned into, kinetic energy. This means that if wave energy has a gravitational charge associated with it, then so must kinetic energy. The law of conservation of mass requires it.

The assertion that kinetic energy has mass is clearly open to experimental testing. If it's correct, a hot object, one which contains much internal molecular kinetic energy, should have more mass than the same object when cold. Similarly, an object which is moving fast will have more mass than the same object travelling slowly. In both cases, the object's mass will have two components. One is the mass of the particles (electrons, protons, etc.) it is made of. This mass it will possess even when stationary and as cold as possible. The other component is the mass of the object due to the kinetic energy of its constituents. To distinguish the two components, the former is called the *rest mass* — i.e. the mass of the object when it is completely still, at rest. The latter will be referred to here as the kinetic mass. To complete the picture, the gravitational charge of waves will be called the wave mass.

Before attempting to measure the increased mass of a hot or rapidly moving object (assuming the prediction that there is an increase is correct) it would be helpful to have some idea of how much of an increase can be expected. It can be predicted mathematically — it is in fact one of the conclusions to emerge from the Special Theory of Relativity — that the equation relating mass to energy is

$$E = mc^2$$

where E is the amount of kinetic or wave energy, m is the corresponding mass, and c is the speed of light. Hence, the mass of an amount of energy E is E/c^2.

The number c^2 is very large: 9×10^{16} m^2 s^{-2}. In consequence, to make a measurable difference to the overall mass of an object, a very great deal of energy must be given to it. This is why hot objects and those travelling at speed are not noticeably heavier than cold or stationary ones. *Very* hot or fast objects, such as those made in the laboratories of particle physicists, can indeed be observed to have an increased total mass in line with the above equation.

Incidentally, you may be wondering whatever happened to potential energy. Does that also have a mass? The answer is indeed yes. Because the potential energy of two stationary attracting objects is less when they are close together than when they are far apart, their total mass is less in the former case than it is in the latter.

The equating of energy to a quantity of mass is only half the story. Just as the realization that waves are sometimes like particles led to the question being asked as to whether particles are not also sometimes like waves, it can now be similarly asked, given energy has mass, whether mass has energy. The $E = mc^2$ equation clearly implies that it does, as it states that mass — in any form — is directly proportional to an amount of energy. To say that x units of kinetic, wave, or potential energy equal y units of mass means reciprocally that y units of kinetic, wave, or potential mass equal x units of energy. This equality between mass and energy makes it reasonable to combine the two terms and hence to refer to kinetic mass-energy, wave mass-energy, and potential mass-energy. Mass and energy are different aspects of the same thing, much as waves and particles are.

So much for kinetic, wave and potential mass-energy. But what about rest mass? Apparently it stands aloof from all this. It does not seem to resemble the mass-energy of motion, waves and potential at all. Surely if it is correct to talk of rest mass-energy, then rest mass should be as energetic as the other forms. It ought to look and behave like they do. For example, it should be possible to observe rest mass-energy changing into, say, kinetic mass-energy in just the same way that wave mass-energy does. Indeed, all four types of mass-energy should be able to change into each other. Because of the size of c^2 in the $E = mc^2$ equation, a little rest mass should be equivalent to a lot of energy. But wherever experimenters look, that energy seems not to be forthcoming. Rest mass appears decidedly unenergetic.

There is one crucial exception to this. Of the three particles with rest mass, isolated electrons and isolated protons have never been observed to change into any other form of mass-energy. Not so the neutron. In certain special circumstances a neutron may be separated from its nucleus. When this happens it turns into other particles whose combined rest mass is less than that of the neutron. The missing rest mass is duly converted into kinetic mass-energy of the resultant particles (see section 3.6). This behaviour of the neutron shows that rest mass can turn into other forms of mass-energy and that it is correct to talk of rest mass-energy. It does, however, raise the problem of why, when the other forms of mass-energy can turn into each other readily, rest mass-energy only turns into other forms in highly restricted circumstances.

The simplest way of evading this problem is simply to say that that's just the way it is. Put scientifically, a law is formulated which describes the situation. Preferably it should be a conservation law because, recalling Occam's razor, conservation laws amount to statements that change (in whatever is being conserved) = 0, and you can't get much simpler than that.

The first law which would at least partly foot the bill is a law that electric charge is conserved; that is, electric charge cannot appear out of nowhere or disappear into nowhere. Observation confirms that this is indeed the case. It means neither an electron nor a proton can change into one of the other forms of mass-energy because the other forms have no electric charge. There is nowhere for the electric charge to go. If electric charge is conserved then so must be any particle possessing it. The law of conservation of electric charge in a sense locks the mass in the electron and the proton up. (See also comments on conservation of mass in section 1.2. The reasoning there can be applied to electric charge as well as to mass.)

Unfortunately, electric charge conservation is not enough. It fails to forbid an electron and a proton combining together to form a single electrically neutral (i.e. uncharged) object. This neutral object — a hydrogen atom? — could then turn into one of the other forms of mass-energy without violating the electric charge conservation law. Consequently two additional conservation laws are needed. One is that the total number of electrons is conserved. The other is that the total number of protons+neutrons is conserved. These two laws complete between them the task of locking up the rest mass in protons, neutrons and electrons and preventing it from turning into kinetic, wave, or potential mass-energy. They amount to statements that the total number of electrons in the universe is fixed forever, and the total number of protons+neutrons in the universe is fixed forever too.

This unification of mass and energy means that the mass conservation law and the energy conservation law can be combined into one: the total amount of mass-energy never changes. That law, together with the three new conservation laws, completely accounts for the interchange of the four kinds of mass-energy into each other. Any interconversion forbidden by these laws does not happen. Any interconversion permitted by them can be observed in appropriate circumstances. All interconversions are carried out through the agency of the forces of electromagnetism and gravity.

To repeat them, the four conservation laws are:

- the law of conservation of mass-energy
- the law of conservation of electric charge
- the law of conservation of electron number (i.e. the number of electrons)
- the law of conservation of proton+neutron number (i.e. the number of protons and neutrons).

UNCERTAINTY

The theory that energy comes in quanta has so far not seemed all that revolutionary. Mass was found to be quantized without there being any dramatic consequences. Why should the same not be true of energy? Yes, quantizing energy leads to an identity between waves and particles, and also invites the discovery of an equivalence between mass and energy, but nothing in quantum theory encountered yet in this chapter has significantly altered the way terrestrial-scale reality is perceived. Alas that cannot remain so. In seeking a complete solution to the still unsolved many-protons-in-a-nucleus problem which was revealed at the end of section 3.1, one final aspect of quantum theory needs to be explored. It is an aspect which greatly changes the way science views reality.

Consider two charged particles, two electrons say, approaching one another. Their respective kinetic energies have been carefully measured. As they get closer together the electrical repulsion between them slows them down. Their combined kinetic energy decreases, being converted into potential energy. The equation governing potential energy, it will be recalled, is $E = \mathbf{F}\mathbf{H}$ (see section 1.2) where $\mathbf{H}$ is, in this case, the distance between the two electrons. What is the value of $\mathbf{H}$ at any instant? To know that, it is obviously necessary to know exactly where the two electrons are. And that, according to quantum theory, is not so simple.

As has been explained, each electron is located in a volume of space at any instant. Exactly where each electron is in this volume is a matter of probability, the probability being expressed mathematically in the form of a wave function. If it can only be said where each electron probably is, it can in like fashion only be said what their distance apart **H** probably is. There is a limit to how precise it is possible to be about it. Consequently, the value of the potential energy of the two electrons can also only be assigned a probable value. It is possible to say that the value lies somewhere within a range of values, but not which value within that range the electrons actually possess. As with position, the probability of the two electrons having a particular value of potential energy is given by a wave function.

The consequence of quantum theory which is of concern here is thus its implications for the law of conservation of mass-energy. Quantum theory requires that the combined energy of interacting particles is always imprecisely known. To use the theory's terminology, the total energy of two or more particles engaged in an interaction is uncertain.

This conclusion is alarming. It has been seen that mass-energy is conserved. The law to that effect is derived from the making of measurements, measurements that must necessarily be precise. The law of conservation of mass-energy is a precise law with no room for any uncertainties. Yet quantum theory holds that the values assigned to measurements of such things as the positions and total energy of interacting particles cannot be precise at all. There is clearly a conflict here. How can a conservation law resting on precise measurements survive if there's no such thing as precise measurement?

The way to tackle this issue clearly lies in finding out how much imprecision quantum theory implies. In terms of energy uncertainties, an equation is required which relates the uncertainty in the energy of interacting particles — written ΔE — to other measurable quantities.

Consider again the two electrons approaching one another. At any precise moment it is known that the electrons are each within a volume of space. Where in the volume they are is uncertain, and it is this uncertainty which leads to the conclusion that their total energy is also uncertain. To sum the situation up succinctly: at any given precise moment the energy is uncertain.

This can be looked at in an inverted kind of way. It could also be said that the electrons *do* possess a precise amount of energy, and what is uncertain is exactly *when* they possess it; that is, for a precise value of energy it is the moment in time which is uncertain. This argues for a relationship between energy and time. (Recall here Noether's Theorem from section 1.2.) Specifically, when the precise moment is known, it is uncertain how much energy the interacting particles have; and when it is known precisely how much energy the interacting particles have, it is uncertain at what moment they have it.

The complementary uncertainties in energy and time can be combined in an equation

$$\Delta E . \Delta t = \hbar$$

where Δt is the uncertainty about the moment in time. As ΔE goes up, Δt goes down and vice versa. Because the amount by which one changes is exactly counterbalanced by the amount the other changes, the number obtained by multiplying the two values together is a constant. This constant, $\hbar$ (pronounced 'h bar'), has the value of Planck's constant divided by 2π.

The ΔE equation places a value on the amount of imprecision in quantum theory, putting uncertainty on a quantitative footing. It reveals that there is a fundamental limit to the accuracy of measurements, and states how to calculate that limit. (It does so paradoxically in a certain and precise way!) The more accurately the energy of an interaction is measured, the less accurately can it be known at what moment during the interaction the energy had that particular value. Conversely, the more precisely the moment in time is measured, the less accurately can it be known what the energy of the interaction is.

This accuracy limitation, it is important to realize, is not a reflection of some inadequacy on the part of the observer making the measurements. It is built into the very nature of interactions between particles. The outcome of an interaction depends on the combined energy of the interacting particles, and the combined energy has an uncertain value. As a result, the outcome of every particle interaction is governed by probability. It will be one option out of the range of options permitted by the uncertainty in the amount of energy involved. All interactions — of which an observer's measurements are but a special kind — are thus fundamentally governed by probability and by uncertainty. This fundamental uncertainty is accorded the status of a principle known as the *Heisenberg Uncertainty Principle*, Heisenberg being the physicist who first expressed it.

Using the uncertainty equation above, it is possible to look at how the law of conservation of mass-energy and the uncertainties of quantum theory fit together. The size of ΔE represents the range of possible values of the energy of an interaction at any given moment. It is, in effect, the size of the violation of the law of conservation of mass-energy that can occur during the interaction. The bigger ΔE is, the bigger is the possible violation of the mass-energy conservation law.

The size of ΔE depends on the size of Δt. From a human observer's viewpoint, the smallest timescale that can be detected by direct observation is of the order of 10^{-1} s. The value of Δt associated with human experience is therefore 10^{-1} s. Putting that value into the uncertainty equation gives a value for ΔE of 10^{-33} J, which is so small as to be unmeasurable. The kinetic energy of a single air molecule at room temperature is 10^{12} times larger. An observer will always find energy to be conserved because the sort of timescales on which violations of the mass-energy conservation law are large enough to be significant are far, far smaller than the timescales over which observers make their measurements. On a human timescale, energy can be measured extremely precisely.

What about very short timescales? When an interaction is pinned down to a very small range of possible moments — small Δt — the law of mass-energy conservation can be violated quite badly. However, although this breakdown in conservation of mass-energy can affect the outcome of an interaction, the violation of the law can itself never be observed. All observations are necessarily made on a human timescale, and by the time any interaction, no matter how short its duration, is actually perceived, the time will have been inflated to the observer's minimum of 10^{-1} s. Any non-conservation of mass-energy that has taken place will by then have shrunk in inverse proportion, as per $\Delta E = \hbar/\Delta t$. In other words, the outcome of an interaction may lead an observer to infer that a violation of mass-energy conservation has occurred, but the violation cannot actually be witnessed. It simply doesn't last long enough.

Quantum uncertainty is thus generally compatible with the law of conservation of mass-energy. The latter is an observed phenomenon; the former, which violates the latter, is unobservable. In an interaction lasting Δt seconds an amount of energy $\Delta E = \hbar/2\Delta t$ can be created, or an amount of energy $\Delta E = \hbar/2\Delta t$ can be destroyed (making the total uncertainty $\hbar/2\Delta t + \hbar/2\Delta t = \hbar/\Delta t$). But that energy can be created or destroyed only for the duration of the interaction. It cannot observably violate the mass-energy conservation law.

The notion that energy may be temporarily created during interactions brings with it an unexpected theoretical benefit; it enables force to be more closely related to mass-energy. Taking two electrons as an example again, when they begin to interact they do so by means of the electromagnetic force. (The gravitational force is also involved but is too weak to make any difference.) As the interaction proceeds, the electrons are able to create some wave energy, ΔE. This passes between them in the form of photons. The photons convey to the two electrons where each of them is with respect to the other, and bring about the acceleration that each undergoes in the course of the interaction. The exchange of the photons made of created energy is thus equivalent to the passage of the electromagnetic force between the particles. Interaction by means of force — it will be recalled from section 1.1 that all interactions are by means of force — is brought about by the exchange of particles made of created energy. The particles are photons in the case of electromagnetism (and gravitons in the case of gravity?).

Because force-carrying photons are made of created energy they do not last long enough to be observed. Their existence can only be inferred from the very occurrence of an interaction. To distinguish them from observable photons, force-carrying photons are called virtual. As they can't be observed they can't be said to *really* exist, but they do *virtually* exist! Forces are transmitted by *virtual particles*; that is, by virtual waves in the lines of force connecting the force exchanging particles. Thus real particles of mass-energy interact by means of forces conveyed by their exchanging virtual particles of mass-energy.

The closer relating of force to mass-energy increases the consistency of the scientific account of the small-scale world. Thereby it adds yet more to confidence that quantum theory is likely to be a true account of objective reality.

This is just as well, for the way energy is accounted for in quantum theory — and upon which the relating of force to mass-energy depends —

strikes at the heart of one of science's most cherished tenets. Its consequences for reality are significant. It was seen in section 1.2 that conservation of (mass-)energy and the law that every event has a cause are inseparable. Yet if energy can be created in interactions between particles so that mass-energy is not conserved, albeit only briefly, it follows that events do not have to have a cause. Every change is not, after all, always the product of previous change. Sometimes change can happen on its own. Events can occur spontaneously.

Examples of spontaneous events — in the form of particle disintegrations — will be found in sections 3.6 and 4.5.

One important point to make about spontaneous events is that they do not occur in a lawless way. Quantum theory can always predict when a spontaneous event will probably occur, and how many spontaneous events will probably occur in a given period of time. Thus strict cause and effect is replaced not by chaotic spontaneity but by probability.

The impact of quantum uncertainty on reality is worth summing up here. It is that the universe of chapter 1, one of precision laws making the future precisely predictable (in theory), has gone. In its place is a universe in which precision has given way to imprecision, and the future is ruled not by certainty but by law-abiding uncertainty. The philosophical ramifications of this, widespread and contentious as they are, will not be gone into further here.

There is one practical benefit of uncertainty that it would be useful to end this section with. That concerns the question of just how elementary the elementary particles are. It is inevitable, given how atoms turned out to be made of protons, neutrons and electrons, that a speculation arises about what protons, neutrons and electrons are made of in their turn.

The question is particularly pertinent for electrons because they are so small — no more than 10^{-18} m in diameter. Suppose they are made of yet smaller constituents. The time taken for light to travel 10^{-18} m is about 10^{-26} s. Putting this value of Δt into the uncertainty equation, $\Delta E.\Delta t = \hbar$, gives a value for ΔE of 10^{-8} J. The mass which is equivalent to 10^{-8} J of energy is 10^{-25} kg. This is 100,000 times larger than the electron's mass of 10^{-30} kg (all figures rounded to the nearest power of ten). So the uncertainty in the mass-energy of the electron's supposed constituents is vastly greater than the electron's certainly known mass. Superficially this is contradictory. There are various ways to overcome this problem but the simplest — and hence the preferred way, using Occam's razor — is to deny that the electron has constituents. It is genuinely elementary.

The same reasoning doesn't work for protons and neutrons as their diameters are a thousand times larger and their masses two thousand times greater. They could indeed be made of smaller constituents. This possibility will be explored in section 3.5.

3.4 THE STRONG FORCE

The three aspects of the quantum world discussed in the previous section — wave-particles, mass-energy and uncertainty — permit the second clash between atomic physics and electromagnetism to be confronted. To recap: the problem is how there can be mutually repelling protons in the nuclei of nearly all types of atoms. What is it that stops the protons in a nucleus from accelerating apart?

The obvious answer to this question is quite simply that nuclear protons must be stuck together. Using the same reasoning as was employed in section 1.4 about inter-molecular stickiness, the hypothesis can be advanced that a force exists between the protons in a nucleus. Of the forces already encountered, electromagnetism is repulsive and gravity too weak, so the force must be an entirely new third force. Since it is strong enough to overcome the electromagnetic repulsion of the protons for one another, it is appropriately called the *strong force*.

Having put forward this hypothesis, the next step is to turn to observation. Since nuclei are too small to be seen, these observations must be indirect. The method used in making them is to all intents and purposes the only one available when dealing with objects smaller than atoms. It is crude in principle but sophisticated in practice. Basically it involves taking electrically charged particles, accelerating them to very high speeds, and then detecting what happens when they collide with one another.

As the strong force is a nuclear force, the sorts of collisions that will shed light on it are those between different nuclei. Often electron-less helium nuclei are used, being greatly accelerated electromagnetically and then caused to crash into much larger stationary nuclei. Results from experiments of this kind were first mentioned in section 3.1. They show that nuclei are extremely small, ranging in size from about 10^{-15} m to about 10^{-14} m in diameter. (Obviously the more protons and neutrons there are in a nucleus, the larger it is.)

They also show that the impacting helium nuclei are repelled by the electric charges of the protons in the stationary nuclei even when the colliding nuclei come within 10^{-14} m of one another before bouncing apart. At such distances there is no stickiness between the protons in the helium and the protons in the target nucleus to counteract the electrical repulsion. The conclusion to be drawn is that the strong force has a limited range. Over distances of 10^{-15} m or less it is stronger than electromagnetism; at distances greater than 10^{-15} m it is not felt at all.

This characteristic of the strong force can be put into force-carrying particle terms. Just as the electromagnetic force is carried by virtual photons, so the strong force should also be carried by virtual particles. The particles concerned are known as *mesons*.

Strong force-carrying virtual mesons can be assigned a quantity of mass-energy based on the range of the strong force. If it is assumed that the strong force can travel at up to the speed of light, then the largest amount of time that a virtual meson can exist without being able to exceed the dimensions of a nucleus is 10^{-23} s. This is how long it takes to travel the nuclear distance, 10^{-15} m, at the speed of light. The timescale associated with strong force interactions is thus 10^{-23} s, and the uncertainty in the time when a strong force interaction occurs is of the same size; i.e. $\Delta t = 10^{-23}$ s. The corresponding value of ΔE $(= \hbar/\Delta t)$ is 10^{-11} J. Strong force-carrying virtual mesons can therefore be expected to have a created mass-energy of at least 10^{-11} J.

It should be noted that 10^{-11} J is the smallest value of created mass-energy that a virtual meson can possess. A smaller value would enable the virtual meson to last longer and thus travel further than 10^{-15} m, thereby exceeding the inferred maximum range of the strong force.

The next observation that ought to be made which would give support to the emerging theory of the strong force is of a virtual meson. Unfortunately, virtual mesons can only last 10^{-23} s, not remotely long enough to be observed.

All is not lost however. The question to ask is: what is the difference between those electromagnetic force carriers, virtual photons, and their real counterpart? In terms of their two primary properties, there is almost no difference. Virtual photons and real photons are both made of mass-energy and are both produced by electrically charged particles. Their sole distinction is that the former cannot be observed on account of their short lifespan (defined by $\Delta t = \hbar/\Delta E$) while the latter can be observed because they are subject to no such restriction. If a similar situation applies to mesons, it can be expected that there should be real mesons as well as virtual mesons in the world.

The comparison between the strong and electromagnetic forces can be taken a stage further. Electromagnetic wave energy — and thereby the equivalent photons — is readily interchangeable with kinetic energy. Supply the kinetic energy in the form of a collision between electrically charged particles, and real photons are duly produced. A similar thing ought to be true of the strong force: supply sufficient kinetic energy — i.e. 10^{-11} J — in a collision between particles affected by the strong force, and a real meson should be produced.

What will a particle containing 10^{-11} J of mass-energy look like? 10^{-11} J of energy is equivalent to a mass of about 10^{-28} kg, which is approximately one tenth of the mass of a proton. It can therefore be predicted that a real meson will take the form of a perfectly ordinary particle like a small proton or a large electron, though with somewhat different properties.

With some idea of what to look for, appropriate experiments can now be performed. Nuclei can be collided together which have a combined kinetic energy in excess of 10^{-11} J. In the right conditions, that amount of kinetic energy should be converted into an observable real meson if the strong force theory is correct. And experiment confirms the prediction. Real mesons are indeed produced as expected.

Three distinct mesons are found, with rest mass-energies all roughly equivalent to 270 electrons. They are distinguished from one another by their electric charges. One meson has a positive charge exactly equal to that of the proton, one is electrically uncharged, and one has a negative charge exactly equal to that of the electron. The three particles are symbolized as π^+, π^0 and π^-. As other more massive mesons are now also known to exist, these first three are called pi-mesons or just *pions*.

When a real pion is produced, all four conservation laws must be obeyed. It has already been remarked that mass-energy must be supplied to make the pion so there are no problems arising from the first of the laws, that of conservation of mass-energy. To conserve electric charge as per the second of the conservation laws, it is required that protons and neutrons turn into each other when charged pions are produced. For example

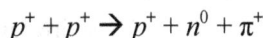

$$p^+ + n^0 \rightarrow p^+ + p^+ + \pi^-$$

$$p^+ + p^+ \rightarrow p^+ + n^0 + \pi^+$$

In these reactions, p^+ and n^0 are of course the proton and the neutron.

To comply with the final two laws, those of conservation of electron number and conservation of proton+neutron number, all that is required is that pions don't count as either electrons or protons or neutrons. As they are quite distinct from electrons, protons and neutrons, that is evidently the case. The strong force thus fits in nicely with the conservation laws as they stand. No new ones need to be invented, nor existing ones amended or scrapped.

It is worth commenting at this juncture on what particle physics reactions such as the two just given mean. They are not like chemical reactions. In chemistry the symbol '$\rightarrow$' implies 'is made of' or 'make'. A $\rightarrow$ B + C implies A is made of B and C. In particle physics '$\rightarrow$' means 'turns into'; no more than that. Each particle should be thought of as a distinct entity in its own right. It would be quite wrong from the first of the two reactions above, for instance, to conclude that a neutron is made of a proton and a π^-. It isn't. But it may turn into those particles when sufficient energy is available to ensure mass-energy is conserved when the transformation occurs. That is the sole import of the reaction.

There is one additional thing which can be inferred from the above reactions, and that is that neutrons as well as protons interact by means of the strong force. The strong force thus not only sticks the protons together in a nucleus; it also sticks the neutrons together as well. Inside a nucleus the picture is one of virtual pions carrying the strong force amongst the protons and neutrons found there. Where the virtual pions are electrically charged, they will transform protons into neutrons and vice versa. Below are two examples of virtual pion interactions inside the nucleus. The first is of two neutrons, n_1 and n_2, feeling each other's strong force. The second is of a neutron, n_1, and a proton, p_2, doing so. In the latter case the two particles swap identities, n_1 and p_2 becoming p_1 and n_2.

$$n_1^0 \rightarrow n_1^0 + \pi^0 \text{ --- then --- } \pi^0 + n_2^0 \rightarrow n_2^0$$

$$n_1^0 \rightarrow p_1^+ + \pi^- \text{ --- then --- } \pi^- + p_2^+ \rightarrow n_2^0$$

In both these cases the exchange of the virtual pion is the significant event. The electric charges involved are irrelevant as far as the strong force is concerned.

The theory of the workings of the strong force brings with it a bonus: the kind of bonus which, as has repeatedly been found before, good theories usually bring. It enables the number of types of atom to be explained. As was seen in section 3.1, the type of atom is determined by the number of protons in the nucleus. The largest nuclei have about 100 protons in them, plus a somewhat greater number of neutrons. The electromagnetic force tries to break up nuclei because all the positively charged protons repel each other. The strength of the repulsion is proportional to the number of protons in the nucleus: the greater the number of positive charges in the nucleus, the greater the repulsion felt by each proton. The strong force opposes this repulsion, but because of its short range it is only felt between neighbouring protons and neutrons. The attractive strong force between individual protons and neutrons is thus fairly constant in value regardless of the size of the nucleus. Consequently, as nuclei get bigger the strong force a proton or neutron feels changes very little, while the electromagnetic force increases. The maximum size of a nucleus will be reached when the growing repulsive electromagnetic force has risen to a value at which it equals the attractive strong force. Any further increase in the number of protons in the nucleus will produce a net repulsive force which will render the nucleus prone to breaking apart. As the repulsive and attractive forces balance when there are about 100 protons in the nucleus, it can be estimated that the strong force is roughly 100 times stronger than the electromagnetic force. The number of types of atom is thus determined by the relative strengths of the strong and electromagnetic forces.

This completes the resolution of the second clash between

electromagnetism and atomic physics. The many-protons-in-a-nucleus problem has been solved. But it has not been a cost-free exercise. The number of fundamental forces has been increased by 50% as the strong force is added to electromagnetism and gravity. Also the number of particles made of rest mass-energy has doubled from three to six: e^-, p^+, n^0, π^+, π^0, π^-. As it has restored consistency to the theories of physics, it seems a price worth paying. Unfortunately the price does not include a number of extras, which the next section will reveal.

3.5 LEPTONS AND QUARKS

ANTIMATTER

The story of pions does not end with the discovery of real pions in high energy nuclear collisions; that is just the beginning. Collisions only create pions. What happens to them after that? Observation reveals that these particles, unlike electrons and protons are not stable; they don't persist.

The fate of the π^0 is the easiest to account for. Being electrically neutral and neither an electron nor a proton nor a neutron, it is protected by no conservation laws except that of mass-energy, which allows it to be freely converted amongst mass-energy's various other forms. As a result, a π^0 turns into electromagnetic wave mass-energy within about 10^{-16} s of being formed.

The charged pions are more of a problem. The law of conservation of electric charge prevents them from turning entirely into waves, for an electrically charged particle must always be left whatever happens. But suppose two pions with opposite electric charges collide together. Their two charges then cancel each other out, leaving nothing to stop them from turning, like the π^0, into electromagnetic wave mass-energy. The reaction between two oppositely charged pions can be written

$$\pi^+ + \pi^- \to \gamma$$

where γ is the symbol for photons.

Aside from mass-energy, all the characteristics of the π^- disappear when it meets a π^+. The one particle exactly cancels out the other.

113

Technically it is said that one is the *antiparticle* of the other and that on meeting they annihilate. All that is left after the encounter are photons.

(A note for enthusiasts: particle annihilations always result in the production of at least two photons which head off in different directions. That is why I have defined γ as the symbol for photons (plural).)

So two oppositely charged pions, when combined, are able to turn into wave mass-energy. But the process is not a one-way affair. Permitted interchanges between the various forms of mass-energy are almost always reversible. It is to be expected therefore that in certain circumstances photons of sufficient mass-energy will be able to turn, even if only briefly, into two pions:

$$\gamma \rightarrow \pi^+ + \pi^-.$$

Such events can indeed be observed. And of course, by the same reasoning, π^0s can also be produced in the same way.

These discoveries would be of limited importance except that they invite speculation about possible similar reactions involving protons, neutrons and electrons. It would be another marvellous piece of encouraging consistency if the conversion of electromagnetic waves into pions was also matched by similar conversions involving these more long-standing particles.

Experiment as ever decides the issue. The following reactions can be observed:

$$\gamma \rightarrow e^- + e^+$$

$$\gamma \rightarrow p^+ + p^-$$

$$\gamma \rightarrow n^0 + \bar{n}^0.$$

In each case it can be seen that electric charge is conserved. Furthermore, to conserve electron number the e^+ of the first reaction must count as -1 electron; that is, it has an electron number of -1. Similarly, the p^- and the $\bar{n}^0$ must each have a proton+neutron number of -1. In fact, the e^+, p^- and $\bar{n}^0$ are the exact opposites of the more familiar e^-, p^+ and n^0 in all respects except that of mass-energy.

Because the universe is made of e^-, p^+ and n^0, these particles are called *matter*, while e^+, p^- and $\bar{n}^0$ are called *antimatter*. (Note that here and throughout, a bar over a particle indicates it is made of antimatter.) As with electrically charged pions, when an e^- or a p^+ or a n^0 reacts with its antimatter counterpart, the two particles annihilate. Everything cancels out except their two rest mass-energies, which are equal rather than opposite. Following annihilation, the mass-energy reappears as photons. The three reactions listed above go from right to left as well as from left to right.

It may be noted here why there is a difference in antiparticle terms between the neutron and the π^0. The π^0 has no antiparticle because it has no characteristics for an antiparticle to be the 'anti' of. The neutron, on the other hand, has a proton+neutron number of $+1$. For all its abstraction, this is a real characteristic of the neutron. Hence the neutron, n^0, has a distinct antiparticle, $\bar{n}^0$, with the opposite characteristic, a proton+neutron number of -1.

One of the extras to be attached to this developing understanding of particle physics is thus a further increase in the number of particles with rest mass-energy. The six have now become nine. And there are still more particles to come.

THE LEPTON FAMILY

The story of electrically charged pions has another ending to it other than annihilation between π^+s and π^-s. Solitary charged pions, which are prevented from turning into electromagnetic waves by the electric charge conservation law, nonetheless have only a short lifespan. After about 10^{-8} s they turn into a smaller particle with a rest mass-energy equal to about 200 electrons, most of the remaining mass-energy appearing as the kinetic energy of this smaller particle. The new particle is called a *muon*. To conserve electric charge there must clearly be two kinds of muon, one with a negative charge that a π^- can turn into, and the other with a positive charge that a π^+ can turn into. The two muons are symbolized μ^- and μ^+. Not surprisingly one is the antiparticle of the other.

Muons themselves only last about 10^{-6} s before they too turn into yet smaller particles. This time the smaller particles are electrons (e^-) or antielectrons (e^+). Pions thus end their brief existence as humble electrons.

But wait! The pion is definitely not protected by the law of electron number conservation, as witness the means of its creation in nuclear collisions. It doesn't count as an electron and so has an electron number of zero. The electron, on the other hand, has an electron number of either +1 (for e^-) or −1 (for e^+). So somewhere between the pion and the electron the law of electron number conservation must be broken.

The way out of this dilemma is getting disconcertingly familiar. Yet another particle must be proposed to exist. The role of this additional new particle is to balance the electron number books. Whenever the electron number appears to go up by 1, one of these new particles must come into being at the same time, counting as −1 electron, and so having an electron number of −1. Similarly if the electron number appears to go down by 1, a new particle with an electron number of +1 must come into being. In this way the law of conservation of electron number can be preserved.

To see how this works, it is helpful to look at the conversion of charged pions into electrons in more detail. Firstly, the pion turns into a muon. Experiment shows muons to be identical to electrons except for being 200 times heavier. Muons, like electrons, therefore have a conserved quantity, muon number, which is +1 for the μ^- and −1 for the μ^+. The conversions can be written:

$$\pi^+ \rightarrow \mu^+ + \nu_\mu$$

$$\pi^- \rightarrow \mu^- + \bar{\nu}_\mu$$

where ν_μ symbolizes the muon number balancing particle, called a *muon-neutrino*. In the first of the above equations it has a muon number of +1; in the second, a muon number of −1. Aside from its muon number, the muon-neutrino is almost non-existent. It has only a tiny gravitational charge and no electric charge.

Muon conversion into an electron also involves neutrinos. The two muon conversion reactions are:

$$\mu^+ \rightarrow \bar{\nu}_\mu + e^+ + \nu_e$$

$$\mu^- \rightarrow \nu_\mu + e^- + \bar{\nu}_e$$

The v_e symbol is that of a particle called an *electron-neutrino*. Electron-neutrinos have the same relationship to electrons as muon-neutrinos have to muons.

The disintegration of single charged pions has revealed several new particles: the muon, which is a heavy electron; the muon-neutrino, which can be thought of as a muon with all its electric charge and almost all its gravitational charge missing; the electron-neutrino, which can be thought of as an electron with all its electric charge and almost all its gravitational charge missing; and there are three matching antiparticles, the antimuon, the antimuon-neutrino and the antielectron-neutrino.

These various particles are given the collective name of *leptons*. Electron number and muon number can be similarly combined as lepton numbers. 'Numbers' rather than 'number' because particle interactions conserve electron number and muon number separately. (Interestingly though, once a neutrino has come into being, its identity — whether it's a muon-neutrino or an electron-neutrino — is subsequently governed by quantum probability.)

To complete the picture of particles having a lepton number, four more have been discovered. One is a really massive electron called a *tauon*, symbolized τ^-. It is as heavy as two protons put together. The second is the tauon's antiparticle, τ^+. The third is the *tauon-neutrino*, v_τ, which is, predictably, like a tauon with all its electric charge and almost all its gravitational charge missing. And the fourth is the antitauon-neutrino. Tauon number is also conserved just as the other two lepton numbers are.

Adding all these particles up, there are now twelve known particles with a lepton number; that is, twelve particles which are affected by the law of conservation of lepton numbers. Of the twelve particles, six are made of matter: τ^-, μ^-, e^-, v_τ, v_μ, v_e; and six are made of antimatter: τ^+, μ^+, e^+, $\bar{v}_\tau$, $\bar{v}_\mu$, $\bar{v}_e$. The six matter leptons each have lepton numbers of $+1$. The six antimatter leptons each have lepton numbers of -1. (See table 3.3.)

QUARKS

If the proliferation of leptons is puzzling — it hardly makes the world a simpler place — the proliferation of non-leptons looks even worse. Collision experiments, which are so useful in investigating the strong force, have produced a whole range of different particles.

With the number of non-leptons exceeding 100, recourse must be had to the classification trail again. Can classification do for non-leptons what it did for chemical elements? Is there anything like a periodic table for particles?

The easiest step to take in attempting a classification is to separate those particles which come within the scope of the law of conservation of proton+neutron number from those which don't. The latter group turns out to consist entirely of mesons; that is, of pions and heavy versions of pions. The former group, collectively known as *baryons*, correspondingly consists entirely of protons and neutrons and heavy versions of these two basic particles, all of which have a proton+neutron number of either +1 or −1.

It can be remarked at this point that since heavy baryons are always made from, and disintegrate into, protons and neutrons, 'proton+neutron' number can be replaced by the more correct 'baryon number'. The law of conservation of proton+neutron number then becomes the law of conservation of baryon number; i.e. it is the number of baryons in the universe which never changes.

The next step in the classification process is to arrange the baryons, and separately the mesons, into groups based on their various properties (mass, electric charge, what they turn into when they disintegrate, etc.). This produces some encouraging results. The patterns which emerge from the particle groupings suggest that baryons and mesons have constituents, three in the case of baryons, two in the case of mesons.

The constituents have been named *quarks*. Six different types of quark, together with six matching types of antiquark, are required to make up all the known baryons and mesons. The six quarks are designated by the letters d, u, s, c, b and t. Just as each charged lepton is paired with a matching lepton neutrino, so the quarks are paired also; d with u, s with c and b with t. The pairs are listed in table 3.3, lightest pair first.

(A note for enthusiasts: you will probably know what d, u, s, c, b and t stand for. I prefer to use the letter designations (i) for brevity and (ii) because the names given to the quarks are, frankly, fatuous.)

As an example of the quark groupings in a baryon, consider the proton. A proton contains three quarks. These are two u-quarks and a d-quark. u-quarks have an electric charge of +2/3 (as do c- and t-quarks), while

d-quarks have an electric charge of $-1/3$ (as do s- and b-quarks). The total electric charge of the three quarks in a proton is thus $(+2/3) + (+2/3) + (-1/3) = +1$, which is the observed electric charge of the proton.

Since all baryons have a baryon number of $+1$, each of the three quarks in a baryon can be assigned a baryon number of $+1/3$. In an antibaryon, made of course of antiquarks, and with a baryon number of -1, each of the three antiquarks can be assigned a baryon number of $-1/3$.

As an example of the quark groupings in a meson, consider the π^+. The π^+ contains two quarks. These are a u-quark and an anti-d-quark. The combined electric charges of the two quarks is $(+2/3) + (+1/3) = +1$, which is the observed charge of the π^+. The total baryon number of the pion — and all other mesons — is $(+1/3) + (-1/3) = 0$. Hence all mesons are unprotected by the baryon conservation law, since the quark they contain is always partnered by an antiquark.

It can be noted here that it is actually the total number of quarks in the universe which is the fundamental conserved quantity. The law that the total number of baryons in the universe never changes is a consequence of quark conservation and follows directly from it.

Of the six quarks, four disintegrate rapidly, usually in no more than 10^{-10} s. The t-, b-, c- and s-quarks end up, by one route or another, as u- and d-quarks. This is only possible, of course, because the t-, b-, c- and s-quarks have more mass-energy than the u- and d-quarks. The four heavy quarks have the same relationship to the two lightest quarks as tauons and muons have to electrons.

THE COLOR FORCE

The presence of three quarks in baryons and two in mesons suggests, if precedent is anything to go by, that a force must be operating inside baryons and mesons which sticks the quarks together. A clue to the nature of this force comes from a property encountered in connection with electrons: no two electrons can be in the same place at the same time. As was explained in section 3.2, place for electrons is defined, not by a precise location, but by four numbers. Something similar applies to protons and neutrons in a nucleus and can be expected to apply to quarks.

An analysis of quarks in baryons leads to the conclusion that in certain baryons all the numbers that specify a quark's place are the same for all three quarks. To avoid the conclusion that several quarks can consequently be in the same place at the same time, and to simultaneously solve the problem of quark stickiness, it is proposed to add a new number to the description of each quark. The new number, which must have three distinct values, enables each quark to be different from the other two in the baryon, thereby allowing each one to be in a different 'place'. The number also provides the source of the stickiness. It amounts to a new kind of charge to add to gravitational and electric charge.

Quark-sticking charge is given the name of *color*. (Color is nothing to do with visual colour. I adopt the American spelling here deliberately on that account.) There are three color charges, one for each quark in a baryon. They are called red, green and blue. There are also three anticolor charges associated with antiquarks. Call these anti-red, anti-green and anti-blue.

The rules of attraction for color charges are as follows. Each color attracts its anticolor. For example red and anti-red attract each other. The three color charges, red, green and blue attract each other. The three anticolor charges anti-red, anti-green and anti-blue attract each other.

Color charges come complete with lines of force, color-force waves and force-carrying particles, which last are called *gluons*. These gluons function in just the same way as the other force carriers described previously. An example of how the quark-sticking color charge works is the following.

$$u_{red} \rightarrow u_{green} + gluon_{red+anti\text{-}green} \text{ --- then --- } gluon_{red+anti\text{-}green} + d_{green} \rightarrow d_{red}.$$

Compare this to the two virtual-pion strong force interactions shown in section 3.4.

Color charges and the accompanying force are not observed in the terrestrial-scale world accessible to human senses. Only gravitational and electric charges have ever been observed. It may be concluded from this that both baryons and mesons, while containing color charged quarks, are overall color-neutral. In suitably chromatic language, the red, the green and the blue color charges inside a baryon add together to make it white; that is, color-neutral = white. In mesons the color charge of the quark and the corresponding anticolor charge of the antiquark cancel each other out more obviously to make the meson white in color.

The five ways to combine color charges to get a color-neutral result can be listed as:

- for a baryon, r + g + b
- for a meson, r + r̄ or g + ḡ or b + b̄
- for an antibaryon, r̄ + ḡ + b̄.

Compare these five to the single way to get an electrically neutral result, which is to combine a positive electric charge and a negative electric charge.

Before declaring quarks and color charges to be another success for particle physics, the theory needs to be tested. Specifically, it predicts the existence of particles a fraction of the size of protons and neutrons, and which possess a fractional electric charge. Can such particles be found?

To date, the answer that has come back consistently from experiment is no. Experimenters have been unable to extract a quark from a baryon or meson. Freely existing quarks, and thereby color charge as well, have remained obstinately undetected. Are ideas about quarks and color wrong after all?

A closer look at the color force shows that pessimism may be premature. In order to make individual quarks and gluons unobservable it is only necessary to take into account that gluons are themselves color charged. (Contrast this with photons, the electromagnetic force carriers, which do not have an electric charge.) Gluons, as a result of being colored, not only stick quarks together; they also stick to each other. The effect of all this stickiness — gluon is really quite a good name! — is that the force that binds quarks and gluons together does not diminish as the distance separating them increases. (Compare this with the effect of distance on gravitational and electromagnetic force.) The result is that color charged particles are permanently stuck together. Hence no lone quark is ever observed, nor is any colored gluon. Only combinations of quarks and gluons having no net color charge, i.e. which are white, can be observed. Only in white groupings can color charged particles escape the unremitting color stickiness.

Thus, providing quarks can be detected indirectly inside baryons and mesons — as indeed they can — the theory of quarks and color charges can survive the absence of lone quarks in the universe.

UNITING THE COLOR AND STRONG FORCES

Gluons, like photons, are a form of quantized wave mass-energy. Real photons (as opposed to virtual ones) with sufficient mass-energy can turn, as was seen earlier in this section, into pairs of particles as long as one of the particles is the antiparticle of the other. Real gluons can do the same.

Now, since the only difference between real particles and virtual force-carrying particles is how long they can last for, it is to be expected that virtual force-carrying photons can turn, like real photons, into pairs of particles, the pairs of particles in this instance, of course, being virtual; that is, they cannot exist for longer than $\Delta t = \hbar/\Delta E$ seconds, where ΔE is the pair of particles' combined mass-energy. Virtual force-carrying gluons are like virtual photons in this respect. They too can use the energy uncertainty equation to turn into pairs of virtual particles.

A significant gluon transformation involves quarks. A gluon may briefly change into a pair of virtual quarks, the pair consisting of a quark and an antiquark. The important thing about this transformation is that a virtual quark and a virtual antiquark together make a virtual meson. Virtual mesons carry the strong force. The strong force can therefore be regarded as due to the color force. The color force gives rise to gluons which give rise to virtual mesons which carry the strong force. The strong force is thus a short-range by-product of the color force. (Compare this situation with the short-range stickiness between molecules in gases and liquids. That was a by-product of the force of electromagnetism; see section 1.5.)

The unification of the color and strong forces keeps the select band of fundamental forces to a mere three in number: gravity, electromagnetism and strong (color). But the number of particles has an unmistakable tendency to increase at every turn. Having reduced the large number of baryons and mesons to a comparatively few six quarks and six antiquarks, the introduction of color charge has added a whole new set of force carriers. To the solitary graviton and photon must be added 8 gluons. The gluons are eight in number because each one has a color charge which is a combination of two quark color charges. One gluon, for instance, has a color charge which is a combination of red and anti-green. All eight combinations will be found listed with table 3.4. Furthermore, it could be argued that since electric charge differentiates one particle from another, color charge should do the same. Hence each quark is actually three quarks, one red color charged, one green and one blue. However, the actual color charge possessed by an individual quark can never be identified, so the color charge distinction is discounted. There are then, practically speaking, six quarks, six antiquarks and eight gluons. The periodic table of baryons and mesons has certainly not proved as economically productive as that for the chemical elements which led to the discovery of protons and neutrons!

3.6 INTERACTIONS: THE WEAK FORCE

One of the key features of particle physics is the interconversions of the various forms of mass-energy into each other. When an interconversion takes place, the different pieces of mass-energy involved must interact with each other. As all interconversions are brought about by means of a force, and as there are three forces, there must be three kinds of interconversion interaction; namely, gravitational, electromagnetic and strong (color).

When an interconversion, of whichever form, takes place it can be classified according to its consequences. Three are possible.

- The interacting particles may find each other attractive or repulsive. Only wave, kinetic and potential mass-energy are interconverted in this case. The rest mass-energy of the interacting particles is unaffected; the particles are either left unchanged or else merely swap identities. (As an example of the latter, recall the exchange of virtual pions by protons and neutrons in a nucleus; see section 3.4.)
- A reaction may occur. In this case, all four forms of mass-energy are interconverted, the interacting particles being detectably changed by the interaction. (For an example of this kind of interaction, again see section 3.4 and the outcome of a collision between a proton and a neutron.)
- A disintegration may occur. In this case, rest mass-energy is

interconverted wholly or partly into one or more of the other forms of mass-energy. (The disintegration of the electrically charged pion is an example; see section 3.5.)

Of these three, the first two kinds of interaction are uncontroversial. But the third is at first sight rather an odd proposition. Surely, when a particle disintegrates it doesn't interact with anything; it just breaks up. This notion is, however, incorrect. The correct way to think of a disintegration is as an interaction between the particle on the one hand and a force of some kind that causes its destruction on the other. Since there are three fundamental forces, it can be expected that disintegrations can be brought about by three fundamental means.

With this realization comes a suspicion; the list of components of reality may not yet be complete. Each force has associated with it a number of characteristics which are part of what distinguishes one force from another. One of those characteristics is an interaction time. Reasonably enough, the stronger the force, the faster the interaction. Applying this to disintegrations, it is to be expected that the speed with which a particle disintegrates will depend on which force causes the disintegration. Here are some typical cases.

Taking the strong (color) force first, the disintegration of a particle called a rho-meson, symbolized ρ^0, is a good example. Rho-mesons are produced in reactions (also due to the strong (color) force) between pions and protons:

$$\pi^- + p^+ \rightarrow \rho^0 + n^0.$$

It takes about 10^{-23} s for a ρ^0 to be produced. This is, as has been stated before, the timescale associated with the strong (color) force. When the strong (color) force brings about the disintegration of the ρ^0 it does so at about the same speed:

$$\rho^0 \rightarrow \pi^+ + \pi^-.$$

The next force to consider, the electromagnetic force, is not as strong as the strong (color) force so the disintegrations it brings about do not take place in quite such a rush. The disintegration of the π^0 into photons that

124

was mentioned in section 3.5 is an example of a disintegration attributable to the electromagnetic force. As remarked at the time, the event occurs in about 10^{-16} s — about ten million times slower than strong (color) force interactions.

Seeing that the characteristic times of strong (color) and electromagnetic interactions are so short, it may be asked what force is responsible for the disintegration of a pion into a muon (in 10^{-8} s), or of a muon into an electron (in 10^{-6} s). These times are too slow to be due to a strong (color) or electromagnetic interaction. But the only other possibility encountered so far, a gravitational interaction, doesn't work. The force of gravity between particles is so weak it is completely negligible. Gravity doesn't disintegrate baryons or leptons.

Accordingly, to account for pion and muon disintegrations it would appear a new interaction is required, an interaction stronger than gravity but sufficiently weak that it has so far escaped notice. Appropriately it is called the weak interaction. It implies there is a *weak force* which must be added to the other three.

On investigation, the weak force turns out to be most unusual. To begin with there is the question of weak charge. The other three forces have a force-specific charge associated with them which is possessed by any particle that feels that particular force. The same must be true of the weak force: there has to be a weak charge. As pions feel the weak force, and pions are made of quarks, it follows that quarks possess a weak charge in addition to color charge, electric charge and gravitational charge. By similar reasoning leptons possess a weak charge in addition to gravitational charge and, in some cases, electric charge. However, whereas gravity produces large aggregates of matter, and electromagnetism produces atoms and molecules, and color produces baryons and mesons, the weak force produces no bound structures at all. Consequently it cannot be said to be sticky, and this is why weak charge is the least tangible of the four types of charge. All it does is enable quarks and leptons to interact in certain characteristic ways.

Then there is the question of range. Gravity and electromagnetism have an infinite range because their force carriers have no rest mass-energy. Gluons similarly have no rest mass-energy, and the color force too would be infinite in range if it wasn't for the law requiring all color charged particles to be in color-neutral bound states, so that color charges cancel each other out. The weak force in contrast is extremely short-range, being effective only over distances of less than 10^{-17} m. To confine it to such short distances, the force must be carried by virtual particles with non-zero rest mass-energy. And as particles go, the rest mass-energy they are required to have makes them veritable giants; the time Δt it takes to travel a distance of 10^{-17} m is 10^{-26} s, which in turn goes with a ΔE of 10^{-8} J. Experimentally, the weak force carriers are found to be three in number, designated W^+, Z^0 and W^-, and they are indeed giants; their rest mass-energy of 10^{-8} J equals around 10^{-25} kg, making them particles which are over 80 times heavier than a proton. (See table 3.4.)

There is a paradox here. The time associated with the weak force on the basis of its range is 10^{-26} s. This stands in stark contrast with the characteristic weak force interaction times of 10^{-8} s or so. What is the cause of this discrepancy?

Because of the size of its force carriers, when a weak force interaction occurs it must do so extremely rapidly, in 10^{-26} s. However, it should not be forgotten that the uncertainty equation which connects the force carrier size and the interaction time — $\Delta E . \Delta t = \hbar$ — is a reflection of probability. It limits the amount of energy that can be created and says that the amount of energy that is actually created lies within a range of values having ΔE as its upper limit. The effect of this on the weak force carriers is, put at its simplest, to make their creation highly improbable. On average, only once in 10^{-8} s does enough energy get created in the vicinity of a pion to make a virtual weak force carrier. So although the interaction is very fast, it has to wait about 10^{-8} s before it gets the chance to happen. For a muon the creation of a virtual weak force carrier in its vicinity is even more improbable; muons must wait on average 10^{-6} s.

There is one weak interaction which is a billion times slower than the 10^{-6} s of muon disintegration. That is the disintegration of neutrons which have become separated from their nuclei. An average non-nuclear neutron takes about 10^3 s to disintegrate. It does so as follows:

$$n^0 \rightarrow p^+ + e^- + \bar{v}_e$$

or in quark terms

$$(udd) \rightarrow (uud) + e^- + \bar{\nu}_e.$$

Despite its slowness this is still a weak interaction (to be described in more detail in section 4.5). The special slowness is due to there being very little difference in the mass-energy configuration before and after the disintegration; almost all the mass-energy remains in a baryon. In a manner of speaking this reduces the incentive that the neutron has to disintegrate.

With the addition of the weak force, the list of forces is complete (with one possible exception to be revealed in section 4.5). The four forces are all that are needed for a description of all the interactions between all components of the universe, from the smallest to the largest, from the lightest to the heaviest, from the slowest to the fastest, that an observer on the Earth could ever possibly detect.

3.7 OBJECTIVE REALITY AT THE END OF CHAPTER 3

At this point, particle physics begins to shade into the physics of the beginning of the universe. Further theories about particles will therefore be left to sections 4.5 and 4.6 where the creation of the universe will be discussed.

For now, it would be worthwhile to summarize how particle physics stands thus far. As what has been described in this chapter is a direct continuation of the process started in the first chapter, you may like to compare what is set out below with the list at the end of chapter 1, seeing thereby how the picture of reality has been modified to reflect the discoveries of particle physics.

- Mass-energy is a fundamental property of everything that exists in the universe. It is found in four forms.
- Rest mass-energy ($E_{rm} = mc^2$) appears in the guise of two types of particles known as quarks and leptons, plus the three weak force carriers.
- Kinetic mass-energy ($E = \frac{1}{2}mv^2$) is possessed by particles of rest mass-energy by virtue of their motion.
- Potential mass-energy ($E = \mathbf{FH}$) is mass-energy in a stored condition.

Objects, including waves, possess potential energy according to their location in respect of sources of force that attract or repel them.

- Wave mass-energy ($E_\lambda = hc/\lambda$) is present in the lines of force emanating from accelerating particles possessing a charge. In quantized form, wave mass-energy appears as a kind of particle (see below).

- Forces mediate the interconversion of mass-energy amongst its four forms. There are four forms of force, which can be found listed in table 3.4.
- Gravitational force affects all forms of mass-energy. All forms of mass-energy are therefore gravitationally charged. The mass aspect of mass-energy (derived from $\mathbf{F} = m\mathbf{a}$, see section 1.1) is identical to gravitational charge.
- Weak force affects those forms of mass-energy which have a weak charge — quarks and leptons, and the W^+, Z^0 and W^- weak force carriers.
- Electromagnetic force affects those forms of mass-energy which have an electric charge — all quarks plus tauons, muons and electrons, and the W^+ and W^- weak force carriers.
- Strong (color) force affects those forms of mass-energy which have a color charge — quarks and gluons.

- Force carriers are virtual particles of wave mass-energy associated with each of the four forces. The virtual particles are the quantum equivalent of virtual waves in the lines of force emanating from each charged object. They transmit forces from one charged object to another. As there are four forces, there are four types of virtual force carrier.
- Gravitons — maybe — carry the gravitational force.
- W^+, Z^0 and W^- carry the weak force.
- Photons carry the electromagnetic force.
- Gluons carry the strong (color) force.
- The three types of force carriers confirmed to exist have real (as opposed to virtual) equivalents which can be detected.

- <u>Laws</u> govern the interconversions amongst the four forms of mass-energy. The laws identified in this chapter (not in fact the only ones) are four in number:
- the law of conservation of mass-energy (= gravitational charge)
- the law of conservation of electric charge
- the law of conservation of baryon number
- the law of conservation of lepton numbers.
- A fifth law can be added to this list; namely, the law of conservation of color charge. Color charge is conserved in the same way that gravitational and electric charges are.

These four parts of reality — mass-energy, forces, force-carrying particles, and conservation laws — all play their part in making the world what it is: mass-energy, in ways constrained by the conservation laws, and by means of force-carrying particles/waves due to the four forces, is engaged in ceaseless interconversions. These interconversions add together to make the universe we live in. It is as complete a description of the basic components of objective reality as has ever been devised.

TABLE 3.1 The periodic table of the elements (numbers 1 to 100).

Hydrogen 1 gas H 1.01									
Lithium 3 Li 6.94	Beryllium 4 Be 9.01								
Sodium 11 Na 23.0	Magnesium 12 Mg 24.3								
Potassium 19 K 39.1	Calcium 20 Ca 40.1	Scandium 21 Sc 45.0	Titanium 22 Ti 47.9	Vanadium 23 V 50.9	Chromium 24 Cr 52.0	Manganese 25 Mn 54.9	Iron 26 Fe 55.8	Cobalt 27 Co 58.9	
Rubidium 37 Rb 85.5	Strontium 38 Sr 87.6	Yttrium 39 Y 88.9	Zirconium 40 Zr 91.2	Niobium 41 Nb 92.9	Molybdenum 42 Mo 95.9	Technetium 43 Tc (98)	Ruthenium 44 Ru 101	Rhodium 45 Rh 103	
Caesium 55 Cs 133	Barium 56 Ba 137	Lanthanum 57 La 139	* Hafnium 72 Hf * 178	Tantalum 73 Ta 181	Tungsten 74 W 184	Rhenium 75 Re 186	Osmium 76 Os 190	Iridium 77 Ir 192	
Francium 87 Fr (223)	Radium 88 Ra 226	Actinium 89 Ac (227) *							

Cerium 58 Ce 140	Praesody- 59 mium Pr 141	Neodymium 60 Nd 144	Promethium 61 Pm (145)	Samarian 62 Sm 150	Europium 63 Eu 152
Thorium 90 Th 232	Protacti- 91 nium Pa 231	Uranium 92 U 238	Neptunium 93 Np (237)	Plutonium 94 Pu (244)	Americium 95 Am (243)

These two rows should be inserted in the positions marked by asterisks above.

key:
element name
atomic number   liquid/gas at room temperature as indicated
atomic symbol
average atomic weight relative to $^{12}C = 12$, to 3 significant figures, or (in brackets) number of protons+neutrons in most stable nucleus

									Helium 2 gas He 4.00
			Boron 5 B 10.8	Carbon 6 C 12.0	Nitrogen 7 gas N 14.0	Oxygen 8 gas O 16.0	Fluorine 9 gas F 19.0	Neon 10 gas Ne 20.2	
			Aluminium 13 Al 27.0	Silicon 14 Si 28.1	Phosphorus 15 P 31.0	Sulfur 16 S 32.1	Chlorine 17 gas Cl 35.5	Argon 18 gas Ar 39.9	
Nickel 28 Ni 58.7	Copper 29 Cu 63.5	Zinc 30 Zn 65.4	Gallium 31 Ga 69.7	Germanium 32 Ge 72.6	Arsenic 33 As 74.9	Selenium 34 Se 79.0	Bromine 35 liquid Br 79.9	Krypton 36 gas Kr 83.8	
Palladium 46 Pd 106	Silver 47 Ag 108	Cadmium 48 Cd 112	Indium 49 In 115	Tin 50 Sn 119	Antimony 51 Sb 122	Tellurium 52 Te 128	Iodine 53 I 127	Xenon 54 gas Xe 131	
Platinum 78 Pt 195	Gold 79 Au 197	Mercury 80 liquid Hg 201	Thallium 81 Tl 204	Lead 82 Pb 207	Bismuth 83 Bi 209	Polonium 84 Po (209)	Astatine 85 At (210)	Radon 86 gas Rn (222)	

Gadolinium 64 Gd 157	Terbium 65 Tb 159	Dysprosium 66 Dy 163	Holmium 67 Ho 165	Erbium 68 Er 167	Thulium 69 Tm 169	Ytterbium 70 Yb 173	Lutetium 71 Lu 175
Curium 96 Cm (247)	Berkelium 97 Bk (247)	Califor- 98 nium Cf (251)	Einstei- 99 nium Es (254)	Fermium 100 Fm (257)			

TABLE 3.2 Electron shells (n) and shapes (l) in order of ascending energy (left to right and top to bottom).

Shape (maximum electrons) →	$l=0$ (2)		$l=2$ (10)										$l=1$ (6)					
Electrons actually present →	1	2	1	2	3	4	5	6	7	8	9	10	1	2	3	4	5	6
$n = 1$																		
$n = 2$															$n = 2$			
$n = 3$															$n = 3$			
$n = 4$						$n = 3$									$n = 4$			
$n = 5$						$n = 4$									$n = 5$			
$n = 6$			*			$n = 5$									$n = 6$			
$n = 7$			*															

Shape (maximum electrons) →	$l=3$ (14)													
Electrons actually present →	1	2	3	4	5	6	7	8	9	10	11	12	13	14
							$n = 4$							
							$n = 5$							

insert
at
asterisks

132

TABLE 3.3 The quarks and leptons.

quarks[a]	symbol and electric charge	rest mass-energy[b,c,f] x 10^{-30} kg		leptons[d]	symbol and electric charge	rest mass-energy[b,e,f] x 10^{-30} kg
d	$d^{-1/3}$	8·3		electron-neutrino	v_e	
u	$u^{+2/3}$	3·8		electron	e^-	0·91
s	$s^{-1/3}$	166		muon-neutrino	v_μ	
c	$c^{+2/3}$	2260		muon	μ^-	188
b	$b^{-1/3}$	7440		tauon-neutrino	v_τ	
t	$t^{+2/3}$	308,000		tauon	τ^-	3160

a) Each quark is 2 particles, one of matter with color charges red, green or blue, and one of antimatter with opposite electric charges and color charges of anti-red, anti-green or anti-blue.
b) It will be noted that the particle rest mass-energies have no pattern to them. Why they have the values they do is a total mystery.
c) To put quark masses in perspective, a proton has a rest mass-energy of 1670 x 10^{-30} kg. Note that 99% of the mass-energy of protons and neutrons is due to the wave, kinetic and potential mass-energy of the mutually interacting quarks and gluons within them.
d) Each lepton is 2 particles, one of matter and one of antimatter with opposite electric charge.
e) Due to their identities being governed by quantum probability (see section 3.5 'The Lepton Family'), the rest mass-energy of each neutrino is not definite. The sum of their combined values is, however, no more than 1/500,000th of the rest mass-energy of the electron.
f) All particle masses are adapted from the Particle Data Group website http://pdg.lbl.gov .

TABLE 3.4 Some properties of the four forces.

force	range	carried by	rest mass-energy[c] x 10^{-30} kg	affects
strong (color)	infinite[a]	8 gluons[b]	0	quarks, gluons
electromagnetic	infinite	1 photon	0	quarks, $e^{\pm}$, $\mu^{\pm}$, $\tau^{\pm}$, W^{+}, W^{-}
weak	$<10^{-17}$ m	W^{+} W^{-} Z^{0}	143,100 162,400	quarks, leptons, W^{+}, W^{-}, Z^{0}
gravity	infinite	1 graviton?	0	everything

a) The strong (color) force range is in practice 10^{-15} m because all baryons and mesons must be white (= color neutral) overall.
b) The 8 gluons are:
red + anti-green
red + anti-blue
green + anti-red
green + anti-blue
blue + anti-red
blue + anti-green
and *two* gluons which are mixture of
red + anti-red
green + anti-green
blue + anti-blue.
c) The W and Z particle masses are adapted from the Particle Data Group website http://pdg.lbl.gov .

4

COSMOLOGY

4.0 OBJECTIVE: To describe the universe, the ultimate objective reality.

We confront in this chapter one of the oldest philosophical questions: 'What is the universe?' To find a scientific answer, it is necessary more than anything else to establish how far away the stars are. Only by this means can some idea be obtained of how big the universe is, size being a key factor in any account of the universe's nature. Starting from the Earth and working outwards, we shall see how astronomers estimate distances using a long chain of connected observations. The knowledge thus acquired will provide the basis for a theory of the structure of the universe — a universe in which billions of galaxies, each containing billions of stars, stretch across space as far as the most powerful modern telescopes can see. This structure is not, however, one which is fixed for all time. We will discover that observation of distant galaxies leads to an additional conclusion that the universe is changing on a scale of billions of years. By finding out what cosmological theories predict about what the universe was like in earlier times, we shall be able to attempt to describe how it began. From that ferociously explosive beginning we shall then travel forward in time, describing how the universe evolved from its initial state to become the one we observe today. Lastly we will see how one of the stars in one of the galaxies, our sun in the Milky Way, acquired its family of planets. Which should bring us firmly back down to earth!

4.1 THE SOLAR SYSTEM

INTRODUCTORY REMARKS

The scientific investigations of dynamics and of particle physics have their roots in people's need to be able to deal predictively with objects encountered in the terrestrial environment. Many, if not most, sciences share this characteristic. Geology, meteorology, botany, and medicine are examples which spring readily to mind. In all these cases, comfort and even survival provide the initial motivation. The same cannot be said of cosmology — the study of the universe as a whole. Cosmology is a product of curiosity for its own sake, as is witnessed by the fact that it has brought no physical benefit at all to humankind.

Yet despite its irrelevance to the imperatives of daily life, it remains a compelling and fascinating subject. Together with its sister science, astronomy, which may loosely be defined as the study of non-man-made lights in the sky, it can claim to be a very ancient subject. Undoubtedly, even far back in the past, humans and pre-humans occasionally found time before going to sleep to look up at a cloudless night sky. At the very least it is a puzzling sight. At best it is an experience of transfixing beauty.

Scientific enquiry into the nature of the universe starts as always with observation. Putting aside such phenomena as auroras, comets and shooting stars, which are not of concern here, naïve observers will identify three types of object in the sky: the sun, the moon and vast numbers of stars. In investigating the last of these three things, the drawback is immediately encountered which sets cosmology (and astronomy) apart from other sciences: it isn't possible to actually get to stars and examine them and do things to them and check they are what and where they appear to be. In consequence, knowledge about the universe rests on only two restricted kinds of observations; namely, the appearance of stars, and measurements indicating how far away they are. Because of these limitations, the observations on which cosmological theories are based are peculiarly vulnerable to reinterpretation. Of all the sciences described in this book, cosmology has the greatest potential for revision and revolution. This should be kept in mind in what follows.

Furthermore, cosmological science is dependent on the assumption that the laws of physics found on the Earth apply both at, and across, astronomical distances. Having regard to the effects on scientific laws of the extremely small distances encountered in the previous chapter, such an assumption needs to be treated warily.

THE MOON AND THE EARTH

The obvious place to start an investigation of the universe is with the easiest object to observe: the moon. Finding the distance to the moon is so straightforward the principle of how to do it was discovered about two thousand years ago.

This earliest attempt at measuring an extra-terrestrial distance starts with a particular observation of a half-moon — what astronomers call a first quarter and a third quarter moon. It is possible, looking at the illuminated half, to picture in what direction the sun must lie with respect to the moon. It can also be seen that the time taken to go from new moon to first quarter is the same as the time taken to go from first quarter to full moon. What these observations indicate is that the sun must lie in the same direction viewed from the moon as it does viewed from the Earth. That in turn means it must be very much further away from either of them than they are from each other. In other words, the sun will look the same size whether seen from the Earth or the moon, or indeed from anywhere in the Earth's vicinity. Specifically, the difference between the direction of one edge of the sun and the direction of the opposite edge, a difference called *angular diameter*, will be the same. Even an observer positioned at the apex of the shadow-cone cast by the Earth will find the sun's angular diameter to be approximately equal to that found by an observer standing on the Earth's surface.

The first step to take in order to find the Earth-moon distance is to measure the *moon*'s angular diameter, which is found to be $0.5°$. Now, because the moon exactly covers the sun during total solar eclipses the sun's angular diameter must also be $0.5°$. And that, since the sun is very far away, is also almost exactly the angular diameter of the sun as seen from the apex of the Earth's shadow-cone.

The next observation to carry out is of a lunar eclipse. The maximum time taken for the moon to enter the Earth's shadow-cone and pass through it enables the diameter of the shadow-cone at the moon's location to be calculated. It turns out that, where the moon is located, the Earth's

138

shadow is about 2·75 times wider than the moon is. In other words, the angular diameter of the Earth's shadow-cone at the moon's distance from the Earth is 2·75 times larger than the angular diameter of the moon.

That leaves one final measurement to be made, namely, the circumference of the Earth. To find this, two identical poles are placed upright in the ground a measured distance apart — say 1000 km — on a north/south line, and the angle each of their shadows makes with the vertical at midday is measured. The two angles will differ, and thereby reveal how far round the curve of the Earth has been travelled going from one identical pole to the other. It's then just a matter of extending the curve arithmetically until it becomes a full circle. In fact a distance of 1000 km corresponds to 9°. 9° is one fortieth of a full circle. Hence the Earth's circumference is 40,000 km. Dividing this figure by π gives a value for the Earth's diameter of about 12,750 km.

Putting these four facts together — the Earth's diameter, sun's angular diameter, moon's angular diameter, and the Earth's shadow-cone's angular diameter where the moon is — reveals the moon's distance to be around 400,000 km. (How the four facts lead to this conclusion I leave to you, if you'd enjoy the challenge, to work out for yourself. Hint: draw an isosceles triangle representing the Earth's shadow-cone and calculate the length of the sides. Then place the moon inside the triangle.)

Since distance holds the key to deciding what any particular astronomical object is, knowing the distance from the Earth to the moon permits evaluation of a number of the moon's properties. For example — now using today's more accurate distance estimating techniques — the moon's diameter can be calculated to be 3475 km. Compare this to the Earth's diameter of 12,756 km. The two values are similar enough to suggest that the moon is like a small Earth, shining by reflecting sunlight, a notion consistent with the phases through which the moon passes. It can also be said reciprocally that in astronomical terms the Earth is like a large moon. It is not a totally separate object from the sun, moon and stars. Astronomers on the moon would regard the Earth in the same way as their counterparts on the Earth regard the moon. So establishing the moon's distance gave astronomers, as a bonus, their first clue to the status of the Earth in the universe.

Knowledge of the distance between the Earth and the moon also enables the masses of the Earth and the moon to be found. Here the laws of dynamics must be applied to the observations. The moon continuously circles the Earth. It can be concluded that since the moon moves in a circle rather than a straight line, it must be constantly accelerating, the acceleration in this case being a change of direction rather than speed. The force that produces the acceleration comes from the Earth and it can be surmised that the force is gravity.

As was seen in section 1.2, the gravitational force between two objects of masses m_1 and m_2 is $\mathbf{F}_g = (Gm_1m_2/r^2).(-\hat{\mathbf{r}})$ where r^2 is the square of their distance apart. Remember also that $\mathbf{F} = m\mathbf{a}$. Putting these two equations together and setting m_1 to be the mass of the Earth, m_e,

$$\frac{Gm_em_2(-\hat{\mathbf{r}})}{r^2} = m_2\mathbf{a}$$

The m_2 values on each side cancel out and the value of G is known. Also, having found the value of $\mathbf{r}$, it is possible to work out the acceleration of the moon towards the Earth, $\mathbf{a}$, from the time it takes the moon to complete an orbit. Putting all these quantities into the equation enables m_e to be calculated. The value obtained is $5\cdot97 \times 10^{24}$ kg. This is the mass of the Earth.

Next, by carefully plotting the Earth's position in space it is possible to evaluate, through the effect of the moon's gravity on the Earth's path, the ratio of the moon's mass to the Earth's. This gives the mass of the moon as $0\cdot073 \times 10^{24}$ kg.

The key point to make regarding this information about the Earth and the moon is that none of it could have been obtained without first determining how far away the moon is. It should be plain from this just how crucial to astronomy distance measurement is.

PLANETS AND THE SUN

Almost all stars appear in the sky night after night, year after year, in the same positions with respect to one another. But there are a few exceptions. These exceptions are called planets and careful, systematic observation of them leads to the conclusion that they orbit the sun. Is it possible to find out how far away the planets are?

There is an interesting mathematical conclusion about orbiting planets which emerges from the theory of gravity: if the time it takes a planet to complete one orbit (T) is squared, and the result is divided by the cube of the relative size (RS) of the planet's orbit, a constant value will be obtained. For every planet, $T^2/(RS)^3$ is the same constant. As far as evaluating the distances to the planets is concerned, this is a step in the right direction. But unfortunately, knowing the relative sizes of the planets' orbits doesn't of itself give any clue to their actual size. Something more is needed.

Enter *parallax*. Parallax entails measuring the position in the sky of an object from two widely separated locations on the Earth's surface. The measurement is made simultaneously at both places. The precise separation of the two locations is also measured. The various measurements can then be combined to give the distance to the object under investigation.

This method of measuring distances works in the following way. There are two observers on the Earth's surface at points A and B. Both observe an astronomically nearby object S. The observer at A notes exactly where S appears to be with respect to a number of background stars. The observer at B does the same. They then compare their results. Because they have been looking at S from different locations, S's position in the sky measured by A will differ from its position measured by B. It is this difference which is known as parallax.

The cause of the difference may not be immediately obvious, but the human brain is well aware of it. Hold a finger at arm's length and look at it with one eye closed. Note how it lines up with things in the background. Then, without moving the finger, look at it with the other eye. The finger will appear to have moved with respect to the background. The brain uses this difference between the two eyes' viewpoints to estimate the distance away of terrestrial objects (up to about 60 m). Astronomers do the same on a much larger scale for planets. (Human eyes are about 7 cm apart; two astronomers may be thousands of kilometres apart.) The end result is that astronomers are able to do for planets using precise measurements and mathematics what the human brain does for terrestrial objects using some clever neurological computing.

The easiest astronomical object to get a parallax value for is Mars. (Venus comes closer, but at its closest it lies in the same direction as the blindingly bright sun.) Simultaneously measuring the position of Mars from different locations on the Earth's surface makes it possible to calculate the distance from the Earth to Mars at their closest approach to each other. This provides a good value for the difference in their orbital radii. If the times the Earth and Mars take to complete one orbit of the sun are written T_E and T_M, and their orbital radii are written R_E and R_M then

$$\frac{T_E^2}{R_E^3} = (\text{constant}) = \frac{T_M^2}{R_M^3} = \frac{T_M^2}{(R_E + x)^3}$$

where x is the closest distance from the Earth to Mars. In this equation T_E and T_M are known and x has been evaluated using parallax, leaving R_E as the only unknown. The value of R_E, which is the distance from the Earth to the sun, can thus be worked out. It may seem a paradox that measuring the parallax of Mars reveals how far away the sun is, but the mathematics is sound.

The Earth-to-sun distance is found to be 150×10^6 km. As with the moon, knowing the sun's distance permits its diameter to be calculated. It has a value of 1,392,000 km. Also, by adapting the $\mathbf{F_g} = (Gm_1m_2/r^2).(-\hat{\mathbf{r}})$ and $\mathbf{F} = m\mathbf{a}$ equations, setting m_1 to be the mass of the sun, m_2 to be the mass of the Earth, and $\mathbf{a}$ to be the acceleration of the Earth towards the sun, the sun's mass can be found. It turns out to be 2×10^{30} kg. Compared to the Earth and the moon, the sun is a million times more massive and a hundred times wider. It is radically different from them.

Consider next the planets. As T_E and R_E are now both known, the 'constant' above can at last be determined, which in turn means that for each T (T_M for Mars, T_J for Jupiter etc.) the corresponding R can be calculated. Thus it becomes possible to say how far each planet is from the sun. Those that have moons of their own, often called satellites, can also have their mass evaluated directly.

(A note for enthusiasts: since the advent of unmanned spaceflight, the planets have had their masses determined by employing spacecraft in the role of (artificial) satellites. The mass of the moon has also been verified by this means.)

The masses of the Earth and the planets and their average distances from the sun can be listed as follows. (All the figures are from http://nssdc.gsfc.nasa.gov/planetary/factsheet .)

Mercury	0·33	x 10^{24} kg	57·9 x 10^6 km
Venus	4·87	x 10^{24} kg	108·2 x 10^6 km
Earth	5·97	x 10^{24} kg	149·6 x 10^6 km
Mars	0·64	x 10^{24} kg	227·9 x 10^6 km
Jupiter	1898	x 10^{24} kg	778·6 x 10^6 km
Saturn	568	x 10^{24} kg	1433·5 x 10^6 km
Uranus	86·8	x 10^{24} kg	2872·5 x 10^6 km
Neptune	102	x 10^{24} kg	4495·1 x 10^6 km
(Pluto	0·01	x 10^{24} kg	5906·4 x 10^6 km)

It can be seen that in terms of mass, the planets are more like the Earth than the sun. Even the largest, Jupiter, is only one thousandth the mass of the sun. As the planets orbit the sun just as the Earth does, and shine by reflecting sunlight just as the Earth would appear to do if it was seen from space, it is justified to classify the Earth as a planet. Thus the sun is orbited by eight planets. (Pluto I have added to the list for historical interest. It is now considered too small to count as a planet, being, in fact, smaller than the moon!) The planets are in turn orbited by a varying number of satellites.

One important point to be made about the solar system — solar system being the collective name given to the sun and its planets and their satellites — concerns sizes and distances. Planets are like tiny dots in a big space. The distance from a planet's centre to its surface is very roughly ten thousand times less than its distance from the sun (a size ratio coincidentally like that between the nucleus and the electron shells of an atom). Well may people talk of outer *space*!

For all astronomers' success in describing the solar system, a glance skyward on a clear night cannot leave any doubt that their task has barely begun. The stars await!

4.2 STARS AND GALAXIES

EXTENDING PARALLAX

Before it is possible to get anywhere in discovering what stars are, some way must be found to determine their distances from the Earth. Only then can it be decided whether they are bright lights far away or dim lights close to. (Far and close here are of course comparative terms.) The trouble is that the only method of measuring distance which does not make unacceptable assumptions in advance is the parallax method, and when that is attempted from different locations on the Earth's surface no parallax can be detected. The stars are simply too far away.

However, all is not lost. It may not be possible to make progress measuring parallax from different points on the Earth's surface, but the same is not true if different points on the Earth's orbit around the sun are used. The diameter of the Earth's orbit is 3×10^8 km. This is 25,000 times wider than the Earth itself is. Points A and B used by astronomers to measure parallax now become points on the Earth's orbit six months apart. Which means that the star whose distance is to be determined can be 25,000 times further away without going outside parallax range.

Because determining stellar distances is so vital to astronomy, astronomers have devoted much effort to refining their parallax techniques. Using telescopes mounted on satellites, and thus unaffected by atmospheric distortion, they can measure the position of a star to better than an astonishing 1/100th of a second of arc; that is, less than 1/360,000th of a degree. This means that any star nearer than 4×10^{18} m away can have its distance established by means of parallax.

Distances to the stars revealed by the extended parallax method begin at a value in excess of 4×10^{16} m and run right up to — and beyond — the maximum range that astronomers are capable of measuring.

To get the distances to the stars into perspective, the distance between neighbouring stars is typically ten million times greater than their diameters. In terms of interstellar space, the solar system is quite crowded!

Despite the general emptiness of space, stars whose distances are determinable by extended parallax can now be numbered potentially at a million. It would not be unreasonable to think that that very nearly wraps it up. It might seem that a million is a fair estimate of the total number of stars. But it is unfortunately wildly inaccurate. Telescopes reveal that stars within reach of extended parallax are a tiny minority. A million barely scratches the surface. The vast majority of stars lie beyond the range of extended parallax.

SPECTROSCOPIC PARALLAX

Parallax makes it possible to evaluate the distances of stars up to about 4×10^{18} m away. For stars further away than this, a different method of distance measurement is needed which overlaps with the parallax method. The overlap is required so that distance values obtained by the new method can be compared with those obtained by parallax. In that way, the assumptions made about any new technique can be tested for validity by checking that it does indeed give the same results as parallax. Providing it consistently passes this test, it can be assumed that it is also reliable far beyond 4×10^{18} m.

The new method must therefore be sought by studying stars within parallax range. A large quantity of data about such stars needs to be amassed in order to see what regularities emerge. If the data can be used to formulate some sort of stellar law, that law may be applied to stars outside parallax range and thereby act as a guide to their distances.

The most obvious data that can be collected consist of measurements of stellar brightness. The brightness of a star is equivalent to the amount of electromagnetic wave energy from the star that a telescope on the Earth receives. By measuring this quantity it is possible to express the brightness in terms of a number of joules per second. The number of joules per second that a telescope receives from a star is a measure of how bright it appears to be when viewed from the Earth, and is called the *apparent brightness* of the star. If the distance of the star from the Earth is known, it is clearly possible to calculate how bright it really is; that is, whether it is a bright light far away or a dimmer one close by. The actual brightness of the star, once allowance has been made for its distance, is called the *real brightness*.

The real brightness of a star can be thought of as the amount of electromagnetic wave energy it emits in all directions. This too can be expressed as a number of joules per second. Amounts vary widely but 10^{26} J s^{-1} is typical. Where this wave energy comes from will be revealed in section 4.6.

The first piece of information which can be tabulated for stars within parallax range is therefore the quantity of wave energy they emit. Although this is not of itself sufficient to formulate a stellar law, it does have one immediate by-product: it reveals that the sun is a star. The sun's electromagnetic wave energy emission is $3·9 \times 10^{26}$ J s^{-1}, which means it's on the large side of being an average star. The reason it doesn't look like a star is simply that it is so close to the Earth. It has an enormous apparent brightness because it is a bright light close to rather than a bright light far away. Its real brightness, taking distance into account, is no different from most stars.

Another piece of data which might be of relevance in constructing a stellar law is the colour of each star; that is, the wavelengths of the electromagnetic waves it emits. When light from a star is passed through a glass prism, it is spread out to form something resembling a rainbow. (In fact droplets of water in rain function in a similar way to glass in a prism.) When this stellar rainbow is examined, a continuous band of colour, called a *spectrum*, running from red to violet is observed. However, close inspection reveals a considerable number of bars in the spectrum. The bars correspond to wavelengths present in below average numbers. To explain this, some atomic physics must be recalled.

It was seen in section 3.2 that electrons emit and absorb electromagnetic wave energy in quanta. When an electron falls from an upper to a lower shell, it emits a wave with a fixed amount of energy. The quantum of wave energy, E_λ, has a specific wavelength, λ, associated with it: $E_\lambda = hc/\lambda$. Conversely, when an electron goes from a lower to an upper shell, it absorbs a quantum of energy which again corresponds to a particular wavelength. The result of the quantum behaviour of atomic electrons is that the electromagnetic waves emitted by a hot body, even a star, have certain wavelengths which are dim because atomic electrons have absorbed some of the waves before they can leave the star and head off into space.

Thus for stars within parallax range it is possible to measure three things in addition to their distance; namely, apparent brightness, colour and spectrum. Knowing the distance as well as the apparent brightness enables the real brightness to be found, and it then becomes clear that with only a few exceptions colour and spectrum and real brightness are closely correlated. In other words, stars with matching colours and spectrums are sufficiently like each other that they have the same real brightness. This is the required stellar law.

Applying this law to stars outside parallax range involves measuring the target star's colour and spectrum, finding similar stars within parallax

146

range, and looking up their real brightness. The target's stars real brightness is then assumed to be the same. Measuring its apparent brightness and knowing its real brightness permits its distance to be calculated. The method is called *spectroscopic parallax* (though parallax is involved only indirectly) and it enables the maximum measurement of stellar distances to be extended 100-fold.

As the distances involved are so huge, astronomers express them not in metres and kilometres, which are hopelessly inadequate, but in light-years or parsecs. A light-year is the distance a photon travels in empty space in one year (about 10^{16} m). A *parsec*, which is the unit of length I shall use from now on, is derived from parallax and is about 3·25 light-years or $3·085 \times 10^{16}$ m. In terms of parsecs, the nearest stars are a little over 1 parsec (1 pc) away. Parallax can evaluate distances to 100 pc. Spectroscopic parallax can evaluate distances to 10,000 pc.

The limitation which stops spectroscopic parallax being used beyond 10,000 pc is that stars further away than this are too indistinct for measurements of their characteristics to be made. However, it is still possible to detect such stars, especially when they occur in groups. For these stars, another way of estimating their distances is needed. Failing that, all that can be said about the structure of the universe is that it exceeds 10,000 pc in size, but it cannot be determined whether it exceeds this value by a little or a lot. It is therefore essential for theories of cosmology that the distances to *all* the stars are known. Another criterion of distance is needed.

CEPHEID VARIABLES

Investigation of stars within the 10,000 pc spectroscopic parallax limit soon reveals that some of them are not normal. Amongst these stellar oddities is one which provides the next distance indicator.

There is a type of star known as a *Cepheid variable*. Its abnormality lies in its brightness, which is not constant but fluctuates. Cepheid brightness changes noticeably, increasing and decreasing in a regular cycle over a timescale of from a day or so up to a month or more. For any one Cepheid, the brightness takes a fixed period of time to go through a complete cycle.

Now, some Cepheids are gravitationally associated with normal stars lying within spectroscopic parallax range. Investigation of these Cepheids leads to a law that those with the greatest real brightness are also those with the greatest time between peaks of brightness. It follows, conversely, that measurement of the time between peaks of brightness enables the real brightness to be calculated. Once again, comparison of real brightness with measured apparent brightness enables distance to be determined.

Because stars having a fluctuating brightness are easier to spot than more nondescript stars, and because Cepheids at maximum real brightness are very bright indeed, they can be seen at distances of up to 20×10^6 pc. Hence the distance to any group of stars which contains amongst its number a Cepheid variable can be established up to this limit.

THE MILKY WAY

With this third criterion of distance it becomes possible for astronomers to put forward their first cosmological theory. That theory is a description of those parts of the universe which lie within Cepheid variable range.

The sun is one unit in a structure which consists of about 10^{11} stars distributed in space in the shape of a slowly rotating disc. The disc is about 30,000 pc across and 500 pc thick. Above and below this disc, and orbiting its centre, are very roughly a further 10^9 stars, about 1% of which are found in spherical groups called globular clusters. Each cluster typically contains between 10^4 and 10^6 stars. Because the clusters are in orbit around the centre of the disc it is possible, by mapping their distribution, to know where the centre of the disc is, and it is this that permits estimates of its size to be given. This is despite most of the disc being hidden from the Earth by interstellar dust, so that only a small fraction of the disc's stars can be seen.

The sun lies about six tenths of the way from the centre to the edge of the disc. From the Earth, looking along the disc as opposed to perpendicularly out of it, an observer sees many stars apparently close together. As the Earth is inside the disc, the band of densely packed stars seems to encircle it. The band of stars is called the *Milky Way*.

GALACTIC DISTANCE MEASUREMENT

Our piece of the universe thus has the structure of a disc surrounded by a spherical halo of stars, some of which are in globular clusters. Does anything lie beyond that? Continuing to concentrate on Cepheid variables, some are found to be so far away they must be completely outside the Milky Way and its halo. Each such Cepheid has many other stars associated with it. Knowing the distance to the Cepheid enables the scale of the star group within which it lies to be determined. It transpires that these other star groups are collections of stars of similar size to the Milky Way. Such star groups are called *galaxies*. Most galaxies contain between 10^7 and 10^{12} stars. The Milky Way is a typical galaxy. The closest comparable galaxy, called Andromeda or M31, is 700,000 pc away.

Even in the largest telescopes all but the nearest galaxies look like fuzzy stars. And all but the nearest ones are outside the reach of Cepheid variable measurements.

Beyond Cepheid range there is one more type of star which can be used to determine distance, a class of star so spectacularly unusual as to make Cepheid variables seem dull by comparison. The stars in question are called *supernovas* and more will be said about them in section 4.6. Considering they are only stars, supernovas produce an enormous burst of light, lasting only a few weeks but so bright they can be seen at a range of 10^9 pc (10^3 Mpc).

In general, the real brightness of supernovas differs from one supernova to another, making them problematic as distance indicators. But there is one type of supernova, called Ia, (pronounced 'one-ay'), for which this is not true. A type Ia supernova begins life as one of a pair of stars in orbit about each other. The more massive of the two stars comes to the end of its life first and its internal fires begin to die down. (The details of this process will be found in sections 4.6 and 10.3.) However, as it cools it also acquires gaseous material from its orbiting partner, becoming thereby even more massive. If its mass eventually exceeds our sun's mass by about 40%, the star explodes catastrophically. This explosion is what is seen on the Earth as a type Ia supernova. As supernovas of this kind all have the same mass at the time they explode, their explosions all involve much the same energy output. Once allowance is made for the variable time-course of the explosion, the real brightness can be evaluated and compared to the measured apparent brightness. The distance can then be calculated.

Unfortunately there are many distant galaxies either outside type Ia supernova range or where no type Ia supernovas have been seen, so a second way to estimate distances on a scale of thousands of megaparsecs is called for. From studying galaxies generally it is found that they are often grouped together in clusters. Discounting the brightest galaxies in such a cluster, it is found that the next brightest galaxies in the cluster tend to have a fairly standard real brightness. As with other standard brightnesses, measuring the apparent brightness of these less-than-brightest members of the cluster, and comparing that with their predicted real brightness, enables their distance to be calculated and thereby gives the distance to all the galaxies in the cluster.

With this new distance criterion — galactic brightness — the task of measuring the distances to all the stars in the universe is nearly complete. There only remain galaxies not belonging to a cluster or that are suspected of being odd in some way. Obviously they cannot have their distances estimated using normal galactic brightness as an indicator. A sixth (and final) distance scale is required.

The sixth distance scale draws once more on atomic spectrums — a new variation on the theme of spectroscopic parallax. Even the spectrum of light from entire galaxies has bars in it, corresponding to the wavelengths emitted and absorbed by atomic electrons in gas distributed within the galaxy. Detection of these wavelengths in the light from distant galaxies does not of itself provide any means of estimating distance. But it is possible to say what the wavelengths should be on the basis of laboratory experiments, and in every case it is found that the wavelengths are too long. The wavelengths are there all right, but it appears they have somehow been stretched on their journey through space. Examination of the data for normal galaxies shows that the amount by which the wavelengths are stretched is proportional to how far away galaxies are. The further away the galaxy is, the greater the stretching. There seems no obvious reason why this should not apply to all galaxies, both normal and abnormal, without exception. It can be concluded, then, that the greater the stretching of the wavelengths from a galaxy, the further away it must be. The technical name given to the stretching of wavelengths is *redshift*, and redshift it is which constitutes the final criterion of distance.

By using the distance scales defined by type Ia supernovas, galactic brightness and redshift, it is possible to determine how far away even the remotest visible objects are. At last, measurements can encompass the entire observable universe. The most distant objects are about 3500 Mpc away, equivalent to 10^{23} km, or 11·6 billion years spent travelling at the speed of light. (More precisely, that's how far away they *were* when the light from them was emitted.) There are thought to be at least 10^{10} galaxies that are potentially observable.

THE SIZE OF THE UNIVERSE

In this section various means of measuring distance have been used to discover the nature of the stars and thereby gain insight into the structure of the universe. These measurements reveal that some of the stars are suns, and that others are collections of thousands of millions of suns. The universe is thus an utterly vast place. The use of exponent notation and units like megaparsecs serves only to obscure the scale of things. But writing the numbers out in full hardly provides any clarification either. To say, for instance, that the number of stars in the universe is at the very least 1,000,000,000,000,000,000,000 doesn't seem greatly to assist comprehension. The universe revealed by science is more awesome than any notion of it devised by non-scientific thinkers.

To further emphasize the point, it can be remarked that observation only fixes the *minimum* size of the universe. It cannot be smaller than the most remote object's distance away, nor younger than the oldest object's age. It could, however, be infinitely larger and older.

Before leaving this section there are two cautionary comments that must be made. Distance estimation is based on a very extended line of reasoning. Parallax can be relied upon. Spectroscopic parallax is slightly less reliable. Cepheid distance cannot be more reliable than spectroscopic parallax distance and is from a practical point of view less so. Similar statements can be made in turn about type Ia supernovas, galactic brightness and the redshift. In other words, by the time redshift distance is being assigned, the actual values must be regarded as approximate.

In a similar vein, all distance scales barring parallax rest on the assumption that the laws of physics on the Earth, particularly those concerned with electromagnetism and atoms, apply throughout the universe. This is especially the case regarding redshift. Sometimes galaxies apparently associated gravitationally with each other (i.e. close together) have very different redshifts. Clearly, if redshift can be caused by some unknown non-distance related process, as such anomalous associations imply, the whole redshift contribution to distance scales becomes unreliable. This is an important point given the central role redshift plays in the current scientific description of the universe.

To conclude this section, here are the main distance scales summarized.

Parallax	to 100 pc	
Spectroscopic parallax	to 10^4 pc	
Cepheid variables		to 20 Mpc
Type Ia supernovas		to 1000 Mpc
Less bright galaxies in cluster		to 3500 Mpc
Redshift		to 3500 Mpc

It should be noted that these are not the only distance scales used by astronomers. Rather, they are six of the most important ones.

4.3 THE STRUCTURE OF THE UNIVERSE

GALACTIC RECESSION

Now that the size of the observable universe has been evaluated, this knowledge can be used as a foundation on which to build a theory of the universe's structure. Aside from the sheer enormity of the place, the single most important factor any theory must take account of is the galactic redshift.

The reason for singling out redshift is its apparently mysterious nature. Of the six distance scales, parallax relies on the geometry of space, while the spectroscopic parallax, Cepheid and type Ia supernova scales use the properties of certain stars. Galactic brightness is based on the properties of normal galactic clusters. The source of the redshift, though, is less obvious. It might at first be thought that stretching of electromagnetic waves is a property of galaxies in the same way that brightness is a property. It is true that the gravity of a galaxy will produce a small redshift in the light it emits, but there is no reason to believe that a galaxy's gravitational redshift will increase the further away the galaxy is.

The same applies to all the other intrinsic properties of galaxies. There are no grounds for believing any of them will produce a redshift which increases with distance.

So what is the cause of the redshift? The accepted answer makes use of a characteristic of electromagnetic waves (and other kinds of waves generally). Specifically, if an object emitting electromagnetic waves moves away from an observer, then the waves continue to be emitted at the same rate, but the peak of each wave has further to go before reaching the observer than its predecessor. This means the observer has to wait longer between the peak of one wave and the next than if the emitter is stationary. In other words, the observer finds the waves have a longer wavelength than would have been the case had the emitter of the waves not been moving away.

Applying this characteristic of waves to cosmology leads to the hypothesis that galaxies in general are moving away from the Earth. The amount of redshift in the light from a galaxy indicates the rate at which the galaxy is receding. In fact, the amount of redshift must be exactly proportional to recession velocity. It then follows from this, since redshift has been found by observation to be proportional to distance, that galactic recession velocity too is proportional to distance. The further away from the Earth a galaxy is, the faster it is receding.

It should be clear now why redshift has been singled out as central to cosmological theories. If galaxies are all receding from the Earth, this must be a property of the whole universe; and it is the whole universe that cosmology seeks to describe.

THE COSMOLOGICAL PRINCIPLE

The relationship amongst distance, redshift and recession velocity is a simple linear one. If the distance is doubled, the redshift is doubled and that means in turn that the recession velocity is doubled. If the distance is trebled, the redshift and recession velocity are also trebled, and so on.

This proportionality is in accord with a proposition known as the *Cosmological Principle*. The Cosmological Principle states that the overall appearance of the universe is the same from every typical point within it. Given that the galaxies are all receding from the Earth, the Cosmological Principle can only be valid if the rate of recession is proportional to distance, as is observed. With simple linear proportion-ality, no matter what galaxy observers are in, they will see all the other galaxies moving away from them with speeds proportional to distance; i.e. the universe will look the same from every galaxy. If distance and recession velocity were not proportional, then the observed motion of other galaxies would vary according to which galaxy an observer was in. The Cosmological Principle would be violated.

The discovery that galactic recession is compatible with the Cosmological Principle is encouraging. This is because the Cosmological Principle implies that the laws of physics derived on the Earth are valid throughout the universe, which was one of the key assumptions drawn attention to in previous sections. It is known that an experiment performed in one place on the Earth will yield the same result as the same experiment performed somewhere else on the Earth. Within reason, it doesn't matter where the experiment takes place. The implication of this is that scientific laws and theories do not simply apply to where they are discovered but apply everywhere. Now, the laws and theories of science amount to a description of the universe. This description is a statement of what the universe looks like. So to assert that the laws and theories of science are valid everywhere is to imply that the universe should look the same wherever an observer happens to be. Which is the Cosmological Principle.

In other words, if the laws of physics are universally valid then the Cosmological Principle should also be valid. The Cosmological Principle predicts that recession velocity is proportional to distance. And observation confirms the prediction, which is gratifyingly consistent. It gives grounds for confidence that the statements made about distant parts of the universe are valid, even if they are based on knowledge obtained purely locally. The Cosmological Principle plays a central role in the scientific account of the structure of the universe. (Though see 'Dark Energy', below, where this role will be questioned.)

THE BIG BANG

Since galaxies in general are receding from the Earth with speeds proportional to their distances, they must also be receding from each other in like fashion. As time passes, the average distance between galaxies is increasing. (Only galaxies which are close to one another tend not to

154

conform to this statement; they are sufficiently attracted by each other's gravity that they are in orbit about one another.) The conclusion to be drawn from the overall universal galactic recession is that the universe is expanding. Each galaxy is gradually occupying a bigger and bigger volume of space as all the other galaxies recede from it.

So, looking towards the future the universe can be seen expanding. It follows that looking towards the past the universe will be seen contracting. While an expansion may not immediately give pause for thought, a contraction surely does, for a contraction cannot proceed indefinitely. Eventually, looking far enough back into the past, a time must be arrived at when the universe has contracted to such a small size that there is nothing left of it. That moment in time is the moment when the universe came into being. To describe what happened then is perhaps the ultimate challenge facing scientific method.

It is crucial to be clear what implication the expansion of the universe has for theories of its structure. The expansion requires the universe to be a changing one. The current structure, consisting of galaxies partially grouped into clusters extending out to 3500 Mpc, and all moving away from each other, is only how it is now. At some time in the future it may come to have a quite different structure. It was certainly radically different in the distant past. A full account of the structure of the universe must describe it at all times, not just as it is at present. A scientific view of reality needs to encompass past and future as well as the present.

Describing the future of the universe will be left until section 10.3. Here attention will be concentrated on the past.

The easiest and most obvious thing to discover about the past is how much of it there is. When did the universe come into being? The speed of galactic recession is the key to this question. Currently the favoured value is 75 ± 3 km s^{-1} Mpc^{-1}. (Plus or minus only 3 is perhaps optimistic.) In simple terms, what this means is that two galaxies 1 Mpc apart will be receding from one another at a speed of no less than 72 km s^{-1}. It also means that with every second back into the past they get at least 72 km closer. Dividing 1 Mpc by 72 km s^{-1} gives about 13,800 million years. This provides a rough idea of how long ago it was that the space between the two galaxies was contracted to nothing.

Because recession velocity is proportional to distance apart, the present separation of any two galaxies makes no difference to the value of

13.8×10^9 years. For example, two galaxies 10 Mpc apart are receding from each other at 720 km s^{-1}. Dividing 10 Mpc by 720 km s^{-1} still gives 13.8×10^9 years.

There is a complication to this simple calculation: just as a stone thrown up from the Earth slows down with the passing of time, so galaxies are now rushing away from each other more slowly than was the case in the past. The gravitational attraction between them will have produced some degree of deceleration. If a larger initial recession velocity is allowed for, it has the effect of reducing the value of 13.8×10^9 years to 9.2×10^9 years.

Unfortunately 9.2×10^9 years is not enough. Stars similar to the sun in the Milky Way have lifespans which can be reliably calculated, and the number of dying stars (white dwarfs, to be described in section 10.3) gives an age for the Milky Way of 10×10^9 years. Obviously the Milky Way cannot be older than the universe. For the moment this problem will be left unresolved. But it clearly cannot be ignored. It will be returned to shortly when discussing Dark Energy.

Whatever the eventually agreed age of the universe, it remains undeniable that at some time in the past the universe must have had a beginning — or at least a transition from a previous state radically different from how it is now. This is the moment when all the matter currently found in galaxies, initially compressed into an incredibly tiny space, began to expand. The beginning of the universe thus seems to have been a kind of explosion of this tiny space, with everything being carried at speed from a starting point of extreme compression. This event is appropriately enough called the *Big Bang*.

The impression must not be formed that the Big Bang took place at some point in the universe and then spread into the surrounding space. Such a picture is incorrect because it would violate the Cosmological Principle. Imagine hypothetical observers able to watch as well as take part in the Big Bang. The Cosmological Principle requires that, wherever they are, they will see the rest of the universe exploding away from them. This would not be the case if they were able to watch the Big Bang from the outside. It follows that it is not possible to be outside the Big Bang. Everything, including space itself, came into being as part of the explosion. 'Everywhere' was contained in the starting point.

THREE MODELS OF THE UNIVERSE

As the explosion proceeded, the universe expanded, as it continues to do today. Space itself got bigger, and is still getting bigger, to accommodate the increasing distances between the receding galaxies that have so far proved to be the Big Bang's end product.

To describe the expansion in terms of the structure of the universe results in something which is very hard to visualize. The best — if not the only — way to picture what is going on is to have recourse to a simplified model of what space looks like. The simplification takes the form of a removal of one of the universe's three space dimensions. In essence this means turning cubes into squares, spheres into discs, and the universe itself into a surface rather like a sheet of paper. Instead of possessing length, width and height, the simplified model only possesses length and width. Height will be completely eliminated from the picture. Stars in normal space will become shining dots on a surface.

The surface that constitutes this model of the universe has only two space dimensions. In other words, to say where a point is on the surface only two numbers need to be given. However, the surface does not have to be perfectly flat for this to be true. As long as it creates no boundaries or edges — which would be a breech of the Cosmological Principle — the surface can take any shape. Fortunately there are only two basic ways to construct a two-dimensional surface which is not flat. (The more complicated ways are ruled out by Occam's razor.) One way is to curve the surface back on itself so that it becomes a sphere. The other is to curve it in the opposite sense to that which produces a sphere. Doing that creates what is known as a *hyperbolic surface*.

There are thus three configurations that the model universe can take up. The model's surface can have the form of a sphere; it can be a completely flat surface; and it can form a hyperbolic surface. These options are illustrated in figure 4.1 where they are labelled model 1, model 2, and model 3, respectively. Note that although models 1 and 3 have now become three-dimensional in appearance, the surface that represents the universe remains two-dimensional in both cases; i.e. you only need two numbers to specify exactly where you are on the model's surface.

How does the Big Bang look in terms of each of the three models?

Model 1
Spherical

Model 2
Flat

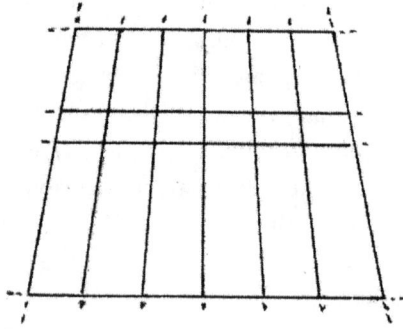

Model 3
Hyperbolic

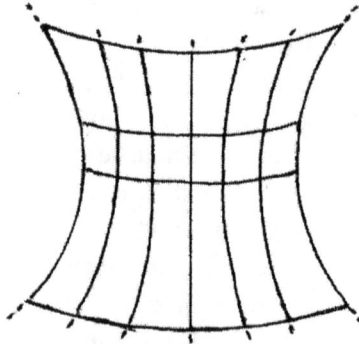

Figure 4.1 Three possible models of space, represented in two dimensions.
Only model 1 can be seen in its entirety. Models 2 and 3 are infinite in extent.

Model 1. The surface begins as a point. Following the Big Bang it expands rather like a balloon being blown up. The galaxies on the surface (which don't expand) recede from each other as the expansion proceeds.

Model 2. The surface is infinite in extent. (It has to be to conform to the Cosmological Principle.) It was so at the time of the Big Bang and always will be. The Big Bang happens everywhere throughout this infinite surface. It remains true that everything was initially compressed into a tiny space; in this infinite model there are an infinite number of tiny spaces containing an infinite number of everythings. As time passes, the surface stretches uniformly in all directions. The galaxies on the surface (which don't stretch) recede from each other as the stretching continues.

Model 3. The surface extends beyond infinity. It did so at the time of the Big Bang and always will do. As time passes, it expands uniformly in all directions. The galaxies on the surface (which don't expand) recede from each other as the expansion proceeds. The state of affairs at the time of the Big Bang with regard to everything being compressed into a tiny space is as for model 2.

These three models look quite different from each other. Ideally a test should be devised to find out which one most closely matches the observed three-dimensional universe. Such a test does in fact exist. To perform it, all that needs to be done is to measure how much mass-energy the universe contains in a given volume; that is, to measure the density of mass-energy in the universe.

The reason why evaluating mass-energy density can decide between the three models of the universe is that it is mass-energy that makes the model universe develop a curve in the first place. If the model contains a lot of mass-energy it curves into a sphere (model 1). If it contains a little mass-energy it curves hyperbolically (model 3). Between these two there is one unique amount of mass-energy which produces no curving at all (model 2). So measuring the mass-energy density of the universe theoretically permits a decision to be made as to whether it has the critical amount of mass-energy that corresponds to model 2, or more (model 1), or less (model 3).

The critical mass-energy density can be evaluated mathematically. It turns out to be 10×10^{-27} kg m^{-3}. What is the measured density? Adding up all the masses of all the stars in all the galaxies gives roughly $0 \cdot 04 \times 10^{-27}$ kg m^{-3}. Interstellar and intergalactic gas is estimated to be about ten times as much as this. Additionally, the way globular clusters orbit galaxies, and the way galaxies in clusters orbit one another, suggests there is much more matter in and around galaxies than can be seen. The form it takes — called *Dark Matter* (see section 4.4) — is currently unknown and a subject of speculation. Even adding in this Dark Matter, highest estimates for mass-energy density in all its forms stand at no more than $3 \cdot 2 \times 10^{-27}$ kg m^{-3}. This is only 32% of the critical density and therefore supports the model 3 universe.

However, various attempts to directly detect curvature in the universe have consistently failed to find evidence for it. It would appear the universe is geometrically flat to within a few per cent. To express this mathematically, cosmologists use a quantity called the density parameter. Written Ω, it can have any positive value. For $\Omega < 1$, the universe is hyperbolic; for $\Omega = 1$, it is flat; and for $\Omega > 1$ it is spherical. The various contributions to the universe's density simply add together. So, for the flat universe indicated by observation, the following equation can be written

$$\Omega_{total} = \Omega_{ql} + \Omega_{dm},$$

where Ω_{ql} is the density of normal matter (quarks and leptons) and Ω_{dm} is the density of Dark Matter. Putting the known values of these things into the equation gives

$$\Omega_{total} \approx 1 = 0 \cdot 05 + 0 \cdot 27!$$

Clearly something is wrong with the theory of cosmology.

DARK ENERGY

The theory of cosmology has so far failed two major tests: its estimate of the age of the universe is too low, and its estimate of the amount of mass-energy in the universe is too low. It also fails a third test; specifically, very distant type Ia supernovas have a redshift indicating one distance, while their brightness indicates a somewhat greater distance. In other words, the two values are incompatible. (N.B. This incompatibility rests on the assumption that 'young universe' type Ia supernovas have the same real brightness as 'current universe' type Ia supernovas.)

As explained in section 2.4, scientists should now reconsider the current theory of cosmology, since it has failed different tests repeatedly. However, as explained in section 2.5, what usually happens first is that ad hoc modifications to the existing theory are attempted. *Dark Energy*— a name which is as good as any other, but doesn't currently mean much — is a modification of this kind.

Here's how it works. Dark Energy addresses the low mass-energy discrepancy directly, by simply supplying the missing 0·68.

$$\Omega_{total} = \Omega_{ql} + \Omega_{dm} + \Omega_{de}$$
$$1 = 0{\cdot}05 + 0{\cdot}27 + 0{\cdot}68$$

Next it solves the age issue by means of a mathematical trick. Cosmologists rewrite the equation

$$\Omega_{total} - \Omega_{de} = \Omega_{ql} + \Omega_{dm}.$$

Placed on the left hand side of the equals sign, Ω_{de} becomes negative, which makes it repulsive — a kind of self-repulsion of space itself. This enables it to counteract gravity and so restore the original satisfactory $13{\cdot}8 \times 10^9$ year age of the universe. Lastly, because Dark Energy is repulsive it will change the relationship between redshift and distance when dealing with very distant objects, thus resolving the discrepancy between the redshift and type Ia supernova distance evaluations.

For one ad hoc modification to solve all three main problems with the theory of cosmology is certainly impressive and quite convincing.

But.... no one knows what Dark Energy is or how it does what it does. No one has ever detected any of it. Is this taking 'ad hoc' too far?

The current theory has a weakness; namely, the Cosmological Principle. This principle exists in two forms. In the first, the universe is the same wherever in space *and time* you are. The Big Bang put paid to that form, since the universe changes over time. Now suppose the second form, that the universe is the same wherever in space (but not time) you are, is not true either. Recall that when the Cosmological Principle was introduced it was described as a proposition; that is, it seems reasonable and convenient, but it isn't a theory. It's more in the nature of an assumption for which some evidence exists. Abandoning the

Cosmological Principle can also solve the three problems listed above.

Here's how it works. Galaxies are not uniformly distributed in space. They occur in clusters, and the clusters occur in super-clusters. The size of the largest cluster structures is 10% or more of the size of the observable universe. Furthermore these galactic structures are arranged in filaments and sheets, with huge empty voids surrounding them. Clearly the reduction in the rate of galactic recession arising from gravity will be greater in the direction of a filament or sheet than perpendicular to it. In terms of the model 1 balloon of figure 4.1, the surface, instead of being smooth, would gradually be made lumpy and distorted by the clustered galaxies on its surface. What this means is that the recession velocity will vary, violating the Cosmological Principle, because it will depend on where in space the velocity is measured.

This has two consequences. Firstly it makes very large redshifts unreliable as an indicator of distance, and thereby removes the conflict between redshift distance and type Ia supernova distance. Secondly it means the recession velocity measured from the Earth, which is used not only to calculate the age of the universe, but also directly determines what the critical density is, will be different if measured elsewhere in the universe. If the recession velocity is variable and so must be averaged and summed across the entire universe to get its proper value, then the uncertainty in its currently assigned value is far above the $\pm\,3$ km s^{-1} Mpc^{-1} given earlier. And a lower overall recession velocity could easily give an age for the universe, even allowing for gravitational deceleration, comfortably above 13×10^9 years, and would also reduce the critical density close to the point where no Ω_{de} is required.

I have presented here the cases for Dark Energy and an alternative to it. Currently most cosmologists favour the former and express no doubts when telling the world about it. What do *you* think?

4.4 COSMOGONY: THE INGREDIENTS

THE PRESENT CONTENTS OF THE UNIVERSE

Fortunately, despite all the uncertainties, a credible picture of how the universe began can still be constructed. The rest of this chapter will be devoted to that end. The branch of cosmology concerned is called *cosmogony*.

The key theoretical requirement is that the present contents of the universe were brought into being during the course of the Big Bang. It follows that in order to know what was present at the Big Bang, it is necessary to know what the universe contains now.

As was made clear in chapters 1 and 3, the universe today contains mass-energy which possesses various charges. Mass-energy and charge are therefore the constituents of today's universe that were produced by the Big Bang. It can be noted here that the proposition that these things came into existence entirely at the start of the Big Bang is justified by the mass-energy and charge conservation laws. These laws imply that neither mass-energy nor charge could have been created since. So how much mass-energy and charge does the universe now contain?

Beginning with the main non-gravitational charges, electric and color, the answer is very helpful. The universe is neutral in them; that is, it contains no net amounts of them. There are exactly as many positive as negative electric charges, and since every piece of mass-energy in the universe is color neutral (white) to outward appearances, the universe as a whole must be color neutral also.

Weak charge is more obscure. Being neither attractive nor repulsive, and serving only to enable weak force carriers to bring about particle transformations, the assignment of weak charge depends on how weak interactions are described in the language of mathematics. What 'works' mathematically is to assign u-, c- and t-quarks and neutrinos a weak charge of $+1/2$, and to assign d-, s- and b-quarks and e^-, μ^- and τ^- leptons a weak charge of $-1/2$. Antimatter quarks and leptons are assigned opposite weak charges to their matter counterparts. Overall this leads to the conclusion that the universe began with, and still has, no net weak charge.

That leaves gravitational charge. The universe is definitely not gravitationally neutral. The quantity of gravitational charge it contains is the same as the quantity of mass-energy. This is because all mass-energy has a gravitational charge and all gravitational charges have mass-energy. So the mass-energy and charge conservation laws require that the universe began with a quantity of mass-energy, but nothing else.

There are four forms of mass-energy: kinetic, wave, rest and potential.

The quantity of the first three in the present universe can be determined by observation. It turns out that kinetic mass-energy is the least significant and can be discounted. The rest mass-energy has already been evaluated as roughly $3 \cdot 2 \times 10^{-27}$ kg m^{-3}. The true figure is unlikely to be less than this but may be more. The wave mass-energy is almost entirely found in its electromagnetic form. Measurement of the amount of electromagnetic wave-energy arriving at the Earth shows that all of space is pervaded by waves of comparatively great length, the wavelengths being of the order of 10^{-3} m. The wave mass-energy density these waves give to the universe is about $4 \cdot 2 \times 10^{-14}$ J m^{-3}. This outweighs the electromagnetic wave mass-energy from more obvious sources such as stars. Finally there's potential energy. Consideration of the amount of that in the universe will be left until discussing the earliest moments of the Big Bang in section 4.6.

The two key directly observable ingredients of the universe that emerged from the Big Bang are thus today's $3 \cdot 2 \times 10^{-27}$ kg m^{-3} rest mass-energy density; and today's $4 \cdot 2 \times 10^{-14}$ J m^{-3} wave mass-energy density. The next move is to see what happens to these ingredients as earlier and earlier times are considered and the universe gets smaller and smaller. What happens to them when they are squeezed into a tiny volume of space?

LOOKING BACKWARDS IN TIME

As the expansion of space is carrying galaxies away from each other, the volume of space that each galaxy has to itself must clearly be increasing. The number of galaxies in a given volume will decrease in inverse proportion; i.e. the density of galaxies will go down. As the expansion of the universe affects all its contents and not just galaxies, the same reasoning applies to mass-energy in all its forms. As time passes, the mass-energy density of the universe is decreasing.

Looking back in time the converse must be true; at earlier times the mass-energy of the universe becomes increasingly dense. The consequences of increasing density for rest mass-energy are different from the consequences for wave mass-energy, so the two must be considered separately.

In the case of rest mass-energy, the relationship between density and the contraction in the volume of the universe is straightforward. When the universe had, say, half its present volume the rest mass-energy was twice as dense. Rest mass-energy density is inversely proportional to volume.

In the case of wave mass-energy the relationship is more complicated. In the first place, whereas rest mass-energy is almost all locked into being by the conservation laws, wave mass-energy can be made and unmade freely. Perhaps the wave mass-energy in the current universe was all made since the Big Bang rather than in it.

The consensus of opinion amongst cosmologists is that the observed wave mass-energy could not have been produced since the Big Bang. Cosmologists hold this view because the bulk of the waves — i.e. those with the 10^{-3} m wavelength — are not found to be coming from particular objects, but come instead equally from all directions in space. (Contrast this with the light from galaxies and stars, which comes only from specific points in the sky.) The directional uniformity of the waves suggests that they were produced at a time when the universe itself was uniform. This is very much how it might be expected to have been initially but is certainly not how it is now, nor how it must have been for most of its existence. The 10^{-3} m waves, then, accompanied the Big Bang rather than followed after it.

Given that electromagnetic waves have been around since the beginning, the key issue is how the wave mass-energy density changes as the volume of the universe gets smaller. Clearly, as the volume decreases, the density of the waves goes up in inverse proportion, in exactly the same way as the rest mass-energy density does. But an additional factor is at work where waves are concerned. As the volume of space shrinks, the lengths of the waves it contains shrink along with it. A reduction in wavelength is equivalent to an increase in energy. (Recall $E_\lambda = hc/\lambda$.) So the waves not only become denser in inverse proportion to the volume, they also become more energetic by virtue of possessing a smaller wavelength. As a result, the wave mass-energy density increases somewhat faster as the universe shrinks than mere considerations of volume alone would predict.

At the current time the rest mass-energy density of the universe is much more significant than the wave mass-energy density. $3 \cdot 2 \times 10^{-27}$ kg m^{-3} of mass is equivalent to $2 \cdot 9 \times 10^{-10}$ J m^{-3} of energy, which is nearly ten thousand times greater than the wave mass-energy density of $4 \cdot 2 \times 10^{-14}$ J m^{-3}. But as earlier and earlier times are considered, the effect of shortening wavelengths is to cause the wave mass-energy density to increase faster than the rest mass-energy density does. The preponderance of rest mass-energy in today's universe is gradually supplanted by a preponderance of wave mass-energy. At the very earliest moments, the rest mass-energy derived from the present-day universe makes so small a contribution to the overall mass-energy that it becomes almost completely negligible.

A description of the Big Bang thus hinges on the properties of wave mass-energy. Two properties are of particular importance. One is that wave mass-energy can be readily converted into kinetic mass-energy. Now, it was seen in section 1.4 that the kinetic mass-energy of a lot of particles moving in random directions is directly related to the temperature of the particles. This means that the wave mass-energy density of the universe can be expressed, via kinetic mass-energy, as a temperature. The denser the wave mass-energy, the hotter the universe.

The second important property is that, subject to the conservation laws, wave mass-energy may be turned into rest mass-energy. (Review section 3.5 with regard to antimatter.) When wavelengths are very short, photons have enough wave mass-energy to turn into particle/antiparticle pairs with rest mass-energy. As a result, the wave mass-energy density is equivalent to a corresponding rest mass-energy density of equal amounts of matter and antimatter as well as to a temperature.

At the moment of the Big Bang then, the universe is envisaged to consist of a colossal amount of wave mass-energy in an unimaginably tiny space: a universe of extreme density and extreme temperature. Within this universe, interconversions between the mass-energy's wave, rest, kinetic and potential forms take place so fast that it becomes impossible to distinguish any one from the others. That is the state of the universe with which the account of the Big Bang and its aftermath must deal.

DARK MATTER

There is one ingredient of the current universe that has been omitted from the Big Bang contents list: Dark Matter. Assuming it actually exists, Dark Matter is one of the great mysteries of cosmology. Gravitational calculations show that for every 18 grams of visible matter in and around a galaxy, there is apparently a further 82 grams of additional matter.

Whatever it is that constitutes this unseen 5/6ths of the mass of galaxies has yet to be identified. Because no one knows what this mass is made of, the topic is rife with untested speculations.

The main things known about Dark Matter are negative; i.e. what it is not.

It would appear not to be made of Jupiter-like bodies inhabiting the vast spaces between stars. Such planets would be invisible — i.e. 'dark' — because there'd be no star near enough to make then shine, but if this is what Dark Matter is, it would mean it consists of ordinary quarks and leptons. Unfortunately, calculations of the amount of $^3\mathrm{He}^{++}$ produced during the Big Bang (see section 4.6), if they take into account the quantity of quarks and leptons in these Dark Matter 'Jupiters', give wrong answers. Furthermore, if Dark Matter formed Jupiter-sized bodies, they would tend to settle into the galactic disc, whereas Dark Matter appears to be distributed universally as a spherical halo in which each galaxy is imbedded. Conclusion: Dark Matter is not made of accretions of atoms and molecules.

Dark Matter can also not be made of 'loose' particles which interact by the strong (color) force and/or the electromagnetic force. That's because color charged and/or electrically charged particles would behave like baryons and leptons. They would be easy to detect. But Dark Matter has so far proved impossible to detect. Conclusion: Dark Matter is not made of particles with color or electric charges.

It also seems doubtful Dark Matter could be made of particles which only interact via the weak force and gravity. The classic weak-force-and-gravity-only particles are the neutrinos, and though they are very hard to detect, they do occasionally interact with a nucleus. Such interactions have been observed. In contrast no Dark Matter interactions of this kind have been found despite various searches being made. Conclusion: Dark Matter probably doesn't have a weak charge.

This leaves two positive possibilities. One is that Dark Matter consists of particles which only interact by the gravitational force; that is, they have only a gravitational charge. That would explain why they can't be detected, since gravitational interactions are extremely slight on terrestrial scales. It would also explain why Dark Matter forms a spherical halo around galaxies. Any Dark Matter particles attempting to form an aggregate body like a sun would have no way to get rid of the kinetic energy generated as the particles fall together; they can't emit electromagnetic waves since they aren't electrically charged. As a result, Dark Matter particles would simply bounce off each other, converting their kinetic energy back into gravitational potential energy in a process that would go on repeating forever.

The other positive possibility as to what Dark Matter is comes down to variations on the theme of not even knowing if Dark Matter consists of particles at all. Could it be a fluid? Could it be a property of space that only manifests itself over distances upwards of a thousand parsecs? Could it simply be that the gravitational equation of section 1.2 needs to be modified when applied to galaxy-sized masses or distances?

Whatever the answers to these questions, from a Big Bang 'ingredients' point of view, Dark Matter would have got denser at earlier and earlier times, probably like rest mass-energy. Other than that, all that can be said is that its lack of color and electric (and weak?) charges means its contribution to the Big Bang is unavoidably going to be a subtle one.

4.5 COSMOGONY: THE PHYSICS

UNIFICATION OF THE FUNDAMENTAL FORCES

There is one particular difficulty which has to be faced in trying to describe the universe at extremely high mass-energy densities: in such conditions it would be unreasonable to expect the laws of physics that describe today's comparatively rarefied universe to be unaltered. To describe the Big Bang it is therefore necessary to formulate some ideas of the way physics changes at increasingly high mass-energy densities. How do the universe's various ingredients interact with one another in such circumstances?

A scientific answer to this question depends on observation. Obviously it is not now possible to observe the early universe directly, so observation must be made of the next best thing: small pieces of the universe in the laboratory.

To simulate the early universe experimentally, very high mass-energy densities need to be created, and this is done, as usual in particle physics, by accelerating electrically charged particles to very high speeds and then causing them to collide. At the instant of collision, all the kinetic and rest mass-energy of the colliding particles is concentrated at the point in space where the collision occurs; the mass-energy density at the collision point is raised, very briefly, to a high value. By observing what then happens, conclusions can be drawn about what the universe was like when the whole of it had a mass-energy density similar to that of the point of collision in the laboratory.

The most significant discovery to come from these high energy experiments is that the weak and electromagnetic forces lose their distinct identities when the mass-energy density exceeds 10^{47} J m^{-3}. Above this value — equivalent to a temperature of 10^{15} K — the two forces become, in effect, a single force which is carried by four force-carrying particles: W^+, Z^0, W^- and the photon. Given the differences between the two forces at more usual temperatures (see table 3.4), this is an astonishing discovery.

In part, the unification comes about because photons at the 10^{47} J m^{-3} electroweak unification density have on average about 10^{-7} J of wave mass-energy. The rest mass-energy of the weak force carriers is also of this size. As a result, subject only to charge conservation, real weak force carriers — real W^+, Z^0 and W^- particles — can be produced. For instance, the fully reversible interaction

$$\gamma \leftrightarrows W^+ + W^-$$

can occur. At densities above 10^{47} J m^{-3} the universe will be populated by real weak force carriers as well as by real photons.

In today's cold universe there are no directly observable real weak force carriers, except momentarily in high energy physics laboratories. In the hot Big Bang universe there were countless numbers of them.

The discovery that when the universe was very young the electromagnetic and weak forces were equal components of a single electroweak force invites a further speculation. Perhaps, right at the beginning, all four forces were but equal components of a single superforce. Can the electroweak trick be repeated for the strong (color) and gravitational forces?

In the case of gravity the answer is at best a heavily guarded maybe. The sort of mass-energy density at which gravity might become incorporated into a unified force is so extreme as to be utterly beyond any possibility of re-creation in an earthbound laboratory, or anywhere else in the universe today. Any observational basis for gravitational unification must accordingly be very indirect. In addition, the mathematics used to describe gravity — the mathematics of the General Theory of Relativity — is not compatible with the mathematics of high energy particle physics, which is quantum based. Consequently, there is as yet no satisfactory theory uniting gravity with any of the other forces.

The strong (color) force is a somewhat different case. The mass-energy density at which it might assume equality with the electroweak force — to make a strong-electroweak force — is again too high to re-create. However, the mathematical description of the strong (color) force is compatible with that of the electroweak force so that the unification can be achieved on paper if nowhere else. And predictions made by the theory of strong-electroweak force unification suggest that the unified force may be observed indirectly, as will be explained shortly.

Strong-electroweak force unification is predicted to occur at a mass-energy density equal to, or greater than, 10^{93} J m^{-3}. At this density, which is equivalent to a temperature of 10^{27} K, individual photons possess about 10^5 J of wave mass-energy on average. To get this amount in proportion, 10^5 J is very roughly the energy of an exploding hand grenade. At a temperature of 10^{27} K *each* photon has this much energy and there are uncountably many of these photons compressed into vanishingly small volumes of space.

With such huge available mass-energy, photons can turn into pairs of real particles, one the antiparticle of the other, even if the real particles are themselves enormously heavy. 10^5 J is equivalent to 10^{-12} kg, which sounds tiny until it's realized a 10^{-12} kg particle would be no bigger than a proton but weigh 10^{15} times as much. Theories which unite the strong (color) and electroweak forces require particles of just such an extraordinary mass to exist. The function of these ponderous particles is to carry a new hybrid sort of force, a force which provides a bridge between the strong (color) and electroweak forces and thereby facilitates their (mathematical) unification.

Because of the new hybrid force, the unified strong-electroweak force has not two components but three. These are the strong (color) component, the electroweak component, and the hybrid component, which last, for want of anything better, will be known as the X force. It should be noted that the X force is not usually regarded as a fifth force to add to the other four. Rather it is a constituent part of the strong-electroweak force out of whose unification it arises.

FORCE CARRIERS AND PARTICLE DISINTEGRATIONS

In the previous section it was seen that at earlier and earlier times the universe gets smaller, denser and hotter, apparently without limit. It would appear then that at a time close enough to the moment when the universe began, photons could well have been hot enough and have possessed sufficient mass-energy to make real 10^5 J X force carriers. What effects would these particles have had on the other constituents of the universe? The strong-electroweak hypothesis must be developed so that it can make predictions about the sort of interactions X force carriers can bring about.

The effects of the X force derive from the way in which force carriers work. To look at this aspect of forces, it would be helpful to start with the better understood weak force.

Consider the disintegration of the neutron. As was seen in section 3.6, this may be written

$$n^0 \rightarrow p^+ + e^- + \bar{v}_e$$

or in quark terms

$$(udd) \rightarrow (uud) + e^- + \bar{v}_e.$$

Overall, the u-quark and one of the d-quarks in the neutron are unaltered by the interaction, so the net reaction is

$$d^{-1/3} \rightarrow u^{+2/3} + e^- + \bar{v}_e.$$

The weak force is involved in this disintegration by way of providing the connection between the left and right hand sides of the reaction. In detail what happens is that the d-quark turns, by means of energy creation permitted by the Heisenberg Uncertainty Principle, into a W$^-$ and a u-quark:

$$d^{-1/3} \rightarrow W^- + u^{+2/3}.$$

The W$^-$ then turns into an electron and an antielectron-neutrino:

$$W^- \rightarrow e^- + \bar{v}_e.$$

The neutron is then observed to decay:

$$(udd) \rightarrow W^- + (uud)$$
$$\downarrow$$
$$e^- + \bar{v}_e$$

Looking at the effect of the W$^-$ on the reacting particles, it can be seen that it has changed a d-quark into a u-quark; conversion of quarks from one type to another is the strict preserve of the weak force. It can also be seen that both steps in the process obey the conservation laws of electric and color charge, and of baryon and lepton numbers.

This example shows how a force carrier sets about causing a particle — in this case a neutron — to disintegrate. The force carrier alters the particle concerned in a way that the conservation laws allow. (Here a d-quark is altered to become a u-quark.) Whatever the difference between the before and after particles is, that difference is absorbed by the force carrier to be released in a new form. Again using the neutron example, the difference between a d-quark and a u-quark is equivalent to an electron plus an antielectron-neutrino. The force carrier thus acts as a kind of translation mechanism. In neutron disintegration the W$^-$ translates $(d^{-1/3} - u^{+2/3})$ into $(e^- + \bar{v}_e)$.

Now look at the X force. There are 4 X force carriers. Each has a gravitational charge of 10^{-12} kg. One has an electric charge of +4/3 and is denoted X$^{+4/3}$; one has an electric charge of +1/3 and is denoted Y$^{+1/3}$; and the remaining two, which are antiparticles of the first two, have exactly opposite electric charges: X$^{-4/3}$ and Y$^{-1/3}$. Each of the four particles has, in addition to mass and electric charge, one of three two-color charges or anticolor charges, so that there are actually twelve variations on the X force-carrier theme. X and Y particles are the only force carriers to possess both color charge and electric charge, and it is this that makes them a bridge between the strong (color) and electroweak forces.

According to the law of baryon conservation, a quark can only disintegrate into another quark. Why? Because quarks have a color charge and color charge is a conserved quantity. In force carrier terms, there is no force-carrying particle which can absorb both a quark's color and electric charges and so permit it to disintegrate into, say, an electron. At least there wasn't until the X force was hypothesized. If the X force really exists — and it must do if the strong (color) and electroweak forces are unifiable — it should be possible to observe quarks changing into leptons in contravention of the laws of conservation of baryon and lepton numbers.

To see how the X force might bring such conversions about, consider how a proton, containing three quarks, would be affected if the proposed X force was to produce an interaction amongst them? There are a number of different options. One is the following. Firstly, utilizing created energy, one of the u-quarks turns into an $X^{+4/3}$ and an anti-u-quark. Here the u-quark has a red color charge:

$$u^{+2/3}_{[red]} \rightarrow X^{+4/3}_{[red+green]} + u^{-2/3}_{[anti\text{-}green]}.$$

The anti-u-quark joins up with the proton's other u-quark — a green one — to form a π^0. Next the $X^{+4/3}$ encounters the proton's one remaining quark, the blue d-quark, and combines with it, turning into an e^+ and thereby abolishing — or repaying — the energy used to make the $X^{+4/3}$ in the first place:

$$X^{+4/3}_{[red+green]} + d^{-1/3}_{[blue]} \rightarrow e^+.$$

Thus the complete X force disintegration is:

$$u + d \rightarrow \bar{u} + e^+.$$

Compare this disintegration with the previously described neutron disintegration. As already remarked, the weak force carrier brings about neutron disintegration by in effect translating $(d - u)$ into $(e^- + \bar{v}_e)$. In a similar fashion the X has been able to translate $(u + d)$ into $(\bar{u} + e^+)$.

Proton decay caused by the X force thus looks like:

$$p^+ = \begin{array}{l} u \quad \rightarrow \quad u \\ \qquad\qquad \downarrow \\ \qquad\qquad\qquad\quad \rightarrow \quad \pi^0 \\ \qquad\qquad \uparrow \\ u \quad \rightarrow \quad \bar{u} \quad + \quad X \\ \qquad\qquad\qquad\qquad \downarrow \\ \qquad\qquad\qquad\qquad\qquad \rightarrow \quad e^+ \\ \qquad\qquad\qquad \uparrow \\ d \quad \rightarrow \quad \rightarrow \quad \rightarrow \quad d \end{array}$$

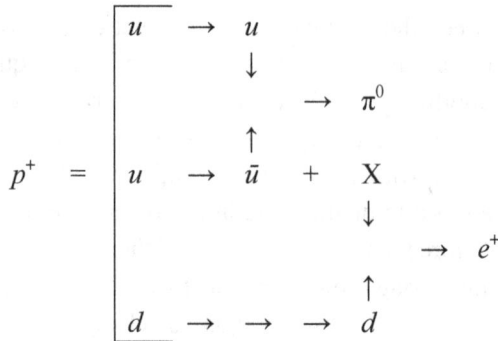

In summary form the reaction can also be written:

$$p^+ \rightarrow \pi^0 + e^+.$$

The X interaction conserves both electric and color charges (remember red + green + blue equals white) throughout. On the other hand, it does not conserve baryon or lepton numbers. This is because, in translating $(u + d)$ into $(\bar{u} + e^+)$, the X force completely eliminates the barrier between quarks and leptons. In a universe containing lots of X and Y particles, the laws of conservation of baryon and lepton numbers are repealed. As these laws were invented in the first place to lock up rest mass-energy (see section 3.3), X and Y particles have the effect of unlocking that form of mass-energy, making it fully interchangeable with the other three forms. The X force puts rest mass-energy on an equal footing with the other types. This is the most significant effect that strong-electroweak hypotheses predict the X force will have on the high density, high temperature Big Bang universe.

What about the effect of X and Y particles on today's universe? Nowadays there are no real X or Y particles anywhere, because the requisite mass-energy density is no longer available. But *virtual* X and Y particles should still exist. This means that if the strong-electroweak force is more than just an entertaining mathematical fiction, it should be possible to observe its X force component bringing about proton and neutron disintegrations in the fashion of the example given above. And yet the laws of conservation of baryon and lepton numbers are a clear statement that no such disintegrations ever take place. Does this mean the strong-electroweak hypothesis must be incorrect?

174

The key to resolving this issue lies in the size of the X and Y particles. As these particles have a rest mass-energy of 10^5 J, that is the amount of energy which must be created via the Heisenberg Uncertainty Principle to make a virtual X or Y. The reaction time associated with such a large amount of energy is 10^{-39} s. Even more so than with the weak force discussed in section 3.6, the creation of this amount of energy is extremely improbable. Only rarely will a virtual X or Y particle be created and an X force interaction occur. Because of the prodigious size of X and Y particles, a u- or d-quark is expected on average to have to wait at least 10^{32} years before one comes into being in its vicinity. And that's the minimum average time. According to some variants of the strong-electroweak hypothesis, it takes considerably longer. It is, in other words, highly unlikely that a proton or neutron would disintegrate by way of the X force while an observer is watching it.

Unlikely does not of course mean impossible. 10^{32} years is how long an average baryon can expect to last before the X force disintegrates it. But some baryons will last much longer than this while others will last much less time. An unlucky few will disintegrate today. An observation of baryon disintegration would be, in effect, an observation of the X component of the strong-electroweak force. As such, observations of baryon disintegrations are being sought by a number of physicists. None has yet succeeded in seeing one. 10^{32} years must be an underestimate!

The hypothesis of the strong-electroweak force can thus be reconciled with present-day baryon and lepton conservation by appealing to probability. Baryon and lepton conservation was introduced originally in a thoroughly arbitrary fashion. The strong-electroweak hypothesis regards the two conservation laws as statements about what is highly improbable rather than what is fundamentally impossible. According to the strong-electroweak hypothesis, the laws of conservation of baryon and lepton numbers are simply a reflection of the consequences of X and Y particles having such a huge rest mass-energy.

In answer to the query raised at the start of this section it can now be stated, on the basis of the foregoing, in what ways it is thought the physics of the Big Bang differed from the physics of today. At the time of the Big Bang at least three of the four forces were but equal components of a unified strong-electroweak force and, as a consequence, the two conservation laws of baryon and lepton numbers were inoperative.

4.6 COSMOGONY: THE EVENTS

WORDS OF CAUTION

The account of the beginning of the universe given in this section deals with times when the universe was very different from how it is today. The densities of mass-energy involved, right at the very start, are far greater than any experiment can ever reproduce. Consequently, the higher the mass-energies concerned, the more speculative must be the description of events pertaining to those mass-energies. Broadly speaking, knowledge of what was going on in the universe at any given time is inversely proportional to the mass-energy density then thought to exist. The lower the density, the more secure the scientific account. The sub-headings of this section reflect that situation by stating whether the following narrative is fancy, speculation or theory.

Fancy is limited to the very first instant. At that moment the universe was so compressed that gravity itself must have behaved in a quantized way. The sort of timescale over which quantum gravity is manifested is determined by a relationship connecting the gravitational constant G, Planck's constant h, and the speed of light c. If the three constants are combined in a particular way, the result is an amount of time. The combination is

$$(Gh/c^5)^{1/2}.$$

The time this expression gives is of the order of 10^{-43} s. So for the first 10^{-43} s or so of the universe's existence, it would have been dominated by gravitational quanta. There is currently no theory, not even a speculative one, that can describe what the universe would have been like in such circumstances. At present, an account of the very beginning has to rest therefore on fancy rather than science.

As a further word of caution, it should be noted that the model 1 universe of section 4.3 will be assumed in what follows to be the true one. This is despite there being no conclusive evidence in its favour. The reasons for preferring it to its rivals are these. Firstly it lacks the infinities of models 2 and 3 and is easier to visualize on that account. Secondly the choice of model has hardly any effect on the early history of the universe, so there is no incentive to select other than the simplest one. And thirdly, as was remarked at the end of section 4.3, the current density of the universe is close enough to the critical value that it's not possible to definitely decide amongst the three models. This means that if the universe isn't exactly flat (model 2), then it must be nearly so. The nearness of the observed universe to having a flat model 2 structure can most easily be explained by blowing the model 1 balloon up to huge dimensions. This gives the relatively small areas of surface visible to any one observer a flat appearance. Such blowing up (see 'The Inflationary Era' below) is a possible consequence of the strong-electroweak hypothesis.

Table 4.1, which can be found at the end of this chapter, shows some of the values associated with important events in the universe's development, and should be studied in conjunction with the text.

IN THE BEGINNING (FANCY)

With no mathematical model for guidance, no observational data, either direct or indirect, and with the universe so small that it could have filled the volume occupied by a proton more than 10^{60} times over, nothing scientific can be said about the moment of the beginning. What follows is compatible with science but is not scientific.

Before the beginning there was no space, no time, no mass-energy. (Use of 'before' in such a context is more or less meaningless but unavoidable — something which points up the fanciful nature of this part of the account.) There was obviously the *potential* for these things to exist, and whatever is was that had this potential included the physical laws. In particular, the laws of quantum physics were in effect which permit spontaneous uncaused events to occur. Thus was a quantity (or perhaps a quantum?) of space and time created. And created moreover in a self-consistent, physically lawful way.

Take the smallest amount of time in which any event — including the beginning event — can occur to be 10^{-43} s. In that time it is possible at the speed of light to travel 10^{-35} m. This distance may be regarded as a guide to how much space was created. In terms of the model 1 analogy, the model 1 balloon would have had a diameter of 10^{-35} m.

The first problem to be faced with this minuscule universe is that if it was able to just start existing in 10^{-43} s, then presumably it ought to have been able to just stop existing every bit as rapidly. Quantum physics is generally symmetric in time; as has been remarked before, quantum reactions can go from right to left as easily as from left to right. (This aspect of physics will be discussed in some detail in chapter 10.) So in a time-symmetric universe the end might be expected to be the exact opposite of the beginning. The universe would end as it began, giving it a lifespan of less than 10^{-42} s.

Obviously this is not what happened. Perhaps it was down to probability. The universe could either have collapsed back to nothing or it could have launched into an expansion. In the event, it did the latter. And an idea arising from the hypothesis of the strong-electroweak force suggests how the expansion might have been driven.

THE INFLATIONARY ERA (SPECULATION)

Mathematical models of inflation exist but there is no direct observational evidence to make it possible to choose amongst them, or indeed to support any of them. The models themselves rest on hypotheses of strong-electroweak force unification for which no observational evidence exists. The size of the universe associated with a time of 10^{-36} s is 10^{-28} m, which is a volume too small to contain sufficient particles to make temperature a meaningful concept. (Recall temperature is a collective property of large numbers of particles.) To extrapolate known physics to the kind of energy densities present at 10^{-36} s after the beginning is at least as risky as using the physics which describes a grain of sand to also describe the sun.

Imagine a completely empty universe: a vacuum. A vacuum is a pure volume of space, unpolluted by mass-energy. Next, apply the Heisenberg Uncertainty Principle. According to quantum theory, a vacuum contains virtual mass-energy which the Uncertainty Principle permits to be created subject to the usual restriction on how long it lasts. A vacuum is thus devoid of real particles but full of unobservable virtual particles made of short-lived created mass-energy. A value can be assigned to the mass-energy density of these virtual particles and this density is usually taken to be very close to zero.

(A note for enthusiasts: the value of the mass-energy density of virtual particles in a vacuum has to be 'assigned' rather than calculated because calculations give a value which is 10^{120} times larger than the maximum possible value obtained from observation. This is far and away the hugest 'wrong answer' in current mathematical physics!)

The important thing about vacuums from a cosmogonical point of view is that it is possible to conceive of a vacuum having a different number of virtual particles to normal. Such a vacuum — known as a *false vacuum* — would have a higher or lower virtual mass-energy density than the 'ordinary' vacuum of today's universe.

Hypotheses of the unified strong-electroweak force permit as a possibility that when the three forces are in their unified state, that is, are equal components of a single force, a vacuum pervaded by the unified force will be a false vacuum. It can be speculated accordingly that the universe was created with the strong-electroweak force in its unified state, and that it therefore consisted of a false vacuum. What would the consequences have been?

One theoretical prediction arising from mathematical models of false vacuums is that they are unstable below the strong-electroweak force unification temperature; that is, when the temperature is less than 10^{27} K. Below that temperature a false vacuum turns into a normal vacuum. The transition takes about 10^{-34} s to complete. While it is taking place, the false vacuum is able to repel itself by way of the force of gravity. In a nutshell this is because an unstable false vacuum exerts a negative pressure on itself. This, according to the General Theory of Relativity, gives rise to a repulsive gravitational force:

negative pressure
$\downarrow$
negative gravitational attraction = repulsion.

The vacuum repulsion causes the vacuum to increase in size exponentially.

('Exponentially' in this context means that if the universe doubles in size in the first unit of time, then in the second unit of time it doubles again, and in the third it doubles again, and so on. After three units of time the universe would be 8 times larger. After ten units it would be 1024 times larger. Suppose, arbitrarily, that the 'unit' is 1/100th of 10^{-34} s; i.e. 10^{-36} s. This would mean that after the final hundredth unit the universe would have increased in size by 2^{100} times; i.e. it would be roughly 1,000,000,000,000,000,000,000,000,000,000 times larger.)

Now to apply these ideas to the Big Bang. Denote the age of the universe by the symbol $\mathbb{T}$. At $\mathbb{T} = 0$ s the universe came into being by way of an uncaused quantum event. This coming-into-being was complete by $\mathbb{T} = 10^{-43}$ s. Whatever processes happened at that time, assume they left the universe with a unified strong-electroweak force, a false vacuum and, by $\mathbb{T} < 10^{-36}$ s, a temperature below 10^{27} K. The false vacuum, unstable by virtue of the low(!) temperature, immediately began to repel itself. The universe started to expand exponentially. (The analogous model 1 balloon can be envisaged undergoing a kind of self-inflation.) By the time 10^{-34} s had passed, the universe had increased in volume by a colossal amount. Depending on the details of the false vacuum — a matter of guesswork currently — the increase in volume can be expressed as 10^x. The 'x' represents the exponential nature of the expansion and could be anything from 100 up to billions and more.

After $\mathbb{T} = 10^{-34}$ s the transition to a normal vacuum was complete and the negative pressure, and hence the gravitational repulsion, disappeared. Separate points in space, which had been accelerating apart at ever increasing speed during the inflation, were left receding from one another at the velocity imparted by the gravitational repulsion during its brief existence. The universe continued to expand, but no longer at an accelerating rate. The expansion continues to this day.

As a result of the tremendous inflation, any real mass-energy which may have been present during the quantum gravity era at $\mathbb{T} = 10^{-43}$ s was dispersed in such a vast amount of space that it effectively disappeared. This makes what went on before the inflationary era irrelevant to the future history of the universe. Ignorance of happenings at $\mathbb{T} = 10^{-43}$ s is not a handicap when it comes to accounting for what happened after that time. Essentially, the Big Bang can be taken to commence when the inflation does.

The strong-electroweak hypothesis's positing of inflation certainly puts the 'Big' into Big Bang. But it has yet to provide the 'Bang'. As the account stands, it has generated a huge universe but an empty one. It is necessary next to fill it with ultra hot, ultra dense mass-energy.

The simplest assumption to make about the universe at the start of the inflationary expansion is that it was indeed empty, containing no mass-energy at all. Kinetic, potential, wave and rest mass-energies were all zero. However, as the false vacuum repelled itself, its gravitational potential energy, initially zero, necessarily became negative. This is because, as will be recalled from section 1.2, whenever anything moves under the influence of a force its potential energy decreases. With some licence the same can be said of the self-repelling false vacuum. So as the false vacuum inflated, its potential energy dropped further and further below zero.

Clearly, to conserve energy some positive non-potential energy had to be created to counterbalance the growing negative potential energy of the vacuum. The strong-electroweak hypothesis enables this non-potential energy — that is, rest, kinetic and wave mass-energy — to be created as the transition to a normal vacuum proceeds. Put very simply, think of the change from false to true vacuum as being like a wave in the fabric of space. Like all other waves, this vacuum wave is a form of energy. And again like all other waves, these waves have a particle equivalent. The particles thus produced — totally hypothetical ones, it must be said — can turn into the other, more usual, forms of mass-energy. In this way, as the vacuum went from the false state to the true one it filled the universe with electromagnetic waves and vast numbers of particles, all at extremely high temperature. (Is this when Dark Matter was created?)

The inflationary universe account of the beginning can thus allow the universe to be filled with (non-potential) mass-energy without necessitating a violation of the mass-energy conservation law. The total mass-energy of the universe was — and remains — zero. The rest, kinetic and wave mass-energy the universe contains is exactly offset by the negative potential energy stored up during the period of inflation. This gives some idea in principle of how the Big Bang could have occurred in a law-abiding fashion.

THE PRODUCTION OF MATTER (FLAWED SPECULATIONS)

The only evidence for the existence of X and Y particles would be non-conservation of baryon and lepton numbers. No violation of these laws has ever been observed. Nor is there any observational justification for asymmetry in X and Y particle decays, assuming X and Y particles actually exist. Nor is the mechanism outlined here capable of producing anywhere near the observed surplus of matter over antimatter. Nor is there any evidence that the other mechanisms suggested below might really have happened. (Though something must have.)

The waves produced as the transition to a normal vacuum ended gave the universe a colossal amount of non-potential mass-energy (a 'Bang' indeed!), raising its mass-energy density to 10^{91} J m^{-3}, equivalent to a temperature of 3 x 10^{26} K. At this temperature an average photon's wave mass-energy is not quite enough to make real pairs of X and Y particles. However, some photons will have mass-energies sufficiently above the average to make possible reactions such as

$$\gamma \leftrightarrows Y^{+1/3} + Y^{-1/3}.$$

Hence after inflation, at $\mathbb{T} = 10^{-33}$ s, the universe still contained a small quantity of real X and Y particles.

Aside from X force carriers, the photon mass-energies were far in excess of those needed to create all the other known particles. The universe at $\mathbb{T} = 10^{-33}$ s accordingly contained the weak force carriers, the six matter/antimatter pairs of leptons, and the six matter/antimatter pairs of quarks in profusion. These various particles, always made in pairs, were produced in such reactions as

$$\gamma \leftrightarrows W^+ + W^-$$

$$\gamma \leftrightarrows \mu^- + \mu^+$$

$$\gamma \leftrightarrows d^{-1/3} + d^{+1/3}.$$

Something to note about reactions occurring at $\mathbb{T} = 10^{-33}$ s is that when waves are converted into rest mass-energy the baryon and lepton number conservation laws are always obeyed. Assuming that the universe was created empty, it must have begun with baryon and lepton numbers of zero. Nothing encountered in the Big Bang so far will have changed that.

The universe continued to contain as much antimatter as matter.

That the universe today is almost entirely made of matter is one of the big mysteries of cosmogony. Some physical process is needed to upset this matter/antimatter symmetry. The decay of real X and Y particles can provide a flawed illustration. It was seen in the previous section how X and Y particles can turn quarks into leptons and antiquarks, and vice versa. This violates the baryon and lepton number conservation laws (whilst respecting the laws of charge conservation). Now, because of this intermediary role of X and Y particles between matter and antimatter, and between quarks and leptons, when a real X or Y particle disintegrates it has two basic options open to it. Take the $X^{+4/3}$ as an example:

$$u^{+2/3} + u^{+2/3} \leftarrow X^{+4/3} \rightarrow e^+ + d^{+1/3}.$$

The $X^{+4/3}$ here can disintegrate into the two quarks to the left, or into the antielectron and the anti-d-quark to the right. Similarly for the $X^{+4/3}$ antiparticle, the $X^{-4/3}$:

$$u^{-2/3} + u^{-2/3} \leftarrow X^{-4/3} \rightarrow e^- + d^{-1/3}.$$

As long as both $X^{+4/3}$ and $X^{-4/3}$ disintegrate in exactly the same way, the quantity of quarks and antiquarks remains precisely equal, as does the quantity of leptons and antileptons; the universe retains its net baryon and lepton numbers of zero. But suppose the $X^{+4/3}$ preferred to disintegrate into the two particles to the left while the $X^{-4/3}$ preferred the two particles to the right. In that case the universe would acquire more quarks and leptons than antiquarks and antileptons. This is such a clever solution to the problem of there being more matter than antimatter in the universe it's a real shame it doesn't work quantitatively.

By $\mathbb{T} = 10^{-32}$ s the post-inflationary expansion of the universe had reduced the mass-energy density below the level at which any photons were likely to have enough mass-energy to make pairs of real X and Y particles. All the X and Y particles then rapidly disintegrated, never to be seen again. It is in their separate guises that the strong (color) and electroweak forces proceeded to fashion the universe.

The universe continued to cool and the mass-energy density fell correspondingly. By $\mathbb{T} = 10^{-11}$ s it had fallen to 10^{47} J m^{-3}, equivalent to a temperature of 3×10^{15} K, at which time the weak and electromagnetic forces lost their equality. Shortly thereafter real weak force carriers disappeared, as had real X and Y particles before them.

It will be noticed that in exponential terms there is a big gap between 10^{-32} s and 10^{-11} s, and therein lies another possibility for creating the surplus of matter. It is known that the weak force has a slight preference for matter as against antimatter, and the weak force would have been as strong as electromagnetism before 10^{-11} s. Perhaps it was the weak force rather than the X force that led to a matter-filled universe.

Perhaps, finally, there were unstable Dark-Matter-like particles created by 10^{-32} s which decayed in a way preferring matter over antimatter.

No one knows.

THE PRE-NUCLEAR ERA (UNCONFIRMED THEORY)

There is no direct observational evidence of what happened before $T = 10^{-2}$ s. However, mathematical models of the separate strong (color), electromagnetic and weak forces are well supported by hard data. If the universe was as described at $T = 10^{-11}$ s, then it will also be as described at $T = 10^{-2}$ s.

With the universe continuing to expand and cool, pairs of heavy quarks and antiquarks (*s*, *c*, *t* and *b*) and pairs of tauons and antitauons ceased to be made as photon mass-energies declined. These heavy particles either annihilated with each other or else disintegrated by means of the weak force into their lighter relatives.

It can be noted here, concerning the number of types of quarks and leptons in the universe at $T = 10^{-11}$ s (and before), that there are not expected to have been more than the six matter/antimatter pairs of quarks and the six matter/antimatter pairs of leptons already discovered. The reasons for this limit are three-fold. Firstly, when real Z^0 weak force carriers are made in high-energy collisions in laboratories, the rate at which the Z^0s break up into quarks and leptons depends in part on how many types of neutrino there are. An answer of three fits the data, whereas an answer of four (or more) does not. Secondly, the number of lepton types affects the ratio of H to He produced during the nuclear era (see shortly below). Again, the observations indicate there are only six leptons (and six antileptons). Thirdly, strong-electroweak hypotheses pair each lepton with a quark type. If there are only six leptons, there must similarly be only six quarks. Consequently the t-quark is thought to be the most massive quark and the tauon the most massive lepton; the huge mass-energy density before $\mathbb{T} = 10^{-11}$ s was not able to make pairs of even more massive quarks and leptons than these because none exist.

By $\mathbb{T} = 10^{-4}$ s muon/antimuon pairs of particles had ceased to be produced, and muons and antimuons mutually annihilated or disintegrated into electrons and antielectrons. At about the same time, the universal expansion had increased the distance between typical surviving quarks to about 10^{-15} m. This is the effective size of baryons. The result is that the boundary between the quarks in neighbouring baryons became distinct where before it had not really existed at all. Mutual annihilation of u/anti-u and d/anti-d pairs of quarks then followed, and the only quarks remaining — those resulting from the surplus of quarks over antiquarks established during the initial stages of the Big Bang — found themselves incorporated into baryons. With this event, the pre-nuclear era was brought to a close.

THE NUCLEAR ERA (FLAWED? THEORY)

If the universe as described at $\mathbb{T} = 10^{-2}$ s is how it actually was, then the fusion of hydrogen to make helium is what would have happened next. The abundance of helium today fits well with predictions, as does the abundance of intermediate products, $^2H^+$ and $^3He^{++}$, but calculations of the tiny amounts of by-product $^7Li^{+++}$ give results which are three times the size of the observed abundance.

By $\mathbb{T} = 10^{-2}$ s the mass-energy density of the universe had decreased to 3×10^{29} J m^{-3}, equivalent to a temperature of 10^{11} K. Its contents were a mixture of protons, neutrons, neutrinos and antineutrinos. The average photon energy was still sufficient to produce electron/antielectron pairs ($\gamma \leftrightarrows e^- + e^+$) so these two particles were also present in vast numbers. The kinetic energy of the protons and neutrons was so large that the strong force between them was quite unable to stick them together. (Compare this to the analogous inability of the atomic stickiness to stick the molecules in a gas together; see section 1.4.) In the same way, electrons were unable to become chemically attached to protons. The contents of

185

the universe were thus in a very simple state with baryons, leptons and antileptons all existing independently of one another.

Independent neutrons, as was described in section 3.6, are unstable and tend to disintegrate ($n^0 \rightarrow p^+ + e^- + \bar{v}_e$). This process would have depleted the number of neutrons as against protons except that at $\mathbb{T} = 10^{-2}$ s the mass-energy density was still so high that neutrino/proton collisions were frequent. These collisions brought about a reverse, neutron forming reaction:

$$p^+ + \bar{v}_e \rightarrow n^0 + e^+$$

which maintained the balance between protons and neutrons and kept their numbers about equal.

By $\mathbb{T} = 1$ s the universe's mass-energy density had fallen to 3×10^{25} J m^{-3}. At this density, neutrinos no longer collided very often with protons to produce neutrons, so the number of neutrons began to decrease, giving rise to a compensatory increase in the number of protons.

By $\mathbb{T} = 200$ s electron/antielectron pairs had long since ceased to be produced in any quantity, average photon mass-energies having fallen well below the rest mass-energy of electrons. Electrons and antielectrons were thereby able to engage in mutual annihilation, leaving behind the surplus of electrons over antielectrons that was created during the first moments of the Big Bang.

By the same time, the gradual decay of neutrons into protons had resulted in a ratio of about 2 neutrons to 14 protons. With the temperature having now fallen below 10^9 K, the strong force was able to stick neutrons and protons together. This it proceeded to do, combining them in groups of four in a net reaction:

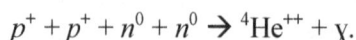

$$p^+ + p^+ + n^0 + n^0 \rightarrow {}^4He^{++} + \gamma.$$

${}^4He^{++}$ (helium-4) is a helium nucleus containing four baryons. Evidently, for every 2 protons and 2 neutrons incorporated into a helium-4 nucleus, 12 protons were left unattached. They remained independent as hydrogen nuclei (${}^1H^+ = p^+$).

At the end of the nuclear era it can be predicted on the basis of the calculated proton to neutron ratio that there was 1 helium nucleus to every

12 hydrogens. This is close to the ratio of helium to hydrogen observed now in interstellar gas and provides this part of the Big Bang theory with observational support.

THE PRE-STELLAR ERA (THEORY)

At last, direct observations are possible. Once electrons are captured by nuclei to form atoms, the universe becomes transparent to light and to electromagnetic waves generally. The current wave mass-energy density of $4\cdot2 \times 10^{-14} J m^{-3}$ (see section 4.4) is what remains of the light bathing the universe when it became transparent, and these waves can be precisely measured on the Earth. This stage of the Big Bang can actually be seen.

After the formation of the helium nuclei, the inexorable expansion continued to reduce the mass-energy density steadily, thereby reducing the temperature as well, but the contents of the universe changed little, apart from getting cooler. Not until about a third of a million years had passed did the temperature drop to a level at which the next change could take place.

Prior to $\mathbb{T} = 300,000$ years, any electrons falling into shells around helium or hydrogen nuclei were rapidly ejected again, either by collisions with other particles or else by absorbing a photon with the requisite energy. After $\mathbb{T} = 300,000$ years, with the temperature dropping below 3000 K, and the corresponding mass-energy density dropping towards 10^{-1} J m^{-3}, electron shells became stable. Fully fledged atoms of helium and hydrogen were able to form. Thereafter the universe was filled with transparent helium and hydrogen gas.

The formation of the universal gas brought gravity, on the side-lines since the end of the inflationary era, back into play. The gas was not perfectly uniform but contained some regions which were slightly denser than others. It is now considered fairly probable that these density variations were formed during the inflationary era and were subsequently enhanced by Dark Matter, thereby creating gravitationally dense regions into which hydrogen and helium gas could fall. This in-fall led to the formation of growing balls of gas, drawing in material from nearby low-density regions of space. The result was a developing hierarchy of gas balls, from small sizes typically of 10^{29} to 10^{32} kg — stars — to aggregates of millions of stars — galaxies — to aggregates of hundreds or thousands of galaxies — galaxy clusters.

Judging by their colour and brightness, the oldest known stars are those in the Milky Way's globular clusters. These stars formed roughly $11·5 \times 10^9$ years ago. Furthermore, the most distant light-emitting point sources of electromagnetic waves have redshifts indicating that they are about 3500 Mpc away, which means stars and galaxies were forming right across the universe more than 10^{10} years ago. From then on the universe at last began to look like the one science confronts today.

THE STELLAR ERA (SECURE THEORY)

Stars in all the various stages of their lives being readily observable, the scientific account is now on firm ground.

With star and galaxy formation well underway, it is time to concentrate attention on just one of the 10^{10} plus galaxies forming within present day observational range. That galaxy is our own, the Milky Way. As the history of the Milky Way is largely the history of the stars it contains, it is necessary to consider here some facets of the physics of stars. This will reveal how the Milky Way has evolved to its present constitution, and in particular to its present chemical composition, bearing in mind that the Big Bang has produced a universe filled solely with hydrogen and helium.

The first point to make is that not all the gas of the Milky Way was incorporated into stars straight away. Formation of star-sized clumps of gas is a protracted business which takes place over billions of years. Stars will still be forming from the galactic gas long after the sun's life is over. So the Milky Way began with a few stars and has been acquiring additional ones ever since. The first of the Milky Way's stars had formed by 10^{10} years ago.

The opening stage in star formation is due to the atoms in a clump of gas trying to fall towards the clump's centre under the force of gravity. In the process of falling, gravitational potential energy is converted into kinetic energy. As a result, an infalling clump of gas becomes hot (and roughly spherical).

What happens next depends on the mass of the particular clump, because the heavier it is the hotter it gets. Clumps of less than 10^{29} kg become hot for a time until eventually the inward force of gravity is matched by an outward repulsive force arising from collisions amongst the many atoms in the clump with each other. Compression then more or less ceases and the clump slowly grows cold as its heat is converted into electromagnetic waves which it emits into space.

What about clumps of gas over 10^{29} kg in size? These clumps become so hot inside that not only are all the electrons given enough kinetic energy to become separate from their nuclei, but the hydrogen nuclei themselves collide so fast that they come within strong (color) force range of one another. This leads to the fusing of groups of four hydrogen nuclei into helium. Since the rest mass-energy of $^4He^{++}$ is 0.5% less than the rest mass-energy of 4 ^1H's, this fusion results in the difference between the two rest mass-energies being converted into kinetic and electromagnetic wave mass-energy. The net reaction is:

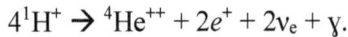

$$4^1H^+ \rightarrow {}^4He^{++} + 2e^+ + 2\nu_e + \gamma.$$

The two antielectrons go on to annihilate with two electrons, producing additional photons. The photons can be interconverted into kinetic energy, thereby making the clump of gas even hotter. That in turn causes even more hydrogen nuclei to fuse, producing yet more photons and thereby heat. Eventually the electromagnetic wave mass-energy emitted by the clump of gas into space balances the heat being generated in its interior by hydrogen fusion. The clump of gas becomes a star.

The importance of the mass of a clump of gas does not end with deciding whether hydrogen fusion will start in its interior. The quantity of mass in a star also determines how long the star will last once hydrogen fusion has commenced. Stars of 2×10^{29} kg last for more than 10^{12} years before their hydrogen runs out. The sun, weighing 2×10^{30} kg, will last 10^{10} years. (And that's despite its currently losing over 4,000,000 tonnes of rest-mass-energy every second through hydrogen fusion!) Stars of 2×10^{31} kg only manage 2×10^7 years. In other words, the more massive the star the shorter its life. This is because, despite having more hydrogen to fuse into helium, bigger stars are so much hotter and gravitationally compressed internally they use the hydrogen up much faster than smaller stars.

When a large star has fused all the hydrogen in its core into helium, its next step is to begin fusing helium nuclei together to make carbon:

$$3\,^{4}\text{He}^{++} \rightarrow\ ^{12}\text{C}^{6+} + \gamma$$

in a net reaction that makes the core of the helium fusing star even hotter. When helium in the core is exhausted, the largest stars turn to fusing carbon nuclei into magnesium, and when the carbon is exhausted they go on to fuse even bigger nuclei. With each increase in nuclear size the temperature inside large stars rises until finally iron nuclei are being produced. This signals the end of the road, for further fusion — of the iron nuclei — doesn't make a star hotter; instead, iron fusion cools a star down by absorbing energy rather than emitting it. At which point, the cycle of ever increasing temperatures fusing larger and larger nuclei comes to an end, as does any star engaged upon it.

When the supply of nuclei to turn into iron at the core of a large star runs out, a star which depends on such nuclear fusion has no alternative to cooling down. As its centre cools, the repulsive force produced by inter-nuclear collisions becomes less strong. Gravity is thereby able to cause a renewed contraction of the star's central parts. This in turn induces an instability in the outer layers of the star which is of veritably explosive magnitude.

The explosion which marks the end of a large star's life turns the star into a supernova. As was mentioned earlier in connection with astronom-ical distance scales, a supernova explosion is of a truly amazing size. As much energy is emitted during the course of the explosion as a star like the sun will emit in the next 5×10^9 years. A supernova will often briefly outshine the entire galaxy in which it resides. Needless to say, most of the contents of the exploding star are hurled out into space at great speed.

Amongst the first stars the Milky Way acquired were a number of very large ones. Within the space of a few tens of millions of years this population of large stars had reached old age and blown up. Their explosions peppered the Milky Way's original mixture of hydrogen and helium gas with all manner of other atoms from lithium up to and beyond uranium in size. As the hundreds of millions of years passed and one supernova explosion followed another, the gas of the Milky Way which had yet to clump into stars became distinctly dusty. Consequently, stars forming well into the life of the Milky Way were not made of the pure Big Bang mixture of hydrogen and helium, but of that mixture with the addition of small amounts of heavier elements derived from supernovas.

Although the large stars that became supernovas have great signifi-cance as the originators of all the elements other than hydrogen and helium, they have always been very much a minority as stars go. Most stars are smaller, quieter and longer lasting. Our sun is a star of this second kind.

The sun began its life some 4.6×10^9 years ago. (The source of this date will be found in section 5.5.) By that time the galactic gas was well contaminated with heavy elements from supernovas. Consequently, as the sun formed, a lot of dust accreted in a disc around it. Gradually, under the force of gravity, this dust came together to make the inner planets. At a greater distance, residual gas and some non-metallic molecules such as methane and water made up the greater part of the outer planets (Jupiter and beyond).

It may be noted here that as the formation of the solar system appears to be a fairly straightforward business it is likely that the Milky Way — and every other similar galaxy — has many such systems. Probably in dusty galaxies more than one star in ten will have planets orbiting it. Because planets are so tiny compared to their parent stars it is technically very difficult to detect them observationally, but this has now been done for a few nearby stars. If these nearby stars and the sun are typical — and it is reasonable to suppose they are — it implies that there may be more than 10^{20} Earth-like planets in the observable universe. You should bear this figure in mind when considering the arguments put forward at the end of section 5.5.

This chapter has looked at the universe as a whole. Observation has revealed how big it is, and hypotheses and theories have described how it may have evolved from its beginning to its present state in a series of events under control of the laws of physics. And now at the end of the process it is time to redirect the gaze of science back to the Earth once more. Having seen how the Earth was formed, the next task must be to look at what has happened here since its formation. That story will be taken up in section 5.5 in the course of beginning to confront what is perhaps as great a marvel as the universe itself; namely, that there are conscious observers to observe it.

TABLE 4.1 The chronology of the Big Bang.

Event	$\mathbb{T}$ seconds	Temperature Kelvins	Energy density[a] $J\ m^{-3}$
Creation!	10^{-43}	?	?
end of inflation	10^{-34}	10^{27}	10^{93}
end of X decays	10^{-32}	10^{26}	10^{89}
electroweak separation	10^{-11}	3×10^{15}	10^{47}
end of μ^+/μ^- production; baryon formation	10^{-4}	10^{12}	10^{33}
$p + n + e^{+/-} + v_e + \bar{v}_e$	10^{-2}	10^{11}	3×10^{29}
end of v/p collisions	1	10^{10}	3×10^{25}
end of e^+/e^- production	10	3×10^9	3×10^{23}
end of $^4He^{++}$ formation	200	7×10^8	3×10^{20}
atom formation	10^{13}	3×10^3	6×10^{-1}
now	$\approx 4 \times 10^{17}$	3	$\sim 10^{-10}$

a) By way of comparison, air at sea level has an order of magnitude rest mass-energy density of $10^{17}\ J\ m^{-3}$ (the energy equivalent to $1 \cdot 25\ kg\ m^{-3}$).

5

BIOCHEMISTRY

5.0 OBJECTIVE: To discuss the form, properties and origin of life on the Earth, this being an essential prerequisite for our next task: to switch our attention from outer space to inner space and investigate the scientific view of the observer.

During the course of chapters 1, 3 and 4 you should have acquired a general feel for the scientific attitude to reality, and how it springs from the belief that all things adhere to simple physical laws, even if those laws are hidden amidst chaos. At first sight, living things appear to stand aloof from this simplicity and to be endowed with some sort of law-breaking 'life essence'. However, any scientific hypothesis about the nature of life must hold that this is not so. Living things must obey the same set of laws that inanimate objects do. (Recall the remarks in section 2.4 about the status of scientific laws, and also about wholes and parts.)

In making the case to support the scientific hypothesis we shall begin by discussing the chemical bonds that form between atoms. The four main classes of biochemicals will then be described, and the way they interact with each other will be outlined. The knowledge thus gained should make the claim reasonably credible that all processes observed in living things at a chemical level (i.e. excluding animal behaviour) can be expected to be accounted for sooner or later without recourse to magical essences.

The next problem to tackle is the very difficult one of how life began. Currently there are no experimentally testable theories of the way this happened. There is, however, no reason to believe that life arose other than by processes occurring under the known natural laws, since these laws do not make the origin of life impossible, but only improbable to a speculative degree.

Having discussed the 'improbability' issue, we will then comment on the process of evolution by natural selection, and the role of chance in the life process. Lastly, the ability of living things to behave purposefully and to make choices will be considered. These two properties, which are central to our perception of whether or not something is alive, will be seen to arise from the sheer complexity of biological systems.

5.1 CHEMICAL BONDS

<u>THE ROLE OF ELECTRON SHELLS</u>

To understand biochemistry the essential first step is to have a good grasp of what chemical bonds are and how they are formed. A *chemical bond* is any kind of short-range force — stickiness — between two or more atoms which enables them to form a stable molecule. A chemical bond is formed or broken whenever a chemical reaction occurs between atoms or molecules.

In section 1.5 chemical reactions were considered in terms of the energy involved. A chemical reaction between two atoms is able to take place if the two atoms collide with so much kinetic energy that their electron shells begin to overlap. A chemical reaction is thus essentially a reaction between the electron shells of the colliding atoms.

This simple picture was modified in section 3.2. There it was seen that all atoms have a number of electron shells. It is the number of electrons in the outermost occupied shell which bestows on an atom its particular chemical properties. In this light, the last sentence of the previous paragraph needs to be revised to read: a chemical reaction is thus essentially a reaction between the outermost occupied electron shells of the colliding atoms.

Now, amongst the chemical properties of any given type of atom are those that determine which particular chemical reactions it will undergo. Atoms are quite selective about which other atoms they will react chemically with, and also about the number of atoms which can join with them to make a molecule (see section 1.3). To take a simple example, when a sodium atom reacts with chlorine to make common salt, it forms a chemical bond with one chlorine atom. Common salt contains one chlorine atom for each sodium atom. A sodium atom never pairs up with two or more chlorines. Observations of this kind have yet to be explained. In section 3.2 it was established *that* the outermost occupied electron shell gives an atom its properties. It has not yet been explained *how* it does so.

The key to solving this puzzle lies in further consideration of electron shell energies. When the outermost occupied electron shell is full, all the electrons in it can swap places with one another without altering the appearance of the shell in any way. In this full state the shell is a bit like someone carrying two equally heavy suitcases, one with each arm. It makes no difference if the suitcases are swapped over. And just as important, the cases are comparatively easy to carry.

If, on the other hand, the electron shell is not full, then it more resembles someone carrying only one heavy suitcase. Transferring the suitcase from one arm to the other makes quite a difference, not least to the side the person has to lean towards whilst walking along, and the suitcase is comparatively hard to carry.

As far as an atom is concerned, carrying an outermost shell which has gaps in it is also hard compared to carrying the same shell when it is full. In terms of dynamics, an atom with a partially full shell possesses more potential energy than one with a full shell. As all objects try to reduce their potential energy to as low a value as possible, the effect of this is to make partially full shells less stable than full ones. Any process which plugs the gaps in a partially full shell reduces the shell's energy. The potential energy well depicted in figure 1.2 is a manifestation of this fact. The conclusion follows that two or more atoms will react together chemically if they can rearrange their electrons in such a way that each atom acquires a full outermost shell.

There are two ways in which a chemical reaction between atoms can lead to the establishment of a full outer shell for each of them. One way involves an exchange of electrons. The other involves the sharing of electrons. Both are equally important and will be described in turn.

IONIC BONDS

When two atoms react by exchanging electrons, one atom adds an electron or two to its outer shell, thereby filling it up, while the other atom loses the electrons concerned and exposes a full, previously inner shell to the world. The two resultant atoms will obviously no longer be electrically neutral. One will have too many electrons and be negatively charged, while the other will have too few and be positively charged. Charged atoms are called *ions*. Charged molecules — which will be encountered shortly — are also called ions. The chemical bond producing two ions, and formed between them, is called an *ionic bond*.

Table 3.2 shows that the outermost shell always contains electrons with shell shape $l = 0$ or $l = 1$. By the time a shell begins to acquire electrons with a shell shape of $l = 2$ or more, it is no longer the outermost shell. Another shell with a higher n value and shell shape $l = 0$ will have been filled with two electrons first. So the fullest possible outermost shell contains eight electrons, regardless of the value of n. The eight electrons are two with an $l = 0$ shell shape and six with an $l = 1$ shell shape. The atoms nearest to achieving a full outer shell therefore lie at the extreme right of the periodic table (table 3.1). Conversely, those that can most easily lose a few electrons from an almost empty outer shell lie at the extreme left.

To illustrate an ionic bond, consider the chemical reaction between sodium and chlorine mentioned above. Sodium is on the left of the periodic table and has in its outer shell one electron which it wants to lose; chlorine is on the right and has in its outer shell seven electrons, so it wants to gain one.

Look at figure 5.1. This diagram shows how a sodium atom's outer shell contains only one electron, and how a chlorine atom's outer shell has one gap in it. After reacting chemically together, both atoms have full outermost shells. One sodium atom is sufficient to supply one chlorine atom with the desired electron. And one chlorine atom is sufficient to relieve one sodium atom of its single undesired electron. The reaction is thus between one sodium atom and one chlorine atom. Additional sodium atoms or additional chlorine atoms have no role to play. That is why the numbers involved in a given chemical reaction are fixed. There are always an exact number of gaps to fill and an exact number of electrons to be relinquished. Those numbers determine how many atoms will combine together in a given molecule.

Ionic reactions can be symbolized in a way which shows the imbalance of electric charge that results from them. The reaction between an atom of sodium and an atom of chlorine may be symbolized as

$$Na + Cl \rightarrow Na^+Cl^-.$$

Incidentally, it can be noted here for future reference that the atoms of gaseous elements kept in isolation from any other element with which they can react often go around in pairs. Chlorine is just such an atom. Hence the reaction above should properly be written

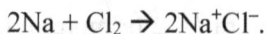

$$2Na + Cl_2 \rightarrow 2Na^+Cl^-.$$

The other di-atomic gases at room temperature and pressure are H_2, N_2, O_2 and F_2. Of these, only fluorine is unimportant from a biochemical point of view.

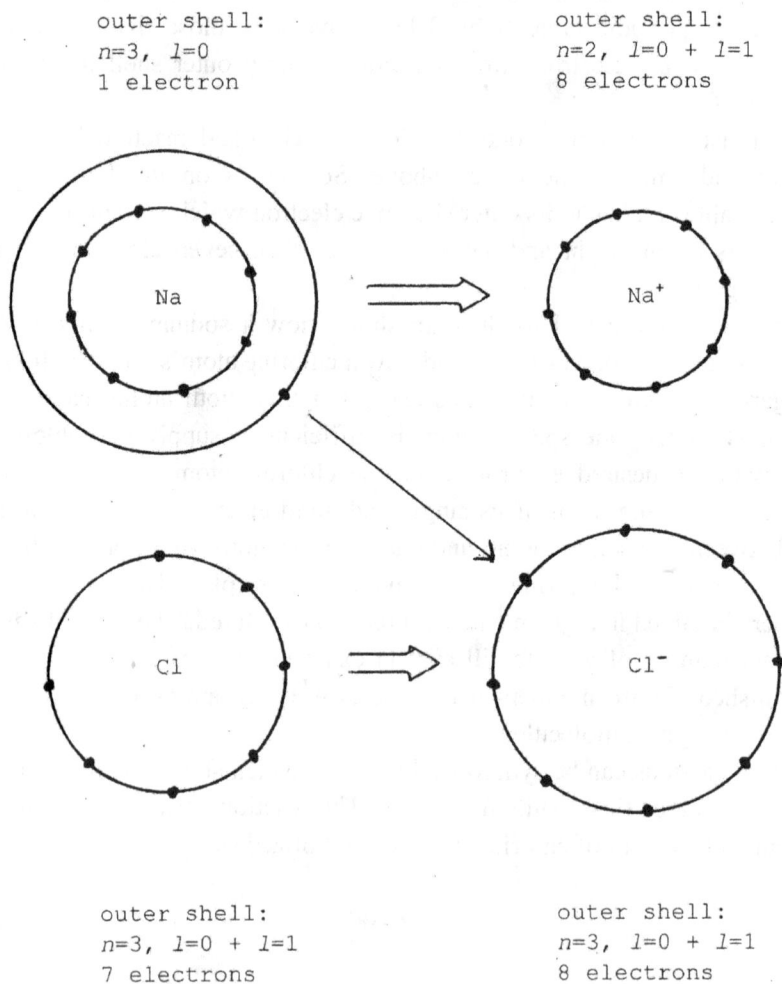

```
outer shell:              outer shell:
n=3,  l=0                 n=2,  l=0 + l=1
1 electron                8 electrons
```

```
outer shell:              outer shell:
n=3,  l=0 + l=1           n=3,  l=0 + l=1
7 electrons               8 electrons
```

Figure 5.1 The chemical reaction between sodium and chlorine in which an ionic bond is formed. Note that for ease of representation the variable shape of the electron shells has not been depicted.

COVALENT BONDS

When the formation of an ionic bond involves the exchange of only one electron, as is the case with sodium chloride — ionic chlorine is always called chloride — the potential well associated with the resultant molecule tends to be quite deep. However, as more electrons get involved, that is, with each step away from the left and right margins of the periodic table inward towards the middle, the potential well gets shallower. This is because of the imbalance of electric charge within the ion. Suppose an atom has two spaces in its outer shell. When the second electron is added to complete the shell, it experiences considerable electrical repulsion from the first of the two electrons which has already taken up its place. The dynamical benefits of a full outer shell are thus offset by the drawback of having to create it in the face of the electrical repulsion between its members, which is no longer fully balanced by the positive electric charge of the nucleus.

For multi-atom molecules and for those atoms needing to gain or lose three or more electrons to achieve a full outer shell or get rid of a nearly empty one, the potential well is often so small as to make unstable any molecules formed ionically. These atoms combine together using the second means available to them: they share their electrons.

When two atoms react by sharing electrons they simply merge their outermost shells. In effect the two shells become a single shell. The electrons in the shared shell arrange themselves in such a way that the stake each atom has in them is equivalent to a full outer shell for that atom. Bonds of this type are called *covalent bonds*.

As an example of a covalently bonded molecule, consider water. Look at figure 5.2. (Note that the electrons in the figure are depicted as originating from one atom or another. This is to facilitate understanding only. In reality the electrons are completely indistinguishable from one another once the covalent bond has formed.) From the figure it can be seen that the upper hydrogen atom has two electrons in its outer shell, both of which are shared with the oxygen atom. It has the equivalent of a full $n = 1$ outer shell. The oxygen atom has a full $n = 2$ outer shell containing eight electrons. In this case two electrons are shared with the upper hydrogen atom and two with the lower hydrogen atom. The lower hydrogen atom is in the same situation, electron-wise, as the upper one. Each of the three atoms thus has the equivalent of a full outer shell.

Water is not in fact completely covalent. At any instant, one molecule in 10^7 in pure water will be ionic:

$$H_2O \rightleftharpoons H^+ + HO^-.$$

When a covalent water molecule becomes ionic, one of the hydrogen atoms separates from the molecule, leaving its electron behind. The resultant HO^- molecule is both ionic (since it is charged) and covalent (since the oxygen atom and the remaining hydrogen atom are sharing their electrons). Notice that both atoms in the HO^- ion still retain the equivalent of a full outer shell.

(A note for enthusiasts: the hydroxide ion HO^- is traditionally written OH^-. However, the surplus electron is mainly associated with the oxygen

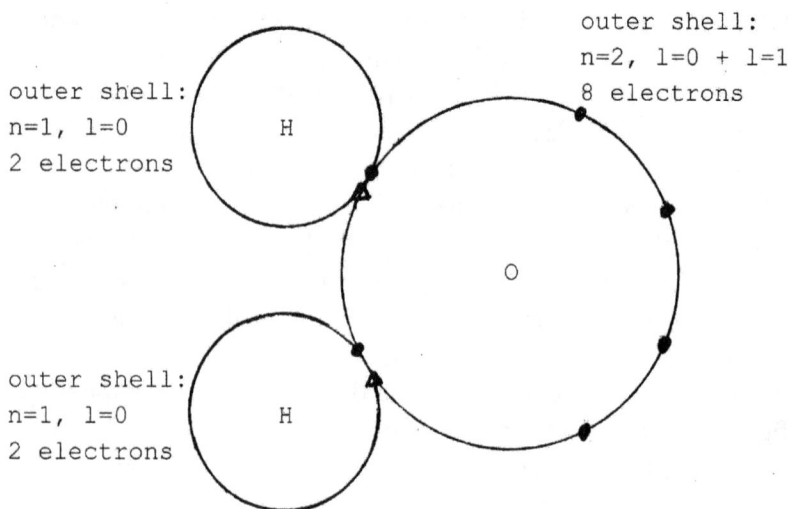

Figure 5.2 The structure of water. Electrons originating with the oxygen are shown as dots; those originating with the hydrogens as triangles.

atom not the hydrogen, so HO^- is a more accurate representation.)

Water is not the only covalent molecule that can be partly ionic. Many others come in this category. They are mainly molecules which contain an atom from well to the left on the periodic table, such as sodium and potassium, and especially hydrogen. When the atom in question is hydrogen, the molecule concerned is generally called an *acid*. One such molecule worth mentioning here because it has a major part to play in biochemistry is phosphoric acid:

$$H_3PO_4 \leftrightharpoons H^+ + H_2PO_4^-.$$

SYMBOLIC REPRESENTATION OF COVALENT MOLECULES

Whereas molecules which are bonded purely ionically contain only small numbers of atoms, those which are covalently bonded can contain large numbers. Of the covalent molecules that will be encountered in the remainder of this chapter, one of the smallest contains twelve atoms and has the formula $C_3H_6O_3$. Many of the others are far larger than that.

The considerable numbers of atoms found in some covalent molecules creates a difficulty in representing their formulas. There are, for instance, many ways in which the thirty-six shared outer shell electrons of $C_3H_6O_3$ can be arranged to give full outer shells for the constituent atoms. The molecules resulting from the various arrangements are found to be chemically distinct despite having the same overall formula. Clearly the bald statement that a molecule has the formula $C_3H_6O_3$ is inadequate to define it.

The solution to the problem is to draw a schematic representation of the molecule, showing the arrangement of its various covalent bonds. When two atoms form a covalent bond in which each atom shares just one of its electrons with the other — called a single bond — the bond is represented by a single line. When two electrons from each atom are contributed to the bond — called a double bond — a double line is used, and so on.

Under this scheme the HO^- ion is

$$H - O^-.$$

Carbon dioxide (CO_2) is

$$O = C = O.$$

But the scheme only really comes into its own with molecules like $C_3H_6O_3$. One arrangement of its atoms and bonds is

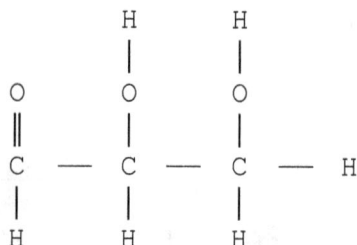

```
        H           H
        |           |
  O     O           O
  ‖     |           |
  C  —  C  —  C  —  H
  |     |     |
  H     H     H
```

With its atoms bonded in this fashion the molecule is called glyceralde-hyde. Its full structure is shown in figure 5.3. Another example of how the atoms of $C_3H_6O_3$ can be arranged will be found in the next section.

It should be noted that many molecular representations of the kind just described — I shall call them formula-diagrams in future — fall short of ideal because they are confined to the flat surface of the paper. To be really accurate, a formula-diagram would need to include perspective. In the case of glyceraldehyde, for example, the left and right carbons, above, should protrude upward from the page, while the middle oxygen and hydrogens should be angled downwards.

The number of lines (symbolizing bonds) that emanate from any given type of atom in a formula-diagram is the same as the number of electrons that each atom needs to gain or lose in order to acquire a full outer shell. Some of the most important examples are H (1 line), C (4), N (3), O (2), S (2) and P (5).

On some occasions the number of lines connected to an atom in a formula-diagram will be too many or too few for a full outer shell. In such cases the atom compensates by ionically losing or gaining, respectively, the necessary number of electrons. This manifests itself as a charge on the atom concerned (as, for instance, H — O⁻ above, and also see proteins below).

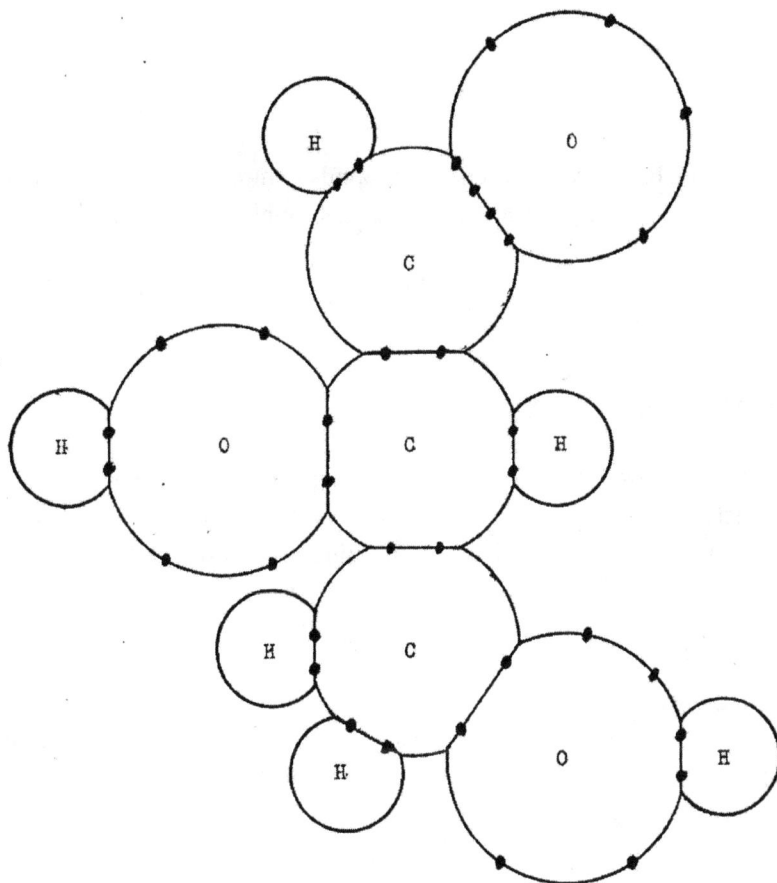

Figure 5.3 The structure of glyceraldehyde. The 3 x 6 (n = 2) electrons from the oxygens, the 3 x 4 (n = 2) electrons from the carbons, and the 6 x 1 (n = 1) electrons from the hydrogens, total 36, are distributed as shown.

In order to limit the size of the formula-diagrams, single shared electrons except between carbon atoms will not generally be shown in future. Under this convention the formula-diagram for glyceraldehyde becomes

```
O            H            H
‖            O            O
C    —    C    —    C
H            H            H₂
```

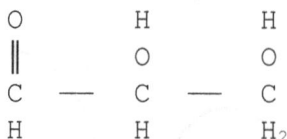

With this brief review of chemical bonds completed, it is now possible to describe the basic chemical processes taking place in living things.

5.2 BIOCHEMICALS

ORGANISMS

In order to construct a theory of what happens inside living things, it is obviously essential to know what chemicals they contain. The first step to establishing this is to define which things are alive. What properties distinguish living things from other objects?

A living thing can be defined as an object:

- which is capable of producing perfect or near perfect replicas of itself, the degree of perfection being sufficient to enable most of the replicas to themselves produce further equally perfect replicas in their turn
- which has a shape and structure that are determined internally; that is, the events that govern the form of a living thing take place inside itself — external events almost always proving destructive rather than constructive.

This definition of living things is wider than it needs to be to encompass all life presently found on the Earth. It is, however, sufficient as a starting point.

It transpires that all known objects which fit the definition given above are made of a distinctive kind of molecules known as *organic molecules*. An organic molecule is a molecule which contains at least one carbon atom and is predominantly covalent.

Living things on the Earth are described as carbon-based on account of this universal prevalence of carbon in the molecules they contain. Given that living things are necessarily complex and that carbon is the only atom capable of forming suitably durable complex molecules, life everywhere else in the universe, if it exists, is also probably carbon-based. The dependence of living things on carbon can be considered to be a third defining characteristic. Thus in addition to the two characteristics listed above, a living thing is an object:

• which has a chemical composition based on carbon.

Objects which qualify as living things are also called *organisms*. Every organism is a living thing and every living thing is an organism.

It should be noted that 'life', in the sense to be understood in this chapter, means 'that which is alive'. The three characteristics of an organism can thus be taken to be a definition of what constitutes life itself.

Now that it has been made clear which things are organisms, attention can turn to examining them with a view to finding out what chemicals they are made of. Part three of the definition has already established that these chemicals are organic (i.e. carbon-based). However, while the number of organic molecules is huge, those found in living things can be expected to be only a subset of the number that could potentially be used.

The way to find out what chemicals an organism contains is to break it into its constituent parts. When this is done, an interesting observation is made which should be remembered for future reference. The observation is that, with the exception of viruses (see section 5.6), organisms are made of units called *cells*. A cell can be thought of as a bit like a balloon. A balloon is a rubber membrane inside which is a gas. A cell is a membrane made of mainly fatty molecules inside which are a collection of other molecules mostly dissolved in water. In some cases the organism consists of a single cell, each such cell being capable of surviving and replicating all on its own. In other cases the organism is composed of vast aggregates of mutually dependent cells.

The relevance of this to investigating the organic chemicals found in living things is that an organism's chemicals are mostly found inside cells. Thus it is cells which need to be broken into their constituents. Since all cells are enclosed by a membrane, breaking a cell simply involves rupturing this membrane and seeing what comes out.

Apart from oddities such as vitamins, minerals and coenzymes, which can be put in a miscellaneous category and (for present purposes anyway) disregarded, the organic chemicals found in cells — chemicals known collectively as *biochemicals* — can be classified into four distinct groups, called sugars, fats, proteins and nucleic acids.

These molecules must somehow enable cells to replicate and to possess the property of having an internally determined structure. To discover how they do this, each group needs to be examined in turn.

SUGARS

Sugars are molecules which consist of a chain of carbon atoms, three or more in number, to which are bonded atoms of hydrogen and oxygen. The ratio of C, H and O is normally 1:2:1. The carbon at one end of the chain forms a double bond (i.e. shares two electrons) with an oxygen. Alternatively this may be done by the penultimate carbon.

The two smallest sugar molecules are glyceraldehyde and dihydroxy acetone. In glyceraldehyde, first encountered in the previous section, the end carbon forms a double bond with oxygen. In dihydroxy acetone it is the penultimate carbon (which in this instance is the middle one) that does so. Writing out their formula-diagrams, it can be seen how the two sugars differ.

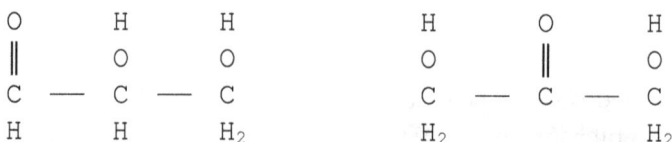

$$
\begin{array}{ccccc}
O & H & H & \qquad H & O & H \\
\parallel & O & O & \qquad O & \parallel & O \\
C - & C - & C & \qquad C - & C - & C \\
H & H & H_2 & \qquad H_2 & & H_2 \\
\end{array}
$$

Glyceraldehyde is on the left; dihydroxy acetone is on the right.

Both these molecules have the same overall formula, $C_3H_6O_3$. They are nonetheless quite distinct from each other chemically, the arrangement of the atoms making a significant contribution to the molecules' chemical properties. The dependence of properties on atomic arrangement is characteristic of biochemicals.

In general most sugar molecules are given a name ending in -ose. One of the most biochemically important sugar molecules is glucose. Glucose is a sugar with six carbons in its chain. It has the formula $C_6H_{12}O_6$. This is its formula-diagram.

```
O        H                    H        H        H
‖        O        H           O        O        O
C   —    C   —    C    —    C   —    C   —    C
H        H        O           H        H        H₂
                  H
```

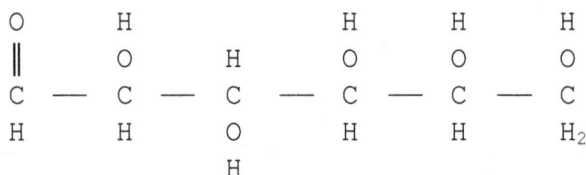

In most sugars with five or six carbons in their chain, including glucose, the fourth or fifth carbon numbering from the left reacts with the $=\!=$O of the first carbon. This is possible because the carbon chain curves into the shape of a 'C', which thereby brings the first and fourth/fifth carbons into proximity with one another. After this reaction has taken place, the sugar molecule has a ring shape. The formula-diagram can be redrawn to reflect this. Glucose is

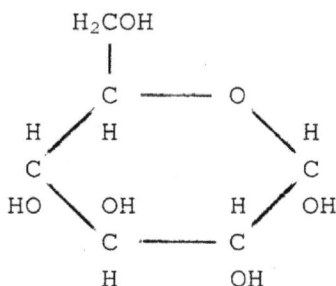

```
                H₂COH
                  |
             C  ——— O
        H  /   H           \  H
          C                   C
       HO  \  OH        H  /  OH
             C  ——  C
             H      OH
```

Note that the reaction does not alter the molecule's formula. Only its shape changes.

Sugar molecules can join together to form molecule chains. Organic molecule chains are called *polymers*. The most well-known sugar polymers are cellulose, which is shown in figure 5.4, and starch. Both polymers consist of chains of sugar molecules bonded covalently. Notice that each link in the chain involves the removal of a water molecule.

FATS

Fats are molecules composed predominantly of carbon and hydrogen. They contain very few oxygen atoms.

Chemically, fat molecules are rather diverse in structure. One of the

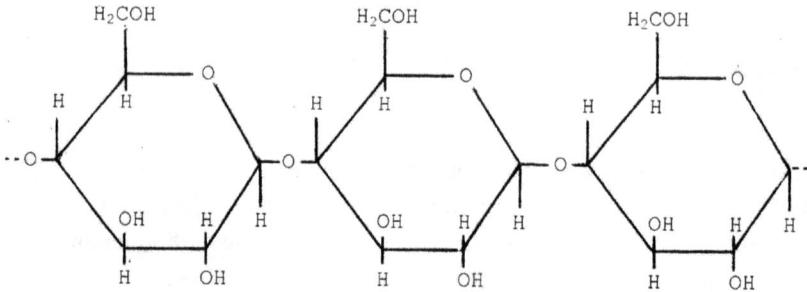

Figure 5.4 The structure of cellulose. The carbons at each of the five points in the glucose rings have been omitted for clarity.

most common atomic groupings found within them is the fatty acid. Fatty acids have the general formula

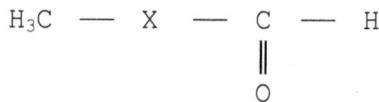

where X is a chain of carbon atoms with attached hydrogens, the number of carbon atoms varying from none to over twenty. The chain may also contain a number of double bonded carbons, as shown in the following example.

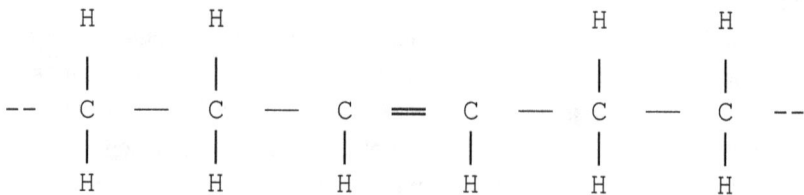

Because chains containing double bonded carbons have fewer hydrogen atoms than those with only single bonded carbons, the fats containing them are known as unsaturated (with hydrogen).

In addition to being found inside cells, fats are also a principal constituent of cell membranes.

PROTEINS

Proteins are chains of molecules called *amino acids*. There are twenty primary amino acids found in proteins, each of which has the general formula

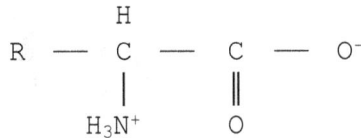

$$
\begin{array}{ccccccc}
 & & H & & & & \\
R & - & C & - & C & - & O^- \\
 & & | & & \| & & \\
 & & H_3N^+ & & O & &
\end{array}
$$

where R is a group of atoms which defines the amino acid. Obviously, since there are twenty different amino acids, R is one of twenty groups of atoms. The twenty amino acids consist of various combinations of H, C, N, O and S.

When amino acids link together to make proteins they do so by forming a single bond between the H_3N^+ of one acid and the COO^- of the neighbouring acid in the chain. This leads to a chain with a zigzag structure.

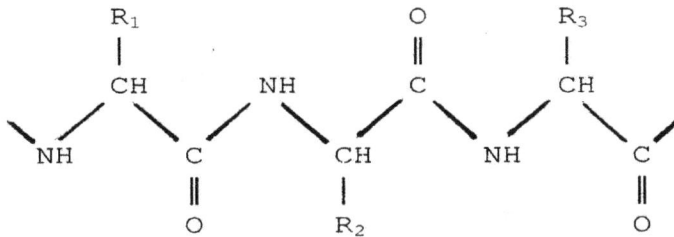

$$
\begin{array}{ccccccc}
R_1 & & & & O & & R_3 \\
| & & & & \| & & | \\
CH & & NH & & C & & CH \\
\diagup\diagdown & \diagup & \diagdown & \diagup & \diagdown & \diagup & \diagdown \\
NH & & C & & CH & & NH & & C \\
 & & \| & & | & & & & \| \\
 & & O & & R_2 & & & & O
\end{array}
$$

At one end of a complete chain of amino acids will be an unattached H_3N^+, while at the other end will be an unattached COO^-. As when sugar molecules polymerize, the formation of each bond in the chain involves the removal of a molecule of H_2O.

The most important property of proteins is the shape they take up. Because the bond between CO and CH, and between CH and NH, is not rigid (study the formula-diagram of an amino acid chain above), amino acid chains can become highly convoluted. Any given sequence of amino acids, however, always becomes convoluted in a fixed way. As a result, two proteins which possess identical sequences of amino acids will also possess identical shapes.

The shapes of proteins can be divided into two categories. In the first category are the filaments. Examples of proteins which take up a filament-like shape are silk and collagen, the latter being found in skin, cartilage and bone. In the second category are the *globular proteins*. In these proteins the amino acid chain is truly convoluted, folding itself up into what often resembles a lumpy ball. There will be more to say about globular proteins shortly.

NUCLEIC ACIDS

Nucleic acids are a kind of polymer. The molecules which polymerize to form nucleic acids are a sugar called ribose, and phosphoric acid. Figure 5.5 shows the formula-diagram for a small section of a nucleic acid polymer. The pentagonal parts of the chain are the ribose molecules, formed into rings in the manner typical of sugars as described above. Each ribose is linked to the next by a phosphoric acid molecule. These latter molecules, when bonded as they are here, are often referred to as phosphate. A nucleic acid is thus formed from a chain of interleaved ribose and phosphate molecules.

Some nucleic acid chains make use of a different kind of sugar. Their sugar resembles ribose except that it has one fewer oxygen atoms. It is sensibly known as deoxyribose. When it is part of a nucleic acid chain, its formula-diagram is as follows. The missing oxygen will (not!) be found on the bottom right.

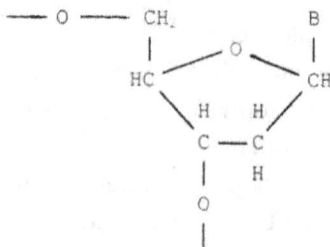

To distinguish nucleic acids made from ribose from those made with deoxyribose, the former are called ribonucleic acids (*RNA*) and the latter are called deoxyribonucleic acids (*DNA*).

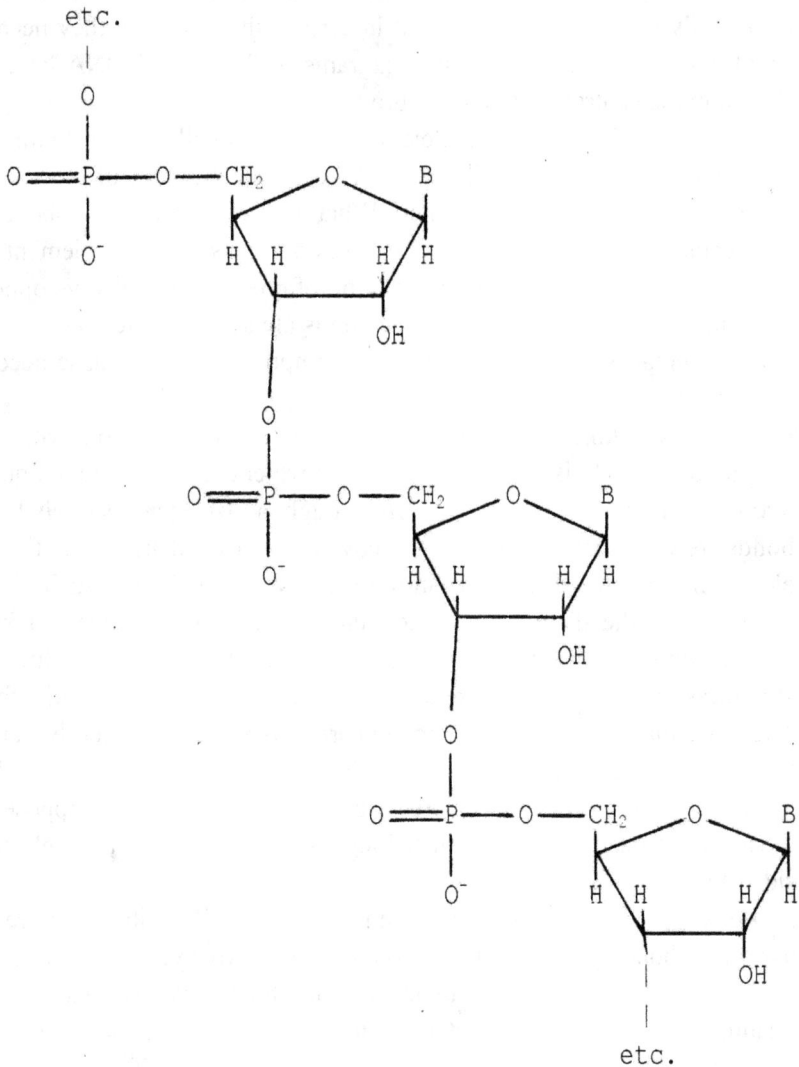

Figure 5.5 RNA. As in figure 5.4, the ring carbons have been omitted.

The formula-diagram shown in figure 5.5 also includes, in addition to ribose and phosphate, other groups of atoms marked B. Each B is one of four possible molecules known collectively as *bases*. In RNA the four bases are molecules called adenine, cytosine, guanine and uracil. In DNA they are adenine, cytosine, guanine and thymine. A number of other chemically similar bases are found in certain situations but they need not be of concern here. The formula-diagrams of the normal DNA bases are shown in the central portion of figure 5.6.

In almost all cases, the nucleic acids found in cells are predominantly DNA rather than RNA. Cellular DNA occurs in the form of two parallel strands twisted round each other. What is most significant about this arrangement is that the sequence of bases on one strand complements the sequence of bases on the other. Adenine of one strand is always opposite thymine of the other. Similarly, cytosine is always opposite guanine. This complementarity turns out to be of vital importance so its cause needs to be explained.

There is a kind of bond that hydrogen atoms can form with their neighbours which is neither ionic nor covalent. It's a bond unique to hydrogen and is called, appropriately enough, a *hydrogen bond*. Hydrogen bonds are weaker than the ionic and covalent types and arise from the way electrons distribute themselves in the vicinity of a hydrogen nucleus. The net result of the distribution is to make the hydrogen appear sticky to nearby atoms. Only two conditions have to be met to produce the stickiness, which is a subtle by-product of the electromagnetic force. These are that the hydrogen atom concerned is already covalently bonded to something, and that the atom to be hydrogen-bonded to it is one which needs to acquire electrons in order to fill its outer shell (as opposed to having electrons to lose); i.e. it belongs to an element on the right-hand side of the periodic table.

Look again at the four bases. It can be seen that, if positioned correctly, hydrogen bonding can occur between them. Two hydrogen bonds will form between adenine and thymine, and three between cytosine and guanine. This is exactly what happens with paired strands of DNA, as shown in figure 5.6. Adenine is not found paired with guanine, nor cytosine with thymine, simply because the two pairs of bases do not fit together in a way that permits the formation of hydrogen bonds.

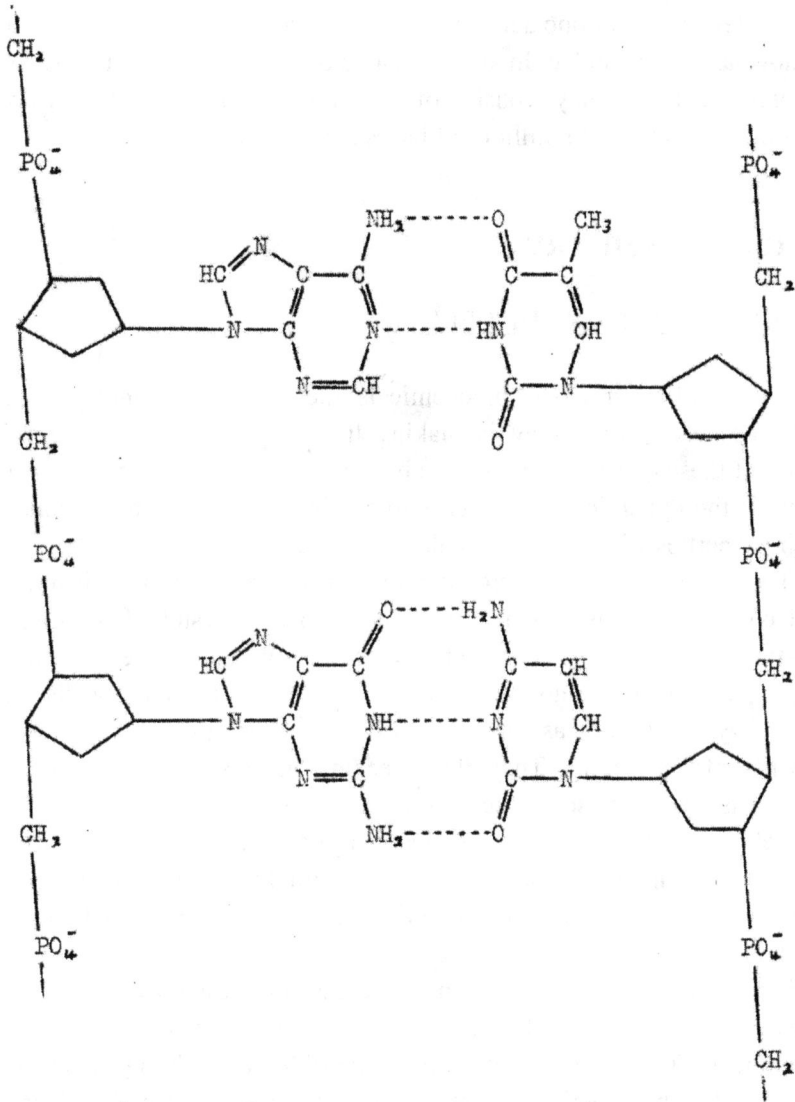

Figure 5.6 Hydrogen bonding between adjacent bases on a double strand of DNA: hydrogen bonds are represented by dashed lines. In this example, adenine (upper left) is bonded with thymine (upper right) and guanine (lower left) is bonded with cytosine (lower right).

213

One other point to make about nucleic acids is that, like chains of sugar molecules and the amino acid chains which form proteins, there is no limit to how large they can be. In all three cases the limit is practical rather than theoretical. DNA may consist of an unbroken pair of deoxyribose-phosphate strands with millions of bases along their length.

5.3 CELL CHEMISTRY

DYNAMICS WITHIN THE CELL

How do the four groups of biochemicals interact with one another? What part does each group play in making living things what they are? To confront these questions, organisms need to be looked at from the point of view of the dynamics of chapter 1 to see if it is possible to account for their properties without violating the known laws of physics.

The observation which provides the starting point is that whenever a cell comes into existence it always does so as a result of an already-existing 'parent' cell dividing into two halves. Where before there was a single parent there are now two offspring half-cells. These half-cells grow until they are as big as their parent was, manufacturing their missing constituents internally. They then become parents in their turn. This process is cyclic and so is able to continue indefinitely.

Cell replication carries the clear implication that the quantity of biochemicals in the world can increase with time; there is no law of conservation of biochemicals. Potentially every cell is able to take part in producing this increase.

Large-scale biochemical manufacture can most easily be observed in plants. As a plant grows, the quantity of biochemicals it contains increases enormously. The environment is quite incapable of supplying such a large amount. The biochemicals must therefore be made inside the plant, as opposed to being assimilated from outside it.

The biochemical manufacturing property of cells in general and plants in particular (plants contain many cells) is a key observation on which a scientific account of organisms can begin to be built. How might biochemical manufacture be accommodated within the theory of dynamics and the laws of physics?

Recalling section 1.5, it can be said that the molecules of which a cell is made are almost entirely of the kind that lie at point 4b on the potential energy curve of figure 1.2. To construct such molecules requires that the constituent molecules at point 1 or point 4a are elevated to this higher energy level; that is, as molecules at 4b have more potential energy than those at 1 or 4a, energy must be added to molecules at 1 or 4a in order to get them to 4b. This is in addition to any energy needed to get them over the potential energy hill at point 3. Biochemicals therefore contain quite a store of potential energy which must be supplied to the constituent molecules whenever a biochemical is made.

Now, all the molecules needed to make a biochemical are available in the environment in which organisms live, and can indeed be seen to be taken into organisms from outside them. Hence the only problem in principle that has to be solved in order to reconcile biochemical manufacture with the theory of dynamics is where the energy comes from to turn those raw-material molecules into biochemicals. The supply and utilization of energy to and within cells is thus the central issue needing to be addressed.

ENERGY SUPPLY

The world is awash with energy, so it should not be difficult to find a source of supply for organisms to tap into. Terrestrial energy comes from two primary sources. The lesser lies within the Earth and is responsible for producing volcanoes and earthquakes. The forms in which this energy leaks out from the Earth's interior make it unsuitable as a source of energy for all but a few organisms. The greater source of energy is electromagnetic wave energy from the sun — sunlight. Being steady and reliable makes this kind of energy eminently suitable and invites the hypothesis that sunlight supplies the energy requirements of cells engaged in biochemical manufacture.

The hypothesis indicates what to do next: take a sample of organisms and keep them in the dark. If the hypothesis is correct, it will be found that cellular growth rapidly comes to a halt.

Experiments yield inconsistent results. Only some organisms, most obviously plants, are affected as predicted; plants do indeed stop growing if kept in the dark. Other organisms, though, are able to function fairly normally in darkness. Thus the hypothesis can only be partially true and needs to be modified. It becomes: organisms which cease growing when kept in the dark obtain their energy requirements directly from sunlight; other organisms obtain their requirements from elsewhere.

Again this revised hypothesis dictates the next experiment to perform. Those organisms largely unaffected by the dark need to be studied with a view to discovering what they take into themselves. Does what they consume supply the energy? It is invariably found that it does. Very nearly every organism which can grow while kept in the dark takes into itself, by one means or another, 4b-type biochemicals derived from other organisms. (The odd exceptions get their energy from 4b-type molecules which are inorganic.)

The picture of energy supply in organisms can be summed up as follows.

- Some organisms utilize sunlight to manufacture their biochemicals. They absorb freely available inorganic molecules at point 4a on the potential energy graph, molecules like CO_2 and H_2O, and use the energy in sunlight to combine them into biochemicals at point 4b.
- Nearly all the remaining organisms get their energy either from absorbing 4b-type biochemicals made by the sunlight-users, or from absorbing 4b-type biochemicals made by other organisms like themselves.

Sunlight thus provides the primary source of energy for living things: directly in the case of sunlight-users; indirectly for everything else.

ENERGY DEMAND

This picture, while dynamically satisfying, is incomplete. Experiments on sunlight-users kept in darkness show that they not only cease growing when deprived of light; they actually start to shrink. Eventually they can even reach a state where they become incapable of ever growing or replicating again. At that moment they do, by definition, cease to be organisms. This step is irreversible, for organisms only ever start functioning as a result of parental division. When an organism ceases functioning, it does so forever. It dies.

216

Non-sunlight-users suffer the same fate if deprived of their source of energy. They too shrink and die.

These observations show that organisms are not internally stable. In dynamical terms their situation is like that of people trying to gain height on a descending escalator. These (somewhat eccentric) people not only need energy to increase their height above the ground (equivalent to growing); they also need energy simply to maintain the height they already have, in order to avoid being carried down the escalator and thrown off the bottom (equivalent to shrinking and dying). The energy they use to maintain their height does not alter their potential energy or their kinetic energy; nor do they emit waves. But the energy is conserved nonetheless. It emerges as molecular kinetic energy. People walking up a down escalator get hot.

This is exactly analogous to what happens to organisms. Their cells consume energy continuously. The energy used just to keep cells going is turned into heat. Functioning organisms are always a little warmer than their surroundings (discounting the effects of surface evaporation — the people on the escalator can cool down even on a hot day by perspiring).

External energy supplies for sunlight users are only available intermittently — the sun disappears at night. Similarly for non-sunlight users, external 4b-type biochemicals are often not readily accessible. Consequently, to meet their cells' continuous moment-by-moment energy demands organisms must have access to an internal store of energy which is immediately available. What supplies this energy?

This question is most easily answered by studying the larger animals. When an animal is deprived of its external energy supply, it is initially the animal's sugar stores — mostly glucose polymers — that diminish in quantity, followed subsequently by fats and some muscle proteins. During this time the animal is able to continue functioning fairly normally and to recover if its external energy supply is restored. It is only when the sugar stores, fats and utilizable muscle proteins have been used up that irreversible degeneration sets in and the animal dies.

These observations lead to a hypothesis for the role of fats and sugars in organisms: their primary function is to meet the organism's moment-by-moment demand for energy. Sunlight-users depend on these substances in the absence of sunlight; non-sunlight-users depend on them all of the time. Proteins, in contrast, are not readily used to provide energy. This indicates they must have a different role to play.

To start investigating this hypothesis about energy, look at figure 5.7 and recall figure 1.2. Sugars and fats are 4b-type molecules. The easiest way to turn them into 4a-type molecules, and thereby get the stored potential energy out of them, is to burn them. This means causing them to react with oxygen. Formulas for the reactions of typical sugars and fats with an unrestricted amount of oxygen are respectively

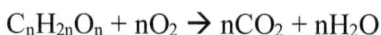

$$C_nH_{2n}O_n + nO_2 \rightarrow nCO_2 + nH_2O$$

and

$$CH_3\!-\!(CH_2)_n\!-\!COOH + \tfrac{1}{2}(4+3n)O_2 \rightarrow (n+2)CO_2 + (n+2)H_2O.$$

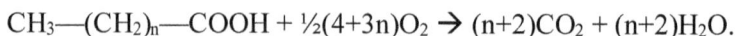

Whenever carbon or hydrogen atoms are bonded to oxygen, energy is released because both CO_2 and H_2O are 4a-type molecules. Fats, having less oxygen in them to start with, supply more energy than do sugars. Clearly sugars and fats are quite viable as a source of internal energy for an organism, providing it has access to a supply of oxygen with which to react them chemically.

Several further observations are relevant here. One is that the requisite supply of oxygen is freely available in the atmosphere. Another is that most organisms deprived of access to oxygen die rapidly, implying that in such circumstances their internal moment-by-moment energy supply is shut down. A third is that sunlight using organisms, when forced to depend on their energy reserves by being kept in the dark, require a supply of oxygen just like non-sunlight-users: without light plants die slowly; without light and oxygen they die much more quickly.

The hypothesis that sugars and fats provide an internal energy source for organisms is thus well on the way to promotion to a theory. All that needs to be done in order to complete the picture is to describe in detail *how* sugars and fats are turned into energy within cells. Could that be what proteins are for?

CATALYSIS

Look at figure 5.7 again. When an organism uses a fat or sugar molecule as a source of energy, it would seem necessary for it to get the molecule over the potential energy hill while reacting it with oxygen. The energy involved in doing this is considerable, presenting the cell with the almost impossible challenge of how to concentrate, contain and control such a large amount of energy in one molecule.

This provides a possible role for proteins. There is no law of dynamics which says that to get a molecule from point 4b to point 4a it *has* to go

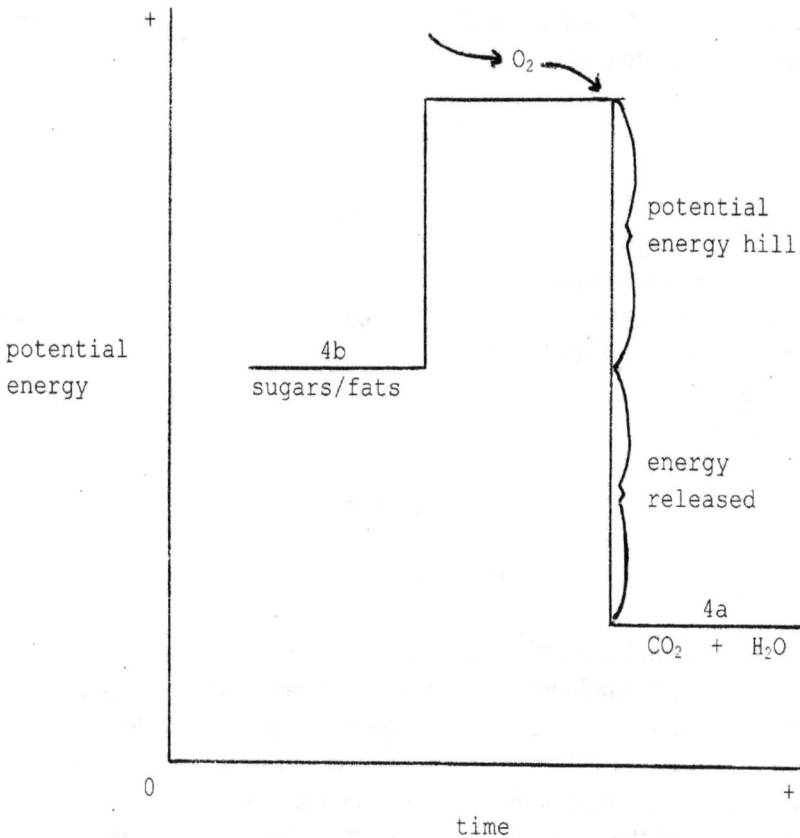

Figure 5.7 A conceptual representation of the potential energy of sugars and fats as they react chemically with oxygen.

over the potential energy hill. All the theory of dynamics requires is that energy is conserved between 4b and 4a. If the biochemicals in a cell were somehow able to tunnel through the hill, they could avoid having to go to the top of it. It can be hypothesized that this is indeed what happens in cells and, furthermore, that it is the role of proteins to do the tunnelling.

Consider the reaction $A \rightarrow B + C$, where A, B and C are biochemicals. Let A be a molecule at level 4b in figure 5.7. Let B and C be molecules at level 4a. E_a is the potential energy difference between level 4b and the top of the potential energy hill, while E_b is the potential energy difference between levels 4a and 4b. In the absence of a tunnel the reaction can be broken down into two stages:

$$A + E_a \rightarrow A*$$

$$A* \rightarrow B + C + E_a + E_b.$$

$A*$ is A at the top of the potential energy hill. Notice that the E_a put into A initially is exactly repaid when $A*$ breaks up into B and C. This leaves a net reaction of $A \rightarrow B + C + E_b$.

The role envisaged for proteins is to replace E_a. The two stage reaction becomes:

$$A + P \rightarrow AP$$

$$AP \rightarrow B + C + P + E_b.$$

P is the protein. By substituting P for E_a, reacting molecules are able to go through the potential energy hill instead of having to climb over it. Any substance which has this energy-replacing property is called a *catalyst*. Thus P is said to catalyse the chemical breakdown of A. Notice that the P entering the first stage of the reaction emerges — like the E_a it replaces — unaltered from the second stage when AP breaks up into B and C. The net reaction is still $A \rightarrow B + C + E_b$.

In order to test this hypothesis about proteins, the question to be answered is: what observable effect can a catalyst be predicted to have on the reactions it takes part in? The effect is fairly easy to deduce. The potential energy hill is what keeps molecules stable. It prevents molecules at 4b breaking up into molecules at 4a. If the hill is by-passed by means of a catalyst, then there's nothing to stop the 4b-molecules breaking up. In other words, supply an appropriate catalyst and a reaction which was previously not taking place will begin to occur.

What needs to be performed, therefore, is an experiment in which a biochemical reaction which was not occurring, or else was occurring only extremely slowly, begins to happen readily when a protein is added to the mixture. It is indeed possible to observe proteins having this effect. Experiment thus supports the hypothesis about the role proteins play.

All the proteins with catalytic properties are of the globular kind. Hence it is globular proteins which catalyse biochemical reactions. To distinguish them from inorganic catalysts, biochemical catalysts are called *enzymes*.

Two points can be made here about enzymes. The first is that, because they are unaffected by the reactions they catalyse, each enzyme molecule is able to catalyse one reaction after another indefinitely.

The second point, which is of crucial importance, is that any given kind of enzyme can only catalyse one specific biochemical reaction. The sequence of the amino acids in the chain that constitutes an enzyme determines the enzyme's shape. And it is the shape of the enzyme which determines the chemical reaction it will catalyse. This comes about because of the way enzymes work. The convoluted amino acid chain gives an enzyme a shape like a lumpy ball. The shape of some part of the enzyme enables it to fit against the chemical whose reaction it catalyses rather like a lock fits a particular key. No other chemical will fit in the lock. When the enzyme encounters the chemical that fits it, a reaction occurs ($A + P \rightarrow AP$ above), and the first step in the catalytic process is complete. So, as a general rule with few exceptions, the cell makes one kind of enzyme for each chemical reaction that occurs within it.

CONTROLLING AND HARNESSING ENERGY

Catalysis removes from cells the need to handle the large E_a energies associated with covalent molecules. (Small E_a energies often remain for the reaction $A + P \rightarrow AP$.) The cell is nevertheless still required to deal with large E_b energies.

To overcome the difficulty they face in this respect, cells break down each chemical reaction into a series of steps. For example, the conversion of sugar into CO_2 and H_2O is achieved by means of about twenty separate reactions, each requiring its own enzyme. With each step along the route, the sugar molecule is altered in such a way that by the end of the twenty steps it emerges as CO_2 and H_2O. With most of the steps a small proportion of the full energy obtained from reacting sugar with oxygen is released. In effect, the cell's enzymes snip away at the sugar molecule a bit at a time. By this means the energy is released in small quantities rather than in an indigestible lump.

By breaking reactions down into steps, cells are able to control the energy released within them. Control, though, is not the object of the exercise. It is no good the cell controlling the release of energy if that energy does no more than make the cell hot. The cell's objective is to harness chemical energy so that the cell can manufacture biochemicals; and more generally so that it can do whatever it needs to do to keep itself alive.

Harnessing intra-cellular energy involves storing it as potential energy. The cell does this by connecting reactions which release energy, such as those involved in converting sugar into CO_2 and H_2O, with reactions which absorb energy.

One energy absorbing reaction that is particularly widely used in cells is the conversion of a molecule called adenosine diphosphate (ADP) into one called adenosine triphosphate (ATP). When ADP is turned into ATP, energy is stored. The cell can then use the stored energy to build biochemicals etc., reconverting the ATP back to ADP as it does so. ATP thus functions in cells like loose change in financial transactions. The cell is like someone who only spends one penny at a time. Bank notes (sugars and fats) are not appropriate currency. They must be converted into small coins (ATP) in order that they can be spent.

Figure 5.8 shows the formula-diagrams for ADP and ATP.

In the course of the twenty steps converting the sugar, glucose, to CO_2 and H_2O, around 30 ATP molecules are produced. The two reactions summarizing the twenty steps are

$$C_6H_{12}O_6 + 6O_2 \rightarrow 6CO_2 + 6H_2O + E_b$$

$$30ADP^{3-} + 30H_2PO_4^- + E_b \rightarrow 30ATP^{4-} + 30H_2O.$$

Overall this gives

$$C_6H_{12}O_6 + 6O_2 + 30ADP^{3-} + 30H_2PO_4^- \rightarrow 6CO_2 + 36H_2O + 30ATP^{4-}.$$

Thus the large E_b energy is shared out amongst 30 ATP molecules. The banknote is converted into loose change.

The dynamics of the processes taking place in cells can be summed up as follows

- Cells get their energy supply from sugars and fats, and in some cases also directly from sunlight.
- They use enzymes (globular proteins) to by-pass the potential energy hill associated with sugar and fat molecules.
- They control their chemical reactions by breaking them down into small steps.
- They harness the energy released from sugars and fats by storing it in molecules such as ATP.

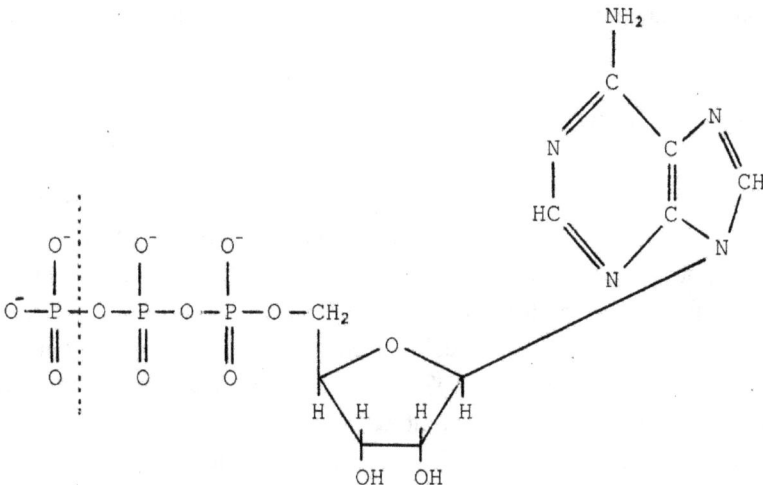

Figure 5.8 ATP. The atoms to the left of the dotted line are lost when ATP is converted to ADP.

It can be noted that all these dynamic activities require the participation of protein enzymes. It is hardly surprising therefore that cells deprived of food avoid using their proteins to supply energy. Proteins are what make all a cell's activities possible. When a cell loses its proteins it loses its life.

Obviously not all proteins are involved with releasing energy from sugars and fats and storing it in ATP. Some are concerned with drawing on the energy in ATP to manufacture biochemicals, including, of course, other enzymes. And these are just two of the functions carried out by proteins. There are many others, some of which will be mentioned below and in chapter 6 and section 8.1.

The picture of cells that has emerged so far is thus centred on enzymes. Without enzymes there could be no energy from fats and sugars. And without energy from fats and sugars (plus other enzymes) there could be no enzymes. A cell's chemistry is thus cyclic in nature. Enzymes release and store energy which other enzymes use to make enzymes. Cyclic processes are very hard to start from scratch. This is because all parts of the cycle must begin working simultaneously. And it is why death is an irreversible event; broken cycles can't restart themselves. It is also why cells always originate from other cells by means of parental division. Only in that way can a cell be born with its chemical cycles intact and working.

5.4 THE ROLE OF NUCLEIC ACIDS

In this account of the biochemical processes taking place inside cells, nucleic acids have so far been conspicuous by their absence. They are involved neither with the supply of energy nor with catalysis. To establish the role they play, a different kind of observation must be made: the quantity of nucleic acids in cells needs to be measured. Under what circumstances does it increase? If it can be discovered when cells manufacture nucleic acids, it might be possible to say what they do it for, and thereby what the biochemical role of nucleic acids is.

It is found that the quantity of nucleic acids in a cell increases prior to cell division. This suggests that nucleic acids play a part in replication. To put forward a hypothesis that reflects this observation, a better idea needs to be formed of what cell replication involves.

When a cell replicates it does so by dividing into two halves. Each half then makes good the loss to itself of those parts taken by its twin. The halves each grow so that they become as large as the original parent cell and also identical to it and to each other. Where there was one cell there are now two. Replication has taken place.

Each of the two cells resulting from the division of the parent has only half the parent's contents. If it is to grow to its parent's size, it needs somehow to sense that it is only half the size it's meant to be. Something must cause the cell to replace what it's missing.

What the cell needs, in effect, is a plan. Any deviation in the form of the cell from the plan — such as would be the case if the cell was only half its proper size — would cause the enzymes of the cell to initiate biochemical reactions to correct the deviation. The plan would give the cell one of the prime characteristics of life, namely a shape and structure determined from within itself.

When a cell divides, each resultant half-cell needs to receive a complete copy of the plan from its parent. Consequently, prior to dividing, the parent must contain two plans of itself. However, unless it is about to divide, the parent doesn't need two copies on its own account. It can be expected then that the second copy of the plan will only be made immediately before cell division takes place. This provides a possible role for nucleic acids. Nucleic acid quantities increase when a cell is about to divide. Could the nucleic acids constitute a plan of the cell?

A look at the structure of deoxyribonucleic acid shows how, if it does contain a plan, cells might make a copy of it. As was seen in section 5.2, DNA occurs as two chains of interleaved deoxyribose and phosphate molecules twisted round one another, the twisting being in the fashion of a very long cylindrical spiral. The bases on one chain complement the bases on the other. See figure 5.6. The result of this complementarity is that if the sequence of bases on one of the chains is known, then it is also known exactly what the base sequence is on the other chain.

Now, suppose some agency (e.g. certain enzymes in the cell) was to separate the two entwined DNA chains from each other. Each unentwined chain could then have formed against it a new complementary chain built from scratch. Having started with one DNA spiral of two entwined chains, there would now be two of them. And most important, because of the complementarity of the bases on each entwined chain, the two spirals would be identical. It can thus be concluded that the chemical nature of DNA is conducive to being duplicated. Given a range of suitable enzymes, the copying of DNA spirals will present no great difficulties to a cell.

From the point of view of replication, the proposed role of DNA is plausible, but replication is only half the story. It is also necessary to suggest how this replicating DNA can contain a plan of the cell in which it resides.

The plan of a cell is essentially a description of the biochemicals it contains. Now, it has been seen that sugars and fats play a largely passive role in deciding the structure of a cell. They do not themselves make biochemicals. That task falls to enzymes. So enzymes it is, and enzymes alone, which build the cell and give it its shape and structure. Consequently a plan of a cell is ultimately a plan of the enzymes the cell contains. The plan of an enzyme is, in turn, a list of the amino acids the enzyme is made of, the amino acids' sequence in the list matching their sequence in the enzyme.

The plan DNA is required to contain is, then, a list of amino acids. Different sections of the list will describe the amino acid sequences of particular enzymes. Those enzymes, once built (by other enzymes), will automatically catalyse reactions in such a way that collectively they give the cell its planned form.

The easiest way for a DNA chain to contain a list of amino acids is for it to have along the length of the chain a series of molecules, each of which corresponds to a particular amino acid. As there are twenty different amino acids used in proteins, there would need to be twenty different molecules. Passing along the DNA chain from one molecule to another would be tantamount to reading a list of amino acids in the sequence needed to build one or more proteins.

Unfortunately, the only molecules found distributed along a DNA chain are the four bases, adenine, cytosine, guanine and thymine. Four bases can only be used to list four amino acids if each base is to correspond to a specific acid. The four bases are clearly inadequate to the task.

Unless, that is, the bases are taken in groups. If the DNA bases are paired, there are $4 \times 4 = 16$ different combinations of bases in a pair. This is still four short of the desired twenty. If the bases are grouped in triplets, the number of combinations becomes $4 \times 4 \times 4 = 64$. It can therefore be concluded that a series of three DNA bases is needed to specify a particular amino acid.

In some respects the DNA bases function analogously to the letters of an alphabet. It may be helpful to pause for a moment and consider some points of comparison between them.

- There are 26 different letters in the written English alphabet; there are 4 different letters in the DNA alphabet, represented by the four bases.
- The letters of the written alphabet can be grouped together to make words; the letters of the DNA alphabet can be grouped together to make DNA-words, each word being exactly three letters long.
- There are many words spelt by the written alphabet; there are 64 DNA-words spelt by the letters of the DNA alphabet.
- Different written words have different meanings, though some are synonyms; different DNA-words signify different amino acids, though because there are only 20 amino acids for 64 words, many DNA-words are synonyms.
- Written words combine to make sentences; DNA-words combine to signify proteins.

There is thus a DNA alphabet of 4 letters (bases) and a DNA vocabulary of 64 words (triplets of bases). These words signify each of twenty amino acids. It can be added that at least one of the triplets of bases must signify the equivalent of a full stop. (Actually three of them do.) The full-stop word is needed to mark the end of a sentence where a protein's amino acid chain comes to an end.

To develop this hypothesis about DNA into a theory, it must be explained in detail how a sequence of three DNA bases comes to be turned into an amino acid. The DNA-words are like a code. They need to be translated in order for their significance — an amino acid — to be realized. The theory must show how this translation comes about.

The theory that has been devised, following much experimentation, is extremely complicated (like most theories in biochemistry). Here it will be reduced to its bare essentials. To turn a DNA base sequence into a protein, a number of steps are needed. The steps are these, described in turn.

1. Most cells contain two entwined DNA chains. One chain is the code chain. The other has complementary bases on it, as already explained. The complementary chain comes away from the code chain at a point where a DNA sentence starts. This exposes the DNA bases to the contents of the cell.

2. An RNA chain forms against the exposed DNA. Aside from substituting uracil for thymine, the RNA chain's base sequence is in the usual complementary form.

3. The RNA chain separates from the DNA. It contains one 'sentence' of bases. As it is carrying a message away from the DNA, this kind of RNA is called *messenger RNA* (mRNA).

4. The mRNA is attached to a large molecule made of proteins and RNA combined. These protein-RNA molecules are called *ribosomes*, and a typical cell has about a million of them. The mRNA is positioned so that the first three bases of its chain are exposed within the ribosome.

5. In parallel with steps 1 to 4, another kind of RNA, called *transfer RNA* (tRNA), has an amino acid attached to it. In principle there are 64 tRNAs, one for each possible three-base sequence. (This is not true in practice.) Each of the tRNAs has three exposed bases. The amino acid it picks up will be the one that those three particular bases signify.

6. The tRNA whose three exposed bases match in complementary form the three mRNA bases exposed within the ribosome enters the latter molecule. Its three exposed bases then form hydrogen bonds with the mRNA's exposed bases.

7. The appropriate tRNA having joined the ribosome, the ribosome moves along the mRNA until it exposes the mRNA's next three bases. These are then ready to receive their own tRNA molecule.

8. The next tRNA joins the ribosome, bringing with it the required amino acid, attached to it as per step 5. See figure 5.9(A). The amino acid attached to the earlier tRNA (on the left in the figure) is moved across to the latest tRNA to arrive and joined to the amino acid found there.

9. The ribosome moves along the mRNA chain for another three bases. This exposes them. At the same time the old, and now amino-acid-less, tRNA is ejected from the ribosome. See figure 5.9(B).

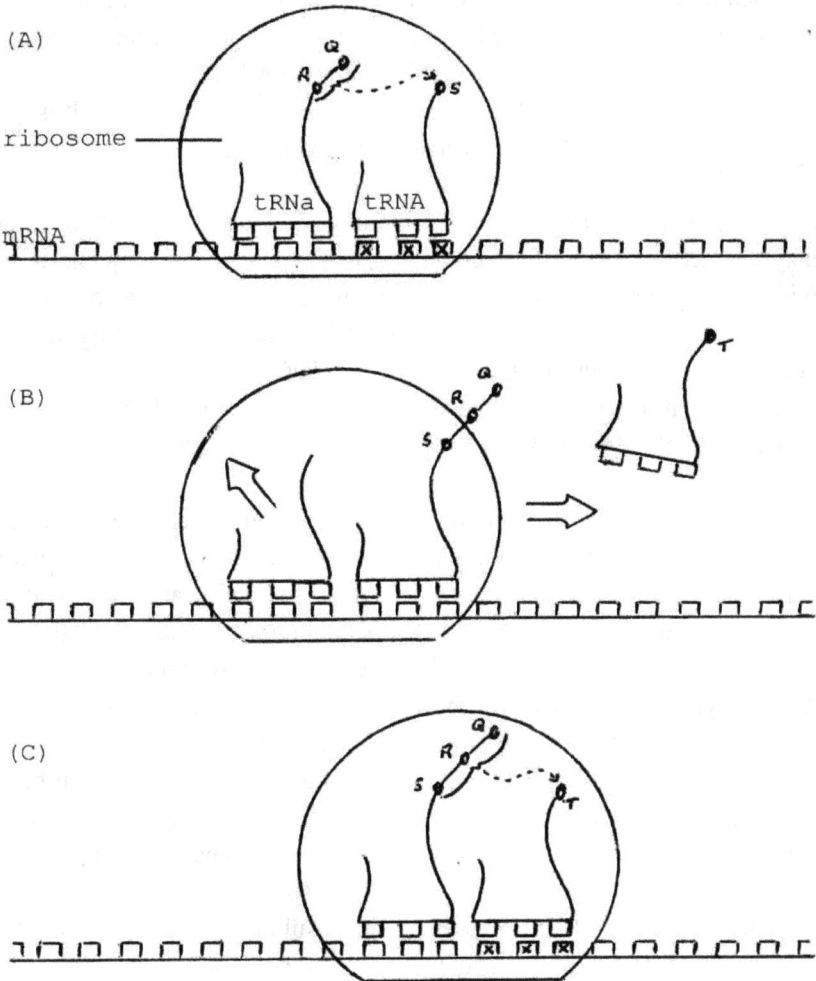

Figure 5.9 A schematic representation of the building of a protein. (A) tRNA with amino acid S attaches to the exposed bases (xxx) within the ribosome. The previously coded amino acid R is joined to S. (B) The old tRNA for amino acid R is released from the ribosome, which moves three mRNA bases to the right. (C) tRNA with amino acid T attaches to the new exposed bases. The cycle begins again and continues until three exposed bases signal a full stop.

10. This step is a repeat of step 8 and is depicted in figure 5.9(C). With each repeat of step 8, followed by a repeat of step 9, the amino acid chain being moved to the latest tRNA to arrive gets one amino acid longer. The cycle continues until three mRNA bases signifying a full stop are encountered. The amino acid chain is then complete. It is released into the cell to play its part as a working protein.

By these ten steps, proteins, including enzymes, are built by other enzymes acting on the code contained in the DNA base sequence. Pre-existing enzymes are heavily involved throughout. They not only release energy for the various steps, but also play a direct role in bringing each step about. For instance, the amino acid that goes with each tRNA is attached to it by means of enzymes. At least some of the time, the enzymes involved in making proteins will be helping to make copies of themselves.

This theory of DNA function reveals a second cycle to add to that found in the previous section. There it was seen that without the energy from fats and sugars enzymes could not be made, and that without enzymes the energy from fats and sugars could not be released. Cells need enzymes to use fats and sugars, and they need to use fats and sugars to make enzymes. In this section, the cycle that has been uncovered involves enzymes and DNA. Without enzymes the DNA base sequence cannot be translated into enzymes. Without the DNA base sequence there can be no enzymes at all. Cells need to translate the DNA base sequence to make enzymes, and they need to have made enzymes to translate the DNA base sequence.

When a cell divides, it passes to the resultant half-cells a complete copy of the DNA plan. It also passes them enough enzymes, fats and sugars to ensure that the two cycles are intact and functioning in its offspring right from the start. The unavoidable conclusion this leads to is that cells never arise by any other means than by division of a parent. The two cycles are vastly too complicated to stand any reasonable chance of arising in (previously) inanimate chemicals.

That organisms only come into being through the process of replication can be considered a law of biochemistry. It is a law that has never been seen to be violated. Yet if it is accepted unconditionally it is totally incompatible with the Big Bang theory of chapter 4. No organism could have existed for millions of years after the Big Bang because the universe was too hot. Furthermore there was no carbon, or nitrogen, or oxygen, or phosphorus in the universe until exploding supernovas made them. So the universe was born lifeless. The Earth too was almost certainly lifeless following its formation.

Life exists today. Given the correctness of the Big Bang theory — you will recall from section 4.6 that all but the first second of the Big Bang theory is well tested observationally — life did not exist in the past. At some time, the law of biochemistry must have been violated. The next step, then, is to return to the Earth left newly-formed at the end of chapter 4, and investigate what was going on there. Somehow life arose from inanimate chemicals. The challenge is to discover how.

5.5 THE ORIGIN OF LIFE

THE EARLY HISTORY OF THE EARTH

The Earth began to form $4 \cdot 6 \times 10^9$ years ago. This estimate of the date rests on nuclear physics. It is found that nuclei of the elements uranium and thorium are almost stable, but not quite. An average uranium or thorium nucleus takes billions of years before it turns via a cascade of nuclear disintegrations into lead. This makes it possible, by comparing the amounts of ^{238}U, ^{235}U and ^{232}Th with their lead end-products, ^{206}Pb, ^{207}Pb and ^{208}Pb respectively, to determine a rock's age even if that age runs to billions of years. Put simply, the more lead end-products there are, the older the rock. As a cross check, other long-lived slightly unstable nuclei can be tested to see what age they give for the same rock. An example is ^{40}K, which turns into ^{40}A or ^{40}Ca. If the age indicated by all these various unstable nuclei comes out the same, it can be taken to be accurate.

Obviously, to find out how old the solar system is requires rocks to be dated which are unchanged since the solar system formed. Meteorites are used for this purpose. Meteorites are lumps of rock that avoided taking part in planet formation and which occasionally fall onto the Earth. Their age is generally found to be the aforementioned $4 \cdot 6 \times 10^9$ years.

The unchanged state of meteorites stands them in stark contrast with rocks on the Earth. Both on the continents and beneath the oceans, the combined actions of weather and geological activity are forever destroying and rebuilding the planetary surface. The oldest rocks on the Earth, dated by similar techniques to those used on meteorites, formed $\approx 4 \times 10^9$ years ago. Rocks of that age are very rare. Most rocks are much younger.

The difference between the age of the Earth and its oldest rocks means that there is no direct evidence of what the Earth was like for its first half billion years. Nevertheless it is still possible to draw a fairly general picture of the Earth during and after its accretion from dust and gas orbiting the newly formed sun.

The accretion of the Earth took about a hundred million years. As the planet grew, large amounts of gravitational potential energy were converted into heat, melting the interior. Having an interior of liquid rock enabled the Earth's heavier elements, particularly iron, to sink downward towards the centre. Lighter elements such as silicon were consequently displaced upwards, floating on the iron. In this way the Earth came to have a metallic core surrounded by a less dense mantle.

It is uncertain how the rate of heat dissipation into space compared to the rate of heat production by infalling material, so the Earth may have begun with a molten surface, or it may have begun with a hot but solid one. Either way, it would have had no appreciable atmosphere and no surface water.

Imagine sending a team of scientists on a visit to the Earth not long after most of the accretion has occurred. They find the planet has no surface water, and no air, but, at least in places, it has solid ground to stand on. They observe the presence of many volcanoes engaged in frequent eruptions and easily conclude that close beneath the surface the heat of accretion has given the Earth a molten interior.

Added to the scientists' panorama of volcanoes, hot rocks, no water and no atmosphere, are meteors. These late arrivals from the accretion process, some of them colossal, are occasionally to be seen falling to the ground, producing craters and, in the largest cases, temporarily melting the surface locally. The sun, radiating with about seven tenths of its present strength, shines from a dark sky perhaps made hazy with volcanic gases and dust, marking out days that are roughly half as long as those of our time. And the moon, about half its present distance from the Earth, is twice as large in the sky and goes through all its phases in approximately ten 24-hour periods (instead of the twenty-nine and a half we're used to).

The scientists might well feel unwelcome on such a planet. Certainly not a place to call home! But let them return a few hundred million years later; say, about $3\cdot8 \times 10^9$ years ago. They now find the sun is a little brighter, the days are a little longer, and the moon is a little further away (all trends which continue to this day). More importantly, the Earth is markedly less hostile to life (and to scientists!). Meteor impacts are fewer in number and the size of meteors is considerably diminished. The Earth's surface is solid everywhere except where volcanoes are erupting, and well below the temperature at which water boils.

And there is another major change to take note of. The cumulative effect of the countless volcanic eruptions of the past half billion years has been to vent some of the gases that were trapped inside the Earth when it formed. These gases have risen, like bubbles in treacle, to be volcanically released to the outside. Among their number are methane (CH_4), carbon monoxide (CO), carbon dioxide (CO_2), ammonia (NH_3), nitrogen (N_2), hydrogen (H_2) and steam (H_2O). Over hundreds of millions of years they have given the Earth an appreciable atmosphere.

But the scientists do not find the atmosphere consists simply of these various volcanically vented gases. This is because the composition of the atmosphere, right from the start, was subject to modification by chemical reactions. Of great importance are reactions between sunlight and gases containing hydrogen. Electromagnetic wave energy from the sun is strong enough to break up molecules of some of these gases. The hydrogen atoms released as a result are able, providing they make their way into the upper atmosphere, to escape from the Earth and drift away into space. The amount of hydrogen in the atmosphere is thus subject to a slow decrease. Nowadays the loss is about 10^7 kg per year. On the early Earth, which had more hydrogen to lose, the rate of loss would have been higher. The net result is that the visiting scientists find an atmosphere which is gradually tending towards consisting mainly of N_2 and CO_2, with an amount of water vapour also.

The scientists notice one other big change: most of the steam introduced into the atmosphere by the volcanoes has condensed as water. Furthermore, many meteors and comets have large amounts of water in them, and their impacts with the Earth have also greatly increased the amount of water on the Earth's surface.

The status of all this water is uncertain. That's because the temperature of the Earth is decisive, and it is not known for sure what that temperature was. As the sun was dimmer than it is today, the Earth should have been cooler. But the gases in the atmosphere were also more effective at retaining heat than the present-day gases, and that should have made it warmer. If the Earth was about as warm as it is now, or a little warmer, the water would have formed oceans. If cooler, there would have been much snow and ice in addition. Possibly over a timescale of tens of millions of years the early Earth may have oscillated between the two states: some of the time ice-bound and some of the time temperate.

Keeping options open on the temperature, the scientists can be pictured walking along the shores of an ocean. The Earth's weather patterns are fairly similar to those of today. Above them, clouds, and dust from volcanoes, shroud the sky. Sometimes it rains (or snows). They can feel the wind in their faces. The land is an endless panorama of rock (and ice), scoured by wind and rain (and shattered by frost). Only two things strike them as incongruous. If they try to breathe the air, it suffocates them. And nowhere is there even a hint of greenery, no trace of purposive movement, no sign of life. The Earth is utterly barren.

OCEAN CHEMISTRY BEFORE LIFE

The oceans of the Earth at this early time would soon have acquired a range of molecules other than H_2O in them. Some of the molecules would have come from rocks, both beneath the sea and from those on land washed by rain. That's because many rocks contain molecules which react chemically with water. What happens is that these molecules break up in the presence of water into ions. The individual ions mix freely with the water and, in a sense, become part of it. The molecules are said to have dissolved.

The most familiar example of a dissolving molecule is sodium chloride, NaCl. In water, the ionically bonded sodium and chloride ions become free of each other. Water in which sodium chloride is dissolved is a mixture of H_2O and separate Na^+ and Cl^- ions.

Solids from the Earth's rocks which dissolve in water contribute many ions to the oceans, including HSO_4^-, SO_4^{--}, $H_2PO_4^-$, Cl^-, NO_3^-, HCO_3^-, Na^+, K^+, Mg^{++}, Ca^{++} and Fe^{++}. What is significant about this, and why it is worth mentioning, is that a lot of these ions have come to be utilized in various roles by living things, including in the human brain, where some of them are vital to its functioning (see section 6.3).

Another source of dissolved molecules in the oceans is the atmosphere. Chemical reactions brought about by energy from lightning and from sunlight would have produced a variety of molecules in the atmosphere which subsequently ended up in the sea. These reactions can be simulated in the laboratory by filling a flask with the hypothesized early atmospheric gases, N_2, CO_2 and water vapour, and then exposing them to artificial lightning and sunlight. NH_4^+ and NO_2^- are found to be produced by this means (mostly the latter) plus a range of organic chemicals, including some amino acids.

Finally, and of uncertain importance, one other source of ocean chemicals are the aforementioned meteors and comets striking the Earth. These extra-terrestrial objects are nowadays found to contain a surprising range of molecules, including simple organic ones, amino acids once again amongst them. As today's meteors and comets are essentially unchanged since the solar system formed, it can be expected that the meteors and comets of four billion years ago were equally well-endowed chemically.

The oceans, then, would have come to contain a quantity of organic molecules, some of which would have been quite large. An observer sampling the water might expect to find a range of amino acids including some of those used in proteins today and some which are not used today, a few amino acid chains, a small range of nucleic acid bases and perhaps some simple sugars and fats. None of these molecules would have had any cyclic function with respect to any of the others. A few of the amino acid chains may conceivably have had catalytic abilities but these would have been exercised purely randomly. There would have been no sequence of bases on DNA or RNA chains. There would have been no DNA or RNA 'words' corresponding to a protein.

It can thus be speculated that, by chance and by chemistry, the oceans acquired within them some of the basic components utilized by carbon-based living things. There would have been a complete absence of order amongst these various molecules. Their relationship to a living thing would be as a scrambled jigsaw puzzle with a lot of pieces missing to a completed one.

Unfortunately, this knowledge of the likely chemical state of the oceans represents the only available data on which to build a scientific account of how life arose on the Earth. Bearing in mind sections 5.2, 5.3 and 5.4, it is not remotely enough. There is an enormous gap between a laboratory flask containing amino acids, or an ocean containing a range of organic chemicals, on the one hand, and a fully operational living thing on the other. Impartial reflection on the sheer complexity of even the most basic biochemical processes will force even the most unrestrained optimist to admit that the transition from organic chemistry to biochemistry is formidably hard to account for.

THE PROBABILITY OF LIFE

So what is the scientific account? The honest answer is that there isn't one. There are no observations, no measurements, no experiments which cast even a faint light on how the first living, replicating entity arose. Nor, since it has left no trace in the fossil record, is there any evidence of what the first living thing actually was — what form it took, what structure it possessed, not even what it might have been made of. The only answer science is currently able to give to the question of how life arose on the Earth is: 'Don't know.'

There are two mutually exclusive views that can be taken of this state of ignorance. In the absence of observational evidence, both are viable, but neither can qualify as scientific. Expressed in their most extreme forms, they are the following:

- EITHER all the theoretical blind spots associated with the formation of the first self-replicating system will be filled in as knowledge advances, with the first organisms proving to have used some means to replicate which was so simple as to make their appearance on the Earth inevitable
- OR the many difficulties associated with forming the first organisms by chance (most of which I haven't mentioned) make the emergence of life an extremely improbable occurrence.

Advocates of the first view argue that our position now is akin to an alien from outer space visiting a human car assembly plant. The alien can see that humans haven't the strength to make so much as one component of their cars but do so using machine tools. But it's equally clear humans haven't the strength to make machine tools either. Those are made by other machine tools. The uncomfortable conclusion, it would appear, is that the process must have been started when a human with a lot of sense came by accident on a perfectly formed set of machine tools fashioned by some freak chance during a volcanic eruption. Yet the visiting alien would be wrong. Machine-based manufacturing is the culmination of a process that began many thousands of years ago, back in the Stone Age, using the bones of dead animals, sharp stones and muscle power, none of which are used today.

This view of life's beginning is best illustrated by an ingenious speculation put forward by Cairns-Smith. (See Bibliography: Cairns-Smith.) He argues that life today is 'high-tech' (his words) — that DNA and proteins are the biological equivalent of machine-based manufacturing. He proposes that the first replicating molecules were not organic at all but were 'low-tech' clay crystals — the equivalent of bones, stones and muscles.

Advocates of the second view argue that life is rare in the universe or possibly even unique to the Earth. Any improbable event will occur somewhere at some time. Even if the emergence of life is so improbable that it is unlikely to occur more than once in the entire universe in, say, 10^{10} years (or 10^{50} years or 10^{250} years!), then if it does happen it has to be somewhere at some time. The 'somewhere' was the Earth. The 'some time' was about $3 \cdot 8 \times 10^9$ years ago.

We are here today. Living things did appear on our initially dead planet. But it cannot be said at present whether that emergence of life was a probable or an improbable event. Science is in the impossible position in this instance of having only one object to observe — and far too late in the day, at that. To settle the issue, systematic observations need to be made. What is required ideally is access to many Earth-like planets, with or without life. Needless to say, with the hopeful exception of Mars, that is not a practical proposition for the foreseeable future.

Given this observational impasse, there are two arguments that can be advanced to take the debate forwards, one on either side. In favour of life being common is the reasoning, at root philosophical, that science hates miracles. Suppose you were sitting beside a pond one day, and by sheer statistical fluke all the water molecules, moving about at random, just happened by chance to find themselves heading towards the centre of the pond simultaneously. The result, when they all collide, would be a water-spout. Nothing in the laws of physics says this can't happen. All the conservation laws, including that of mass-energy, are obeyed in the process. But would you, seeing this water-spout, say to yourself that you'd just witnessed an amazing against-all-the-odds occurrence? Or would you look for a non-statistical 'natural' cause for the observation? Every scientist, without exception, would choose the latter. Extreme statistical rarity is too close to miraculous for science to be comfortable with it as an explanation. And that applies to the origin of life on the Earth every bit as much as to mysterious water-spouts in ponds.

And the argument for life being rare? This is the only piece of observational evidence which has a direct bearing on the question. If life is probable, then it will have arisen on many of the other Earth-like planets that almost certainly exist throughout the universe's more than 10^{10} detectable galaxies. But looking out into space, using equipment suitable to monitor any chosen electromagnetic wavelength, from the longest to the shortest waves, the universe appears lifeless. It looks *exactly* how it would look if it was totally dead. If life is common, why can no sign of it be discerned? Where is it? Could it really be the case that the Earth is unique? An extreme statistical fluke?

5.6 EVOLUTION

NATURAL SELECTION

There are three stages in the emergence of life:

1. the formation of the Earth and its atmosphere and oceans, and therein the development of an environment in which organic chemicals can form — a more or less inevitable step taken easily and simply on any Earth-like planet
2. the transition from organic to biochemical activity, from non-living to living matter — this is the big step whose probability of occurring is completely unknown

3. the elaboration of the replicating entity (taken below to be RNA) — a straightforward process once life has started.

The first stage has been covered fairly fully above, and the second stage — one of the scientific world-view's remaining blank pages — has also been commented on. It remains to look into the third stage. Why didn't the first replicators remain as they were? What made them become more elaborate as the millions of years passed?

To answer the first of these two questions, think of the environment the first carbon-based replicating molecule would have found itself facing. Whenever the raw materials were available, it would make copies of itself systematically. The copies, self-replicating in their turn, could expect to become steadily more numerous in their environment.

This gives rise to a problem. The world is not infinite. Orderly replication runs up against resource limits. For instance, there is a limited amount of space that molecules can occupy; a limited supply of the energy needed by molecules to copy themselves; a limited availability of sugar and nucleic acid bases (or the raw materials to make them); and so on. Thus a situation arises where competition becomes a factor in RNA replication.

Competition on its own is trivial. As long as RNA replication is perfect, the competing RNA replicators will all be identical, and the competition will be a perfunctory affair. But the more RNA chains there are, the greater is the likelihood of replicative errors. When replication takes place today there are error correction mechanisms in use to enhance fidelity, but these would not have been present to assist the first replicating molecules. It can therefore be expected that error rates for early replicators would have been much higher than is the case for the sophisticated life forms on the Earth now. Not too high, obviously, or the whole process would have descended into chaos, but high enough to introduce considerable variety into RNA chains as the millions of generations passed.

Once replication errors are introduced into the picture, the situation regarding competition changes. Not all RNA chains will be identical. Some will differ in ways that make them better or worse at producing copies of themselves. In the competition for limited resources there will be winners and losers.

These two factors — chance modifications to RNA base sequences, and inter-RNA competition for resources — provide the variations and the mechanism respectively for the process called *evolution*. Differences in RNA base sequences are, if you like, the raw material on which evolution works, while limited resources provide the driving force. Competition is able to select from amongst differing RNA chains the ones to survive. This selection-by-competition, which is an entirely natural process (i.e. it conforms to the laws of physics), is called, appropriately enough, *natural selection*. RNA chains which are unsuccessful in the competition are not selected and become extinct; RNA chains which are successful do get selected and go on to replicate themselves.

What is of crucial importance to evolution is that errors in replication are happening continually. Successful RNA chains, as a result of errors arising during the process of replicating, will occasionally pass altered RNA to their offspring. This altered RNA will make the offspring winners or losers in their turn. In this way, natural selection (of winners and losers) causes replicating systems to change as the generations pass; that is, they evolve. All this is so obvious it is practically self-evident.

The answer to the second question above — why did life become more elaborate as the millions of years passed? — is not so clear. It is not inevitable that changes must be in the direction of increasing elaboration. A particular type of organism may well find that becoming simpler benefits its ability to replicate. (The viruses mentioned below may be an example of this.) However, examination of the fossil record shows very clearly that following a trend towards simplicity is the exception. Most organisms evolve in the direction of ever-increasing complexity as their way of enhancing their replicative success. Whatever the reasons, when it comes to competition for resources, it seems the more sophisticated your biochemistry, the more successful you're likely to be.

EVOLUTION AT WORK: CELLS AND PROTEINS

Having seen how fundamentally simple the mechanism which produces evolution is, the next thing to do is to look at what several billion years of the progressive elaboration of living things has led to.

It may seem optimistic to say so, but once evolution is up and running, all the subsequent nucleic acid modifications needed to create the many forms of life that have been known to exist for the past $3 \cdot 5 \times 10^9$ years can be accounted for, at least in principle. The key point is that the only criterion by which a modification is (naturally) selected is whether it improves replicative ability. *Any* modification permitted by the laws of physics and chemistry is allowed.

One of the most important modifications evolution enabled RNA to produce is the cell membrane. How it did this, as with most developments in early life forms, is entirely unclear. Assume though that the process required proteins. If that is the case, it means cells first appeared after, or at about the same time as, protein enzymes started being used as the catalysts of choice. These first hypothetical RNA-protein built cells are called ribocytes.

Aside from keeping the molecules of the replicating system together and forcing replication to be orderly, the emergence of cells has three other important effects. Firstly, cells inevitably have an internally determined structure — one of the three definitive characteristics of living things listed in section 5.2. Secondly, cell membranes keep 'foreign' molecules away from those they enclose. Any one group of RNA molecules, residing inside their cell, will only have access to each other and to the unique intra-cellular micro-environment they have created for themselves. They will not have access to any molecules outside the cell. Nor conversely will anything outside the cell have access to the benefits provided by its interior. And thirdly, possessing a cell membrane enables the living thing to become mobile without risking dissipating. It can go wherever there is water.

Mobility is particularly significant when considering the rate at which evolution can proceed. Evolution depends on chance modifications of nucleic acid chains. Most of these modifications will not be beneficial; they will almost certainly produce changes to the internal form of the affected organism that will be of no account in respect of its competitiveness or will make it less competitive and set it on course for extinction. On the face of it, the accumulation of changes that give cells advantages in the competition amongst them ought to be a very protracted business.

But such a reckoning fails to take the size of the Earth into account. Suppose that when cells first became mobile the oceans only supported organisms to a depth of 100 m. This would provide about 10^{16} m^3 of ocean for those organisms to live in. Suppose further there was only one organism per cubic metre — the number of organisms in an average present day pond is billions of times higher than that — and that a replicating cell has only a one-in-a-million chance of making an error copying its RNA. Even with these limits, there would still be 10^{10} errors occurring globally with each new generation. Viewed in this light it does not seem unreasonable that in 10^{9} years or so a great number of useful changes could occur, producing a large number of useful proteins.

'Useful', in the context of living things is, as already implied, a matter of having an impact on replication; when a new or altered protein comes into existence, its usefulness to the organism making it is always determined by how much it facilitates replication. Facilitation, though, need not be direct. It may for example take the form of simply helping the organism to survive for longer, and thereby have more time to replicate in. Nor may the nature of the facilitation be obvious. For example, one of the early useful roles acquired by proteins was that of making deoxyribose for use in nucleic acid chains in preference to ribose. It is not clear what advantage this gave to replicating systems, but whatever it was it led to the almost complete conversion of RNA to DNA as the nucleic acid used by organisms to store the plan of their structure.

Other ways proteins can be useful in making cells more replicatively successful are limited only by imagination. There are proteins which enable the cell membrane to be selectively permeable to substances the cell needs that are outside itself. (This will be discussed further in the next chapter.) There are proteins which give cells the ability to obtain energy from sugar and store it in ATP. There are proteins which enable cells to harness the energy in sunlight. There are proteins which enable a cell to obtain the biochemical raw materials it needs by breaking up — killing — other cells. There are proteins which give cells the capacity to fend off such lethal attacks. There are proteins which enable virtually all organisms other than the simplest to engage in sexual reproduction; that is, reproduction in which the DNA from two nearly identical cells is merged in one offspring. (Organisms that are identical enough to join their DNA in this way, and in so doing produce viable offspring which can replicate by the same means in turn, are said to belong to the same species.) And so on, and so on. If a protein can arise by chance to do something replicatively useful, then it probably will.

SEVEN KINGDOMS

The DNA from one species of cell never gets mixed with the DNA of any other species of cell. (This rule is broken by bacteria.) In consequence, changes to the DNA of one member of a species can be passed to its offspring, which belong to the same species, but not to the members of any other species. The result is an increasing diversity amongst living things, with each species becoming gradually more different from every other species as accumulating changes to their separate DNA chains occur.

The effect of this diversification, after more than three and a half billion years of replication, has been to produce seven distinct forms of life, each form being known as a *kingdom*. The seven kingdoms represent the extremes of DNA diversity that have arisen over the billions of years of divergent base sequence modifications. In describing each kingdom it will be seen how differently it is possible for the DNA-protein system to become elaborated.

The kingdom containing the simplest organisms is that of the *viruses*. Viruses consist of a DNA chain, or sometimes an RNA chain, covered in a protein coat. The base sequence on the DNA or RNA chain specifies the amino acid sequence of the protein coat, as well as a number of other proteins. Viruses do not possess a self-enclosing membrane, and so are not cellular; nor do they possess any molecules capable of translating nucleic acid bases, nor any source of energy. It is possible they are former single-celled organisms which have simplified their structure to become the ultimate parasites. Alternatively they may be descendants of organisms which never underwent even quite primitive elaborations of the RNA-protein or DNA-protein system. If the latter is correct, they represent a stage in the emergence of life before cell membranes had evolved. Perhaps they indicate that the first tRNA molecules led a physically separate existence from the RNA chains they helped translate into proteins. Nowadays the only places where viruses can find tRNAs are inside the cells of more complicated organisms. When a virus invades an appropriate host cell, the virus's DNA or RNA is readily translated and produces proteins which facilitate replication of the virus.

A more complex group of organisms than viruses, though still comparatively simple in biological terms, are the members of the two prokaryote kingdoms. *Prokaryotes* possess a self-enclosing membrane and are therefore cellular. Each prokaryote cell is a completely independent living entity. Prokaryotes don't indulge in sexual reproduction, but interestingly this doesn't prevent them from acquiring each other's DNA; they have asexual ways of swapping it. Is that how it was done universally before sexual reproduction evolved?

The first of the two prokaryote kingdoms is the *archaebacteria*, a group of primitive single-celled creatures which are functionally living the same life as the first prokaryotes of $3 \cdot 8 \times 10^9$ years ago: an oxygen-free existence, often coping with extreme conditions — hot acidic water for example — which would kill any other life form.

The second prokaryote kingdom is the *eubacteria*. These are typical bacteria. Some obtain their energy from sunlight, while others obtain it from other organisms or from organic detritus.

Archaebacteria and eubacteria, and presumably viruses too, were present on the Earth by around $3 \cdot 5 \times 10^9$ years ago, and for a long time they had the planet to themselves. Well over one and a half billion years passed before they were joined by a fourth kingdom.

The fourth kingdom consists of considerably more complex organisms called *protists*. Like the prokaryotes, protists are cellular, with each cell being an independent entity. Also like the prokaryotes, some protists get their energy from sunlight, while others get it from other organisms. Protists differ from prokaryotes in that their biochemical structure — a structure called eukaryotic — is far more elaborate. Protists include amongst their number the famous amoeba and some forms of algae.

It fell to species possessing eukaryotic cells to be the founders of the remaining three kingdoms. What presumably happened was that following cell replication the two replicas, instead of parting, stayed together. Further cell division then led to the formation of a cell colony. Eventually the species concerned would have been represented not by single cells dispersed in the sea but by many colonies of cells. The irrevocable step in the process occurred when individual members of a cell colony lost the ability to survive out on their own as independent cells. From then on, the species would only be found as a mutually dependent collection of cells. Each member of the species would be a multi-cellular organism.

The simplest of the multi-cellular kingdoms is that of the *fungi*. The fungal kingdom includes amongst its members mushrooms, toadstools and moulds. Fungi generally behave like plants in that when mature they do not have the ability to move around the landscape. In most other respects they are quite different from plants. In particular they obtain their energy from other organisms.

The sixth kingdom is the *plant* kingdom. Its members are immobile and obtain their energy from sunlight. They range from seaweed, through lowly mosses and horsetails, to trees 50 m high and to flowers of exquisite delicacy.

Finally there is the *animal* kingdom. Its members are generally active and obtain their energy from other organisms. They range from the sedate sponge to the 90 km hr^{-1} cheetah, from the microscopic mite to the 100 tonne blue whale (human wrecking of the marine environment notwithstanding), and from the brainless jellyfish to human beings.

The fifth, sixth and seventh kingdoms all seem to have been well established by 650,000,000 years ago.

(A note for enthusiasts: the classification of living things is not entirely settled. Some authorities classify prokaryotes as one kingdom, not two; some authorities classify eukaryotic organisms as two kingdoms, not one; and most authorities regard viruses as not being alive at all. Thus the number of biological kingdoms can be considered to be anything from five up to eight. The classification given here is accordingly open to disputation.)

PHOTOSYNTHESIS AND THE ATMOSPHERE

It will be apparent from the description of the seven kingdoms that species may be divided into those that get their energy from sunlight and those that get it from other species. The former have, in the course of obtaining their energy, affected not just their immediate vicinity but the entire Earth. The changes they have wrought involve the Earth's atmosphere and were brought about in the following way.

As has been mentioned, the most important biochemical energy sources are sugars and fats. These biochemicals are made in the first place by those organisms which can harness the electromagnetic wave energy in sunlight, reacting CO_2 and H_2O together to make sugar:

$$6CO_2 + 6H_2O \rightarrow C_6H_{12}O_6 + 6O_2.$$

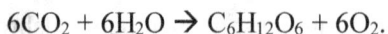

The conversion of electromagnetic wave energy into biochemical potential energy is called *photosynthesis*.

Photosynthesis supplies the sugars, and indirectly the fats, that all organisms need in order to stay alive. It also produces as a by-product oxygen (O_2). Now, if all the sugar made by photosynthesis was eventually consumed by organisms using it as a source of energy, then the amount of O_2 released during photosynthesis would be balanced by the amount used in turning sugar back into CO_2 and H_2O (see section 5.3). However, some of the carbon incorporated into sugar eventually ends up contributing to the hard inedible parts that some organisms possess, and is thereby removed well-nigh permanently from the atmosphere it came from. Furthermore, organisms are sometimes buried, which also locks up the carbon they contain. Photosynthesis thus gradually brought about an increase in the amount of oxygen in the atmosphere.

The result of several billion years of this natural oxygen 'pollution' was the conversion of the atmosphere from its original composition of N_2 and CO_2 to a mixture of N_2 and O_2. The missing carbon now resides in rocks partly made of carbon (e.g. chalk — $CaCO_3$), and in coal and oil, substances made from buried organic matter.

(A note for enthusiasts: most of the CO_2 initially present in the early Earth's atmosphere was removed inorganically by water-based chemical reactions between CO_2 and metal ions, the resulting insoluble substances ending up compressed into ocean-floor sedimentary rocks.)

Life has thus played a major role in making the Earth what it is today. And by enriching the atmosphere with oxygen, it has also made possible its own emergence from the sea onto the land following suitable DNA modifications. The colonization of the land began about 430,000,000 years ago.

If the planet-visiting scientists of section 5.5 were to return three and a half billion years after life began to work its changes, they would find the two earlier incongruities gone. The air would now be breathable. And everywhere there would be living things. Thus have the descendants of a microscopic self-replicating system that first came into being $3\cdot8$ billion years ago made the Earth into their ideal home.

5.7 BIOLOGY AND PHYSICS

CHANCE AND COMPLEXITY

The approach that has been taken to living things in this chapter is essentially one of treating them as physical systems; that is, as collections of atoms obeying the usual physical laws. And yet when organisms are regarded as complete entities there seems to be more to them than that.

In previous chapters it was found that accurate knowledge of the forces, charges and types of mass and energy in the universe can be combined to give a description of everything that goes on in the terrestrial world. But the objects being looked at in those chapters were essentially simple; they were made of aggregates of simple parts combined in simple ways. From the point of view of living things it raises the question of what exactly 'simple' implies.

The implication in the word 'simple' is this: each possible arrangement of the parts of a simple system is indistinguishable from every other arrangement. Nothing needs to be known beyond the laws of science, a few physical constants, and the properties of elementary particles, in order to predict what any such aggregate object will be like. It is possible, for example, to predict the properties of a nucleus from knowledge of its constituent quarks, because quarks can be rearranged inside the nucleus quite freely without making any difference to it. Similarly, to give another example, predictions can be made of what a star will be like from knowledge of the nuclei it contains. The precise distribution of the nuclei has no effect on the star's properties.

In contrast, when it comes to living things the arrangement of their parts is crucial to their identity; every different arrangement is unique. The sequence of bases on a DNA chain perfectly illustrates this point. If that sequence is altered, the proteins being built will change and thereby the properties of the resultant organism will change too. This makes the sequence impossible to describe in simple terms; any description is unavoidably complicated. Thus living things are not 'simply' physical systems. If physics alone is used to describe them, the results will prove inadequate. Organisms are indeed more than collections of atoms obeying the usual physical laws.

The essential complexity of living things has as a consequence that the sciences of biology and biochemistry cannot be derived from knowledge of physics alone. And that is something, in a strictly limited sense, which makes life greater than the sum of its parts. More than physics is needed. Physicists may be able to predict the potential for life in the universe, but they cannot predict the forms it has taken. In order to describe life, the outcomes of several billion years' worth of the random chance events which have decided the DNA base sequence and produced today's huge range of organisms must be added to the physical properties of participating atoms. It is organisms' complexity and the chance events that gave rise to it that underlie their appearing to have more to them than being merely a peculiar form of inanimate matter.

CHOICE AND PURPOSE

The distinction between living and non-living objects can be formalized by proposing that there are two *levels of complexity* associated with arrangements of matter. The first, which I shall call the *physical level*, is that with which chapters 1, 3 and 4 are concerned. Arrangements of matter are indistinguishable from each other; there is an underlying simplicity to be discovered. The second, which I shall call the *biological level*, is that of living things. Arrangements of matter are significant; they are fundamentally complex.

If the only consequence of passing from the physical to the biological level of complexity is that matter becomes arranged in more complicated ways, there would be little point in coining the terms since they are self-evident. But there is more to the issue of complexity than that. Aggregates of matter can always be viewed as collections of parts or as integral wholes. At the physical level of complexity the choice of viewpoint has no dramatic consequences, for the parts determine the whole. But at the biological level, because of the need to add the outcomes of complexity-producing chance events to the laws of physics, the parts do not of themselves determine the whole. The choice of viewpoint makes a difference to what is observed.

If organisms are viewed from the perspective of the biological level of complexity — that is, as integral objects — they acquire two properties which make no sense at all viewed from the physical level. Those properties are choice and purpose. Organisms seem able to choose what goals to pursue. They also seem able to pursue a course of action; that is, their activities are purposive. How do these properties manifest themselves at the physical level of complexity? How, in other words, can they be accommodated within the laws of physics?

To answer these questions, consider the coming together of two organisms initially apart, and compare it with the coming together of two freely moving magnets. The former occurs at the biological level of complexity and can be viewed either from the biological level or from the physical level. The latter occurs at the physical level and can be viewed in only one way.

Taking the magnets first, the way they come together is a straightforward matter of force-and-energy dynamics. If the position of one magnet is altered, the other responds by direct detection of the change in magnetic force from the moved magnet. The magnets approach each other in a way that is describable in simple terms.

In the organisms' case, the process is not a simple mechanical one. Suppose the two organisms approach each other following detection of a chemical each emits — a scent. If the position of one organism is altered, the other will not change its direction of movement immediately. It first needs to interpret the chemical message it is receiving. This act of interpretation makes a description of the two organisms' approach unavoidably complicated. It also — and this is crucial — disengages the message passing between them from their reaction to it. It makes their reaction potentially variable. That is why the coming together of the two organisms is not a simple mechanical process.

The difference between the two magnets and the two organisms can be put this way. The attraction between the two magnets was also the means by which they came together — magnetic force in both cases. The attraction between the two organisms was a scent, but this did not provide their means of approach. That had to be brought about by enzyme-assisted chemical reactions inside the organisms. The separation between attraction and means of approach makes it possible for the organisms to respond variably to the attraction. If their energy reserves are depleted for example, they may not move towards each other despite picking up each other's scents. It also makes it possible for organisms to have more than one means by which approach can be initiated. By contrast, the magnets have neither an option-to-ignore nor a range of options by which they can detect one another.

Viewed in terms of the physical level of complexity there is no difference in principle between the magnets and the organisms. Like the magnets, the organisms approach each other as a result of an unbroken series of inevitable caused events. There is nothing mysterious about it. No laws are broken or need to be invented. The organisms possess no properties which do not arise naturally from the laws of physics.

But the existence of a series of events between the initial cause (receipt of a scent in the example given above) and the final effect (movement towards the source of the scent) produces a *qualitative* difference between events occurring at the biological level and those occurring at the physical level. Seen from the physical level, events at the biological level involve a series of interactions connecting cause and effect. Seen from the biological level, from which vantage point the complex objects are seen as integral entities and the series of connecting interactions is invisible, events involve choice and purpose.

Two things, then, are essential to the appearance of choice and purpose in living things. One is the invisibility of the internal chain of physical events linking cause and effect; the other is that the chain of events is complex. Complex chains can diverge and they can merge together. Diverging chains gives rise to choice; merging chains gives rise to purpose.

Think again of the two organisms attracted by each other's scent. If the invisible chain linking cause to effect follows its main path the organisms appear to choose to approach each other. This seems to be a choice because observers know from experience that it is possible that the organisms may not approach each other at all. They know that the invisible cause-and-effect chain is able to take an alternative path at some point along its length. Viewed from the biological level of complexity, with the series of internal events being invisible, the organisms choose what to do. In contrast, when viewed from the physical level, the particular path which turns out to be the one taken can be seen to be followed in accordance with the laws of physics, leading to the conclusion that there is no choice.

A similar position arises with regard to purpose. With choice there is one initial event giving rise to more than one possible final event. With purpose it is the other way round. There is more than one possible initial event giving rise to a single final event. Any one final event need not have a unique cause. For example, the two organisms might have several different ways of detecting each other's presence. Any of these non-unique ways might cause them to come together. Thus, the initial cause can differ but the final event remains the same. Coming together thereby seems to be the goal of the two organisms; an observer knows from experience that if they can't do it one way, they'll do it another. Achieving the goal is perceived to be their purpose. Again this is only so when viewed from the perspective of the biological level of complexity. From the biological level the merging of cause-and-effect chains leading to a single final event is invisible; the organism appears to have a purpose. From the physical level the merging series of events can be seen to obey the laws of physics and the conclusion follows that there is no purpose.

The source of the properties of choice and purpose can be summarized as follows.

- In biologically complex structures, series of events tend to diverge and merge.
- One event may lead to more than one possible next event (diverging).
- One event may be initiated by more than one possible causing event (merging).
- A sequence of events which can diverge produces choice.
- A sequence of events which can merge produces purpose.
- The series of events is observed from the physical level of complexity.
- Choice and purpose are observed from the biological level of complexity.

It may seem that there are two mutually exclusive propositions being supported here simultaneously: that living things have no choice or purpose, and that they do have them. But this is not so. The two propositions are incompatible but not exclusive. They are rather like two photographs of the same scene, one with the foreground in sharp focus and the background a fuzzy blur, and the other with the background in focus and the foreground blurred. Only one of the photographs can be taken at a time, but they are both of exactly the same scene.

It cannot realistically be claimed that living things have no choice or purpose, since they can be observed objectively to have them. Yet physics, which applies to everything, does not find any evidence to support the existence of choice or purpose in any of the interacting particles found in biochemical systems. The resolution of the paradox is that, where living things are concerned, the context in which they are observed affects the properties ascribed to them. The physicist sees living things choicelessly and purposelessly obeying physical laws. The biologist sees living things making choices and having purposes. Both are correct in their respective contexts. The existence of choice and purpose in biology does not imply that the laws of physics are inapplicable to living things, but only that they are not being taken into account by the biologist, for whom they are largely irrelevant.

6

NEUROBIOLOGY

6.0 OBJECTIVE: To describe the human brain.

It is generally agreed that if observers are to be found anywhere within a physical structure, that structure is the brain. Hence to find out how science views the observer it is necessary to apply scientific method to the brain. Now that we have some basic knowledge of biochemistry and of how cells function, we are in a position to do this.

The chapter begins with an account of the most important developments in animal biology since DNA modifications first brought animals into existence. It will become clear that in terms of ability to survive and replicate, most animals benefit from possessing sensing and motion systems capable of facilitating rapid reactions to what is going on in the environment. The types of cells within an animal's body which perform the sensing functions, and which govern its responses to what has been sensed, cells known generically as neurones, will be described in some detail. We will see neurones obey the laws of physics, just as every other kind of cell does.

Having looked at neurones individually, we shall next investigate their aggregate properties, starting with a description of the structure of the brain, limiting ourselves to its most important parts. The brain's biggest component, the cerebral cortex, will then be described, being subdivided into a number of lobes. It will be seen that each lobe carries out a range of specialized functions. Finally, the behavioural consequences of damage to the cerebral cortex will be described. This will provide overwhelming evidence for the hypothesis that an identity exists between observers and their brains.

It will be noticed that there is a gradual shift in this chapter from questions of 'how' relating to neurones, to 'what' when the brain as a whole is considered. This reflects the fact that the science of how the brain does what it does is still quite rudimentary. 'What', being descriptive, is an easier question to answer than 'how', which is explanatory. This is not to say that a comprehensive explanation is impossible; it is merely that insufficient observational data exists at the moment for explanation (in most respects) to be advanced.

6.1 THE NEUROMUSCULAR SYSTEM

CELL DIFFERENTIATION

It is reasonable to suppose that the very first multi-cellular organisms consisted of collections of identical cells. However, it is in the nature of cell groups that some cells will be on the outside of the group while others will be inside. The environment inside a body of cells will clearly be different from that at the surface. Such a situation favours DNA modifications which produce proteins enabling individual cells to know where they are in the group. Those inside the organism can then be spared making proteins that will only be of use at the organism's surface, and vice versa.

Cells can only know where in the organism's body they are if they communicate with one another. This process often involves chemicals produced by proteins in one cell being released from that cell and detected by other cells nearby.

Once cells know where they are within the organism, the scene is set for those at particular locations to acquire special functions by way of specialized proteins. These proteins, on being added to the organism's protein repertoire, would be made only at specific sites in its body. Each site would consist of a collection of identically specialized cells not found anywhere else in the organism.

The acquisition of special functions by cells in a multi-cellular organism is known as *cell differentiation*. The many forms of differentiated cells cooperate together to make up the complete organism.

Differentiated cells usually retain all the basic cellular characteristics; that is, they obtain energy from the conversion of sugar to CO_2 and H_2O, they manufacture proteins, and so on. Each differentiated cell's specialization is added to these basic abilities by way of the specialized proteins it makes. Since all cells possess the full complement of DNA required to build the total organism and *all* its proteins, the differentiation process can be regarded as the switching on or off of this or that section of DNA bases.

(Actually there are exceptions to the rule that all cells possess a complete DNA complement. Rules in biology are generalizations rather than laws in the physicist's sense. Most of the all-embracing statements made in this chapter — and in the previous one, come to that — have their exceptions. You should remember this as the existence of exceptions will often not be mentioned.)

In the course of hundreds of millions of years, the multi-cellular organisms of the animal kingdom acquired a range of specialized cells, mostly grouped together to form organs (e.g. kidneys, liver), or located in particular areas (e.g. skin, hair), or serving a well-defined function (e.g. blood). Each type of cell complements the others in forming the complete animal.

Cells in an animal's body need to know more than 'this is where in the body I am'. They also need to know 'this is what we're all doing'. Animals — and multi-cellular organisms generally — are single coordinated entities, and their many cells need to act in concert. Coordination of cellular activity can be achieved in three ways: externally, internally by means of biochemicals, and internally by means of specialized structures.

For those multi-cellular organisms adopting a passive approach to survival — almost all plants and fungi and a few animals such as sponges — the first two means are adequate. Speed is not of the essence, and external environmental influences like temperature, the direction of the force of gravity and the intensity of sunlight can be used to coordinate cellular activity on a daily and seasonal basis. Internal biochemical communication amongst cells runs hand in hand with, and supplements, this system.

But the vast majority of animals also need to employ the third coordination system. Pursuit of an active survival strategy requires rapid escape from predators and sometimes rapid chasing of prey. This cannot be accomplished by the first two systems because they tend to concert actions on timescales of minutes to months rather than fractions of a second. The third system, on the other hand, can produce results at the desired speed because its specialized structures are specialized to achieve exactly that. They do not attempt to communicate with all the cells of the body, as is broadly the case with the biochemical system, but only amongst themselves. The body's other cells are carried along passively. Collectively the cells making up the third communication system are called the *neuromuscular system*.

NEUROBIOLOGY

NEUROMUSCULAR CELL TYPES AND LAYOUT

In all except the most primitive animals such as jellyfishes, the cells specialized to enable an animal to react rapidly can be classified into five types.

- Sensory cells detect what is taking place in the animal's environment, and they also monitor the state of the animal's interior.
- Intermediate cells evaluate what the sensory cells have detected. They determine what the animal will do in response to incoming sensory data.
- Motor cells receive the results of the evaluation performed by the intermediate cells. They directly control muscles.
- Muscle cells are able to change their shape by contracting lengthwise. Not all muscles are concerned with movement about the environment, but those having this function are able to drag the immobile remainder of the cells of the body around with them. They achieve this by being attached to hard, rigid structures either encasing, or else imbedded in, the animal's body.
- Glandular cells carry out internal biochemical coordination, the second kind of coordination mentioned above. In some instances their chemical emitting activity is influenced by what may loosely be regarded as motor cells. Although strictly speaking not part of the neuromuscular system, they are included here because of their close involvement with sensory-motor coordination.

Of the five types of cells, the first three are sufficiently similar to be accorded a common name. They are called *neurones*. Hence there are *sensory neurones*, *intermediate neurones* and *motor neurones*. Neurones collectively constitute the *nervous system*. The nervous system is the neuromuscular system excluding the muscles (and glands).

The neuromuscular system tends to have a fairly standard structure. Some quite wild and weird variations are found — think of anemones, starfishes and limpets as examples — but the most common layout is one which follows from animals being shaped roughly like a tube with various appendages attached to it for use in locomotion. These appendages must be connected to muscle cells. Hence muscle cells need to be widely distributed throughout the body. The same is true for the parts of sensory neurones responsible for detecting, for example, touch and body position; they are also widely distributed. On the other hand, light, sound, smell and taste sensors are located at the front end of the tube — 'front' being defined by the end where food enters the animal's body. The bulk of the intermediate neurones are also found at the front end.

This general body plan results in the formation of a number of clearly defined nervous system organs. Sensory neurones form key parts of the eyes and ears. Intermediate neurones form the brain. And in the group of animals which attain the largest sizes, called the chordates, motor neurones are located in a cord-like organ (hence the group's name) which passes down the back of the animal. Some chordates encase this cord in a bony covering, which is known as the spine. Those that do this are called vertebrates, and human beings are of course included amongst their number.

6.2 GENERAL NEURONE BIOLOGY

THE FUNDAMENTAL HYPOTHESIS OF NEURONE BIOLOGY

Since it is humans who are of the greatest neurobiological interest, possessing as they do a nervous system of unmatched sophistication, it is human neurones, brains, and nervous systems generally which will be under consideration from now on. It can be noted, however, that a lot of what is said about humans also applies to a similar or lesser degree to other animals.

It is possible to establish experimentally that (human) sensory neurones react to particular kinds of events external to themselves; that is, when appropriate external events occur, chemical reactions are brought about within the sensory neurones as a result. It is also possible to establish that movements in muscle cells are brought about by motor neurones. This follows from the observation that the character of muscular activity is profoundly altered if the connection between a muscle and the motor neurones in contact with it is severed. These experimental facts provide strong grounds for thinking that both the sensing of events and the carrying out of movements are purely physical processes. They occur by means of biochemical reactions and as such are subject to the laws of physics.

The biochemical events which take place in people's muscles are obviously related to what is happening in their environment. Now, only sensory neurones are able to react in a systematic way to environmental events external to the body. And with few exceptions (see reflexes below), sensory neurones are not connected to muscles, nor even to motor neurones. Instead, sensory neurones are connected, as explained in the previous section, to intermediate neurones. It follows that muscular activity comes to be related to external events by means of a chain of connections:

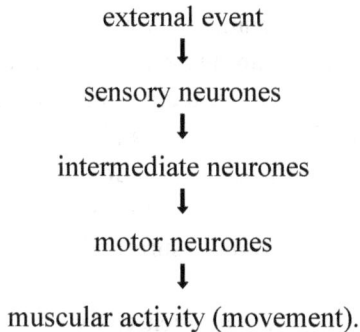

external event
↓
sensory neurones
↓
intermediate neurones
↓
motor neurones
↓
muscular activity (movement).

It has already been established that the first and last links in the chain, those between external events and sensory neurones, and between motor neurones and muscular activity, are biochemical in nature. It is therefore plausible to hypothesize that the other links are also biochemical. This is the fundamental hypothesis of neurone biology: that when people (and this applies to all other animals too) react to an external event they do so purely as a result of a sequence of biochemical reactions. It implies that behaviour — behaviour being the sum total of muscular activity — is due entirely to biochemical processes obeying the laws of physics.

NEURONES AS ORDINARY CELLS

In order to develop this hypothesis further, it is clearly necessary to look more closely at the cells of the nervous system. By doing this it should be possible to discover what sort of biochemical processes are taking place that could meet the requirements of the hypothesis. So how do neurones compare to other cells? In what ways are they the same? In what ways are they different?

Size. This is an obvious place to start an investigation. An average cell is intermediate in size between an atom and a human. Atoms found in living things weigh about 10^{-26} kg, a cell weighs about 10^{-13} kg, and a human weighs about 60 kg. Similarly, in volume terms an atom occupies about 10^{-30} m^3, a cell occupies 10^{-15} m^3, and a human a little over 10^{-1} m^3. A cell contains about 10^{13} atoms; a human has more than 10^{13} cells in his or her body (plus a huge number of bacterial cells). Neurones are fairly unremarkable in terms of these dimensions.

Ease of destruction. Neurones differ from other cells in being fussier about their immediate environment. Direct contact with blood rapidly destroys them. Almost all neurones are surrounded by specialized cells called astrocytes which act as a protective interface between the neurones and the blood that supplies their sugar and oxygen. Astrocytes work by causing the cells which form the walls of blood vessels to seal tightly against each other, so preventing any substances other than desired ones from passing from the blood into the surrounding brain. Astrocytes are the brain's gatekeepers.

Energy supply. Neurones obtain their energy from glucose and oxygen and store it in ATP. This makes them similar to other cells, though with a difference. Almost all other types of cells can survive brief periods of oxygen shortage by converting glucose to lactic acid, which is another arrangement of $C_3H_6O_3$ like glyceraldehyde (see sections 5.1 and 5.2):

$$C_6H_{12}O_6 + 2ADP^{3-} + 2H_2PO_4^- \rightarrow 2H_3C\!-\!(CHOH)\!-\!COOH + 2ATP^{4-}.$$

Neurones are very heavy consumers of energy and need a lot more than 2 molecules of ATP to keep going. (Recall from section 5.3 that complete conversion of glucose to CO_2 and H_2O produces 30 molecules of ATP.) Consequently neurones die within minutes of their oxygen supply being cut off.

Replication. Most cells die and are replaced over and over during an individual's life. Neurones are different once again. They are unable to replicate, not even to repair damage. Sensory neurones responsible for the sense of smell break this rule, dying and being replaced roughly eight times every year, but they are almost the only exception. For nearly all other neurones in the nervous system, replication, which proceeds prodigiously in the months following conception, comes to a halt at around the time of birth and never resumes again during the lifetime of the individual. Consequently, damage to the nervous system is irreparable. This will be discussed further in section 6.7.

Shape. This is another point of comparison which distinguishes neurones from other cells. Most cells are roughly spherical. Not so, neurones. In terms of shape, a typical neurone (there are lots of atypical neurones) can be divided into three parts called the dendrites, the cell body and the axon.

- The *dendrites* are spindly outgrowths of the cell body which sprout from it rather like the branches of a tree from a tree trunk. They usually extend about a millimetre away from the cell body in many directions. Dendrites make intimate contact with numerous axons belonging to other neurones.
- The *cell body* is where the neurone's DNA is found, and also where most of its enzymes are located. It is very roughly spherical and is of the order of 10^{-2} mm in diameter.
- The *axon* is another outgrowth of the cell body. Axon lengths vary widely from neurone to neurone, but 10 mm would not be out of the ordinary; this makes an axon a thousand times longer than the distance across the cell body from which it comes. The diameter of axons is upwards of 1×10^{-4} mm, making an axon a very long, very thin tube. Because an axon is so long, it enables the neurone to potentially affect other neurones or muscle cells or glands far from where the neurone is based (i.e. where its cell body is located).

Neurones acquire their shape by way of filament proteins (see section 5.2) made internally by specialized enzymes not possessed by other types of cells.

MEMBRANE PROTEINS

Of all the differences between neurones and other cells mentioned so far, shape is the most significant for the specialized role neurones have in the body. But there is one other specialization which is equally important, and it concerns some of the proteins that neurones make within themselves. These proteins do not catalyse chemical reactions inside the neurone; instead they are imbedded in the neurone's cell membrane. Not surprisingly, such proteins are called *membrane proteins*.

Of course, all cells have proteins imbedded in their membranes. As implied earlier, it is by means of membrane proteins that separate cells communicate with each other. But some of the membrane proteins found in neurones are unique, or at least occur in combinations not found in any other kinds of cell.

There are five classes of membrane proteins. They are:

- structural proteins
- external enzymes
- receptor proteins
- pumps
- channels.

Structural proteins attach cells to other cells or body structures. They are used, for instance, to attach muscle cells to bones (indirectly). Structural proteins are thus of great importance for the muscular side of the neuromuscular system. In contrast they have little part to play in neurone functioning and so will not be considered further.

External enzymes, whilst made inside the cell, are fixed to the cell membrane in such a way that they can catalyse reactions outside it.

Receptor proteins react with molecules located outside the membrane. The reacting molecule causes the part of the receptor protein on the inside of the membrane to change shape, which in turn brings about chemical reactions within the cell. (Shape, remember, is fundamental to enzyme function.) This property makes receptor proteins in effect a cell's sensory system. They enable a cell to react to substances outside it without permitting those substances to enter it by crossing the membrane. Cells sense their surroundings by means of their receptor proteins.

Pumps take substances from one side of the cell membrane to the other. They do so in the opposite direction to that in which the substance would move if the membrane had holes in it rather than membrane proteins. This means that the substances being taken across the membrane by the pumps are either being moved up a potential energy hill or else are more concentrated on the side they are being moved to than on the side they are being moved from. In either case the molecules do not move willingly across the membrane, but must be pushed. That, of course, is why the proteins concerned are called pumps, for the name exactly describes what they do. Like every other kind of pump, membrane protein pumps need a supply of energy — often ATP — in order to work.

Channels allow specific substances to pass through the membrane in the same direction as the substance would move if the membrane was full of holes. Because substances passing through a channel are travelling down a potential energy hill or from a region of high concentration to a region of low concentration, no energy is involved in transporting them across the membrane.

Between them, pumps and channels enable a cell to control its internal chemical composition. Molecules that the cell needs can be pumped in, and those that it doesn't can be pumped out. Excess concentrations of molecules on either side of the membrane can be reduced by channels which permit the particular molecules concerned to pass in or out of the cell.

External enzymes, receptor proteins, pumps and channels all play a part in the functioning of neurones.

6.3 SIGNALS

THE SODIUM PUMP

Anything which travels from one place to another in such a way as to be able to cause at its destination an event which is related to the event at its source is called a *signal*. In the case of neurones, when the sensory neurones react to an external event, they need to send a signal through the intermediate and motor neurones to the muscles. The muscles will then react in a way that is related appropriately to the originating external event.

In recent decades it has become possible to make measurements of individual functioning neurones, and it transpires that the most crucial component of neurones in their role as transmitters of signals is a

263

membrane protein which pumps sodium ions (Na^+). Sensibly enough, these proteins are called *sodium pumps*, and each neurone has about a million of them, all working continuously and collectively capable of pumping 200 million sodium ions every second.

Sodium pumps operate in a cyclic fashion, transporting sodium ions across the membrane so that the neurone has less sodium inside than outside. To describe a complete cycle, it would be simplest to start at the moment the pump protein is ready to begin expelling some sodium from the neurone. An enzyme inside the neurone attaches three sodium ions to the pump in a reaction which consumes energy — some ATP is converted to ADP. The attachment of the sodium ions causes the pump to alter its shape in such a way that the ions, while still attached to the pump, are moved to the outside of the membrane. There a reaction occurs between the three sodium ions, the pump and two potassium ions (K^+) — potassium ions being found all over the body, including on the outside of neurones. The two potassium ions exchange places with the sodium ions in a reaction which produces rather than consumes energy. This means it happens readily, with no energy supply being needed for it. The attachment of the potassium ions causes the pump to change its shape once more, this time transporting the potassium ions into the neurone. Once inside, the potassium ions are released from the pump by the same enzyme which attaches the next three sodium ions to it. This completes the cycle, which continues indefinitely.

From this description it can be seen that sodium pumps do three things. Firstly they expel sodium ions from the neurone. Secondly they concentrate potassium ions on the inside of the neurone's cell membrane. And thirdly they cause an imbalance of electric charge. This third feat is achieved because three sodium ions are transported out of the neurone for every two potassium ions that are transported in. With each cycle of the pump, one positive charge is pumped out of the neurone.

Of the three consequences of sodium pump operation, the one the neurone is least keen on is the inflow of potassium ions. The neurone has another membrane protein which attempts to deal with this. It is a potassium channel. Potassium channels allow the potassium ions pumped into the neurone by the sodium pumps to leave again. The neurone's attempts to get rid of its potassium ions are not completely successful, however, because of the large number of positively charged sodium ions on the outside of the membrane. Their presence causes the equally positively charged potassium ions to be repelled back into the neurone when they attempt to leave it. There are as a result many more potassium ions on the inside of the membrane than on the outside.

The combined effect of the various ion movements is to leave the inside of the membrane relatively sodium-less and negatively charged. If a hole was made in the membrane, sodium ions would flow into the neurone, not out of it. Similarly, any other electrically charged particle in such a hole would accelerate towards one side or the other of the membrane according to which charge the particle possessed. Of course, the acceleration of the charged particle means that it is being affected by the electromagnetic force. The units in which electromagnetic force is expressed are volts (see below). The voltage across the neurone membrane is usually around 70 millivolts (70 mV) though some types of neurone depart significantly from this value.

The maintenance of the voltage and the imbalance in sodium ion concentration across the membrane requires a continuous expenditure of energy. (Recall section 5.3 and comments about maintaining height on a down escalator. In many kinds of cells, including neurones, keeping sodium pumps running is what some of that energy is used for.) If the sodium pumps were switched off by some means, sodium and potassium ions would eventually leak into and out of the neurone until there were as many ions inside as outside. The membrane voltage would fall towards zero. With its sodium pumps working, the neurone is thus in an unstable state. Unstable states are easy to upset. Signals in neurones capitalize on this instability.

A BRIEF ASIDE ON VOLTAGE

If you are unclear what voltage is, it can best be thought of as being to electromagnetism as difference-in-height is to gravity. A ball rolls down a slope because different parts of the slope are at different heights, and the ball is in its most stable position when its height is as low as possible. The force of gravity acts on the ball because of the difference in height of the slope. In an analogous way, the force of electromagnetism acts on a charged particle if there is a difference in the 'height' of the electromagnetic slope — a voltage.

This is the simplest explanation of voltage, but it is actually a little more complicated than that. Gravitational difference-in-height obviously has to be in respect of a massive body, usually the Earth. Thus it implies a rate of acceleration which in the Earth's case is $9 \cdot 8$ m s^{-2}. Voltage, on the other hand, rather than implying some rate of acceleration, explicitly incorporates it. Hence electromagnetic voltage is really equivalent in comparable gravitational terms to

difference-in-height x rate-of-acceleration.

This means that the faster the rate of acceleration produced by the source of the electromagnetic force, the shorter the distance an attracted object must cover in order to fall through a voltage of one volt.

To give an idea of the size of a volt, a proton accelerating at $9 \cdot 8$ m s^{-2} due to gravity would need to fall a distance of 10,000 km in order to acquire as much kinetic energy as it does falling through 1 volt. The equivalent distances for other objects depend on the relative sizes of their gravitational and electric charges.

SIGNAL TRANSMISSION WITHIN NEURONES

If a neurone is left entirely to itself, its pumps and channels maintain a constant voltage across the membrane and a constant concentration of sodium ions on the outside. For all the activity taking place, nothing actually changes.

Now clearly, if a neurone is to transmit a signal, something has to change. Biochemical reactions that were not taking place before must occur as the signal passes. Indeed, those biochemical reactions *are* the signal. So the question to ask is: in what way could a neurone change? The obvious answer is that it is the membrane voltage and sodium ion concentration which change and, in changing, which constitute the signal.

Given what was said about receptor proteins in the previous section, the simplest way to start the voltage and sodium ion concentration changes off would be to place a chemical in a neurone's vicinity which reacts with a receptor protein. The receptor protein, 'sensing' the presence of the chemical, would convey the information to the interior of the neurone and cause the neurone to alter its membrane voltage and sodium ion concentration. If this is indeed what happens, a signal can be hypothesized to pass from one neurone to another by means of a chemical emitted by the sending neurone which reacts with the receiving neurone's receptor proteins.

Experiment confirms this hypothesis; neurones do indeed 'talk' to each other using chemicals.

When a receptor protein reacts with the chemical emitted by a sending neurone, how might it produce the expected changes to the receiving neurone's membrane voltage and sodium ion concentration? There are two principal ways. One is comparatively simple; the other is more complicated. The simpler way will be described first in some detail.

In this comparatively simple case — 'comparatively' is an important word here! — the function of the receptor protein is not only to detect an incoming chemical signal, but also to directly permit ions to enter the neurone. It thus functions as a channel as well as a receptor.

The key point about receptor proteins which double as channels is that the channels are not always open. Indeed, whenever the neurone is not receiving a signal the channels are firmly closed. This variable configuration which some channel proteins are able to take up is called *gating*. Whether the gate is open or closed is determined by the receptor part of the protein. When the receptor protein receives an incoming chemical signal, its shape changes such that a gate which is otherwise closed opens.

Receptor protein gates are not merely like holes in the neurone's cell membrane. They are highly specific for electric charge. One kind of gate will only permit positive ions to pass through it; another kind of gate will only permit negative ions, principally chloride (Cl^-) to pass. (Like potassium ions, chloride ions too are found all over the body.)

The positive-or-negative nature of the gate is crucial to the role it plays. If the gate in the receptor protein is a negative ion gate, it will facilitate the entry of chloride ions into the neurone. Since chloride ions bring a negative charge with them, and since there is already a surplus of negative charge on the inside of the neurone, opening a negative ion gate makes the neurone more resistant to disturbance. That's because open negative ion gates tend to lock in, or even increase, the -70 mV difference in electromagnetic 'height' across the cell membrane. (The minus sign, '-70 mV', arises because the inside of the neurone is electrically negative compared to the outside. Note that when I refer to increases/decreases in membrane voltage, below, I will always be referring to the *size* of the number, without reference to the minus sign.)

To further strengthen the stabilizing effect of chloride ions entering the neurone, many neurones have a chloride pump that expels chloride ions to the outside. This ensures that when a negative ion gate opens, there are plenty of chloride ions on the outside available to pass to the inside.

So much for negative ion gates. What about the opening of positive ion gates? The effect in their case is far more dramatic. There is a small flow of potassium ions out of the neurone but this is swamped by the sodium ion inflow. Because there are many more sodium ions outside the neurone than inside, sodium ions pour inward through the open gate. In addition, the surplus of positive ions on the outside also pushes sodium ions through the gate. (Contrast this with the movement of potassium ions and chloride ions, where the flow from the high concentration side of the membrane to the low concentration side is opposed by the surplus of positive ions on the outside.) Thus, opening a positive ion gate can, by way of sodium ion inflow, potentially cause a large disturbance to the membrane voltage.

These two opposite effects that ion gates can have when opened in response to an incoming signal are referred to as inhibitory (where chloride ions are involved) or excitatory (where sodium ions are involved). A neurone can thus be in receipt of an *inhibitory signal* which stabilizes its membrane voltage, or an *excitatory signal* which destabilizes it. In practice almost all neurones are in receipt of both kinds of signal most of the time. Some sending neurones will be passing an inhibitory signal to the receiving neurone, while other sending neurones will be passing an excitatory signal. A typical neurone has to add up all the inhibitory signals and all the excitatory signals and choose whether it's going to respond by being inhibited or excited in its turn.

How, in chemical terms, does the neurone make this choice? The first consideration here is that an incoming chemical signal does not affect the entire neurone but only the very localized part of one small piece of its membrane where the chemical reacts with the receptor protein. The flow of chloride ions through their gates stabilizes or increases the local membrane voltage of -70 mV. The flow of sodium ions through their gates reduces the local -70 mV membrane voltage. The part of the neurone membrane where these ion flows are occurring combines the two effects, and providing the result is a membrane voltage that remains above −55 mV, it does nothing. If, on the other hand, the membrane voltage falls below −55 mV, what is in cellular terms a spectacular sequence of events gets initiated.

The neurone cell membrane has, in addition to gated channels built into receptor proteins, a number of other stand-alone gated channels. As with ion gates and channels generally, these stand-alone channels are selective. Some permit only sodium ions to pass through, and some permit only potassium ions to. But they also have another vital property: their shape is determined by the voltage of the membrane in which they are imbedded. They only open if the voltage drops below −55 mV. For this reason they are called *voltage-gated channels*.

Voltage-gated channels for sodium ions have two gates in their structure. Gate 1 is open when the membrane voltage is above −55 mV, while gate 2 is closed. When the voltage drops below −55 mV, gate 2 opens, doing so in less than a millisecond. This means that both gates are now open, permitting sodium ions to flood into the neurone. The sodium ion influx has the effect of changing the local membrane voltage from −55 mV to +20 mV; i.e. it actually creates a surplus of positive ions on the inside of the neurone's cell membrane. Over the space of another couple of milliseconds gate 1 responds to the +20 mV voltage by closing, shutting off the flow of sodium ions. Meanwhile slower voltage-gated potassium ion channels open, typically taking about 2 ms to do so. By that time the membrane voltage has reached its +20 mV value and potassium ions, being positively charged, are repelled by the positive inside voltage to the outside. As they leave the neurone they take their positive charges with them so that the membrane voltage becomes negative again and in fact usually briefly overshoots the normal −70 mV value. Restoration of the normal voltage closes the voltage-gated potassium channels, closes gate 2 in the sodium channels and reopens the sodium channels' gate 1. The standard sodium pumps quickly put the migrated sodium and potassium ions back where they belong, and the membrane returns to how it was before.

It can thus be seen that when the combined incoming excitatory and inhibitory signals produce a sufficiently large drop in membrane voltage — to below −55 mV — the membrane locally flips, just for a couple of milliseconds, to a completely different, excited +20 mV state.

The important thing about this voltage flip is that, although it is localized to a small part of the neurone's cell membrane, it leaks to neighbouring pieces of membrane. The sudden change from −55 mV to +20 mV triggers nearby voltage-gated sodium channels to open so that the voltage flip spreads from one piece of membrane to the next until it arrives at the initial segment of the axon. This part of the neurone is especially sensitive to voltage drops. If and when the voltage flips there, the voltage disturbance becomes more pronounced, inevitably spreads over the entire cell body and dendrites and travels down the axon to its very tip. The neurone is said to have 'fired'.

Before looking at how neurones use firing to pass signals to each other, mention needs to be made of the other, more complicated, way in which an incoming chemical signal can affect a neurone's membrane voltage. In this alternative case the receptor protein does not double as a gated channel. Instead, when it receives a signal it changes its shape on the inside of the neurone and this causes another protein attached to the inside of the cell membrane to react chemically with the receptor. This second protein, called a G-protein, then changes shape itself, harnessing a molecule called guanosine triphosphate, GTP, as a source of energy. (GTP is closely related chemically to ATP; see section 5.3.) The GTP/G-protein combination activates yet another membrane protein, enabling it to catalyse chemical reactions within the neurone. Two such reactions, though by no means the only two, are those which produce chemicals that bind to open potassium ion channels and block them and those that bind to closed potassium ion channels and open them. Blocking the outflow of potassium ions by closing open channels increases the number of positive ions on the inside of the cell membrane, and so reduces the voltage from −70 mV to nearer −60 mV. This doesn't cause the neurone to fire, but it does increase the likelihood that it will do so in response to excitatory signals received by receptor proteins doubling as gated channels. Conversely, opening closed potassium ion channels increases the membrane voltage above −70 mV and makes the neurone resistant to firing. Importantly, these G-protein mediated changes in membrane voltage, far from being over in the milliseconds that it takes a neurone to fire, can last for a minute or two or even longer.

This more complicated way by which a neurone can process an

incoming signal is referred to as *second-messenger* and plays a role in memory and learning, as will be described in section 8.1.

SIGNAL TRANSMISSION BETWEEN NEURONES

Observation reveals that most of a neurone's gated-channel receptor proteins are located on its dendrites and cell body. It can be concluded that when a neurone receives a signal it does so through these two of its three parts. What role, then, does the axon have? Recall from the previous section that axons are very long tubular outgrowths of neurones. This makes them ideally shaped to carry signals to other neurones (and eventually to muscle cells and glands). Thus, signals enter a neurone through its dendrites and cell body and leave along its axon. And it is from a neurone's axon that signals pass to the next neurones down the line. It has already been established that signals between neurones are chemical. But neurones' dendrites and cell bodies have converted an incoming chemical signal into an electrical voltage flip. To pass the signal on, their axons need to convert the voltage flip back into the required chemical one. How does this come about?

At many points along its length, particularly near its end, a neurone's axon will come into close contact with the dendrites and cell bodies of other neurones. The points of close contact are marked by the development of a knob-like structure on the surface of the axon, one knob at each point of close contact. The knob is separated from the receiving neurone by a gap, typically of less than 1×10^{-4} mm. This gap, across which the signal passes, is called a *synapse*. An axon may make synapses with hundreds or thousands of other neurones. Similarly, a receiving neurone's dendrites and cell body will have hundreds or thousands of synapses on them from many different axons.

The conversion of the axon's electrical signal into a chemical one happens at these synapses. When the voltage disturbance gets to the synaptic knob of an axon, voltage-gated calcium channels there open, allowing calcium ions (Ca^{++}) — another type of extra-cellular ion found throughout the body — into the synaptic knob. The influx of calcium ions causes tiny bubbles stored within the axon to fuse with the axon's surface membrane and then to discharge their contents — the chemical signal — into the synaptic gap. From arrival of the voltage disturbance to release of the chemical signal takes less than half a millisecond.

Each bubble in a synaptic knob contains about 5000 molecules of the signalling chemical, and perhaps ten bubbles may fuse with the membrane with each voltage flip. The released chemical crosses the synapse in about one tenth of a millisecond and reacts with receptor proteins which double as gates for positive or negative ions, or else with receptor proteins which activate the second-messenger system. How the receiving neurone responds when this signal arrives has already been described.

For obvious reasons, the various chemicals used to transmit signals across synapses are known as *neurotransmitters*.

Because of the way a signal passes down an axon, it can be regarded as a pulse — an electrical one. The pulse, on arrival at a synaptic knob, switches on the release of the neurotransmitter for the duration of the pulse, typically about one millisecond.

Neurotransmitters, in contrast to axon pulses, are chemical rather than electrical in nature, so they could potentially linger in the synapse for a long time. A few do indeed do this, diffusing into nearby parts of the brain. But most neurotransmitters need to be deactivated rapidly to make way for the next electrical pulse travelling down the signalling axon. This enables the receiving neurone to distinguish each release of neurotransmitter from the preceding and following releases.

As a high-speed alternative to letting neurotransmitters diffuse away, there are two ways to remove them from the synapse. The most widespread one is reuptake. In this case the cell membrane in the synaptic knob contains imbedded within it proteins whose function is to react chemically with the neurotransmitter and pump it back into the axon it came from. The axon then recycles it.

The other means is degradation. External enzymes catalyse the chemical breakdown or modification of the neurotransmitter so that it can no longer react chemically with the receiving neurone's receptor proteins. This switches the neurotransmitter off. The products of the degradation are then reabsorbed by the sending axon. Once this has happened, the neurotransmitter can be resynthesized ready for future re-release.

There is no standard chemical used for transmission across the synaptic gap. The most prominent chemicals, known as *classical neurotransmitters*, are ten or so in number. They include adrenalin and two closely related molecules (noradrenalin and dopamine), a less closely related molecule called serotonin, a few amino acids, and acetylcholine. Three will be described as examples here.

Acetylcholine. One of acetylcholine's main roles is to transmit signals from motor neurones to muscle cells. It has the formula:

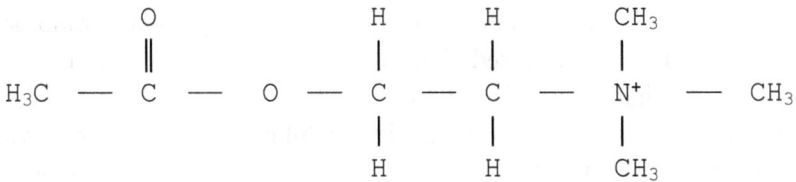

$$
\begin{array}{ccccccccccccc}
& & \text{O} & & & & \text{H} & & \text{H} & & \text{CH}_3 & & \\
& & \parallel & & & & | & & | & & | & & \\
\text{H}_3\text{C} & - & \text{C} & - & \text{O} & - & \text{C} & - & \text{C} & - & \text{N}^+ & - & \text{CH}_3 \\
& & & & & & | & & | & & | & & \\
& & & & & & \text{H} & & \text{H} & & \text{CH}_3 & & \\
\end{array}
$$

GABA. One of the main inhibitory classical neurotransmitters in the brain is gamma-amino butyric acid (GABA):

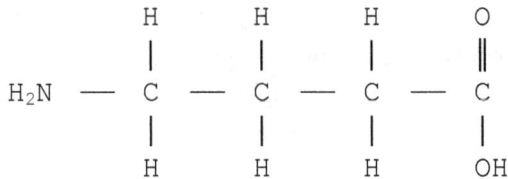

$$
\begin{array}{ccccccccc}
& & \text{H} & & \text{H} & & \text{H} & & \text{O} \\
& & | & & | & & | & & \parallel \\
\text{H}_2\text{N} & - & \text{C} & - & \text{C} & - & \text{C} & - & \text{C} \\
& & | & & | & & | & & | \\
& & \text{H} & & \text{H} & & \text{H} & & \text{OH} \\
\end{array}
$$

Glutamate. This is one of the main excitatory classical neurotransmitters:

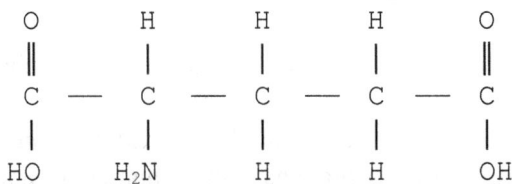

$$
\begin{array}{ccccccccc}
\text{O} & & \text{H} & & \text{H} & & \text{H} & & \text{O} \\
\parallel & & | & & | & & | & & \parallel \\
\text{C} & - & \text{C} & - & \text{C} & - & \text{C} & - & \text{C} \\
| & & | & & | & & | & & | \\
\text{HO} & & \text{H}_2\text{N} & & \text{H} & & \text{H} & & \text{OH} \\
\end{array}
$$

It will be noticed how similar GABA and glutamate are. The source of these two neurotransmitters reveals that biochemistry is often economical as well as complicated. One of the molecules formed in the course of converting glucose to CO_2 and H_2O is alpha-ketoglutarate. A single enzyme can divert the alpha-ketoglutarate away from the glucose to CO_2-and-H_2O path and convert it to glutamate, which is one of the twenty amino acids used to make proteins as well as being a neurotransmitter. And only one more enzyme is needed to convert glutamate to GABA.

Hence a neurone which possesses this final enzyme inhibits other neurones by using GABA to signal with, and a neurone which doesn't possess the enzyme excites other neurones by using glutamate.

In addition to the classical neurotransmitters there are dozens of non-classical neurotransmitters, though proof of neurotransmitter activity in these cases can be difficult to obtain. Most are fairly small amino acid chains. They are often released alongside a classical neurotransmitter and can have either inhibitory or excitatory effects.

Whether a particular neurotransmitter inhibits or excites a receiving neurone depends not only on which neurotransmitter it is, but on the specific receptor protein with which it reacts. Any given receptor protein will only receive one neurotransmitter, but one given neurotransmitter may be received by many more than one type of receptor protein.

This variability creates the potential for confusion. Fortunately there are some rules which are generally obeyed.

Receptor proteins which double as gated channels are few in number, so the rule for them is quite straightforward: neurotransmitters always have the same effect. The following are examples.

- Acetylcholine is excitatory.
- GABA is inhibitory.
- Glutamate is excitatory.
- Serotonin is excitatory.

In addition, neurotransmitters which react with gated-channel receptors are always classical ones.

For receptors which operate via the second-messenger system, the number of different receptor proteins is more numerous, and the rule — if it can be called a rule — is that the effect of a given neurotransmitter varies. There are various second-messenger receptors for the classical neurotransmitters. Furthermore, all the non-classical neurotransmitters use the second-messenger system.

SIGNAL CHARACTERISTICS

An electrical signal passing down an axon travels at as much as 120 m s^{-1}. It can be calculated that a signal travelling from a receiving motor neurone in the spine to a muscle in the calf of an adult human can take less than eight milliseconds.

This high velocity of signal transmission is achieved in two separate ways. One concerns axon diameter. A thin axon may have a signal speed as low as $0·5$ m s^{-1}. By increasing the diameter of the axon the signal speed can be made considerably faster.

The other means of increasing signal speed involves equipping axons with electrical insulation. This is brought about by two kinds of cell: oligodendrocytes inside the brain, and Schwann cells outside it. The shape of both kinds of cell is that of a tube enclosing the axon, and most axons have these insulator cells distributed along their length. Despite this, not all axons end up being insulated. In uninsulated axons the insulator cells fail to develop their insulating potential. In insulated axons, by contrast, the insulator cells wrap themselves around the axon very thoroughly, making a kind of Swiss roll shape with the axon running down the middle. The insulator cell membranes contain a fatty substance called *myelin* which provides the insulation. Myelin brings an almost complete halt to the movement of ions across the membrane of the axon. This effectively seals the various ions in (and out) and insulates the electric charges inside the axon from those outside.

The insulation is not perfect, however. At various points along the length of the axon there are left small uninsulated gaps where one insulator cell ends and the next one begins. These gaps typically occur at 1 mm intervals. The electrical signal passing down the inside of the axon jumps from one uninsulated gap to the next, so giving rise to the high transmission speed.

While signal speed may vary from axon to axon, in all other respects neurone signals differ from one another hardly at all. Once the part of the neurone membrane where the axon joins the cell body has experienced a voltage reduction below -55 mV, the $+20$ mV voltage flip spreads inevitably down the full length of the axon. The course of events is always the same. Because of this, one signal passing down an axon is identical in appearance with any other signal passing down it. Furthermore, signals passing between neurones depend only on the neurotransmitter in use and the receiving neurone's response to it. Neither the location of the neurones concerned (spine, brain, eyes, etc.) nor the cause of the excitation makes any difference to any one neurone's signals.

Neurone signals are thus very nondescript. Indeed the only way single neurones can transmit a variable signal is by sending signals at a variable frequency. In a strongly excited neurone, a pulse of voltage disturbance may pass down its axon 200 times a second. A weakly excited neurone, in contrast, may not send any pulses of voltage disturbance down its axon at all. Thus the signals passing through the nervous system amount to statements by individual neurones of 'this is how much I am being excited'. That is all one neurone can say.

6.4 INPUTS AND OUTPUTS

SENSORY NEURONES

Sensory neurones are much like the typical neurones described above. They transmit signals through themselves in the form of membrane voltage reductions and pass those signals to intermediate neurones by means of neurotransmitters. What distinguishes them is the source of the signals. In all except sensory neurones, the source is receipt of neurotransmitter molecules; in sensory neurones it is some event external to the nervous system. It is of course the external source of the signal that provides the defining characteristic of sensory neurones.

By this definition, sensory neurones include not only neurones that react to events outside the body but also those that react to internal events. Neurones which initiate signals in response to a decrease in the amount of water in the blood, for example, can be considered sensory.

The means whereby external events cause sensory neurones to begin signalling are mostly physical or chemical in nature. An example of the former is sound. In neurones which are sensitive to sound, very rapid oscillatory air movement (which is what sound is) brings about deformation of the neurones' membranes. The way this occurs is very complicated. The deformation — which is purely physical — causes ions to cross the membrane, reducing the membrane voltage, and this results in signal transmission in the usual way.

With chemically initiated signals — taste and smell primarily — it can be assumed that the chemicals concerned react with receptor proteins in the appropriate sensory neurones' membranes, thereby altering the membrane voltage in much the same way as neurotransmitters do.

Light sensing neurones are an interesting example of signal initiation which is neither physical nor chemical, and of how some kinds of neurones work quite differently from the 'standard' neurones of sections 6.2 and 6.3. Light sensing neurones have membranes with a voltage of only -40 mV, which means they are signalling continually. They also contain a chemical which absorbs photons. When a photon possesses sufficient energy, the chemical undergoes a number of reactions, during the course of which various enzymes convert it back to its original form. As these reactions take place they bring about an *increase* in the neurone membrane voltage. This stops the neurone signalling. In other words, light sensitive neurones release more neurotransmitter when it is dark than when it is light. They send signals when nothing is happening and stop sending signals when struck by incoming photons. This is the opposite of the way neurones usually work.

INTERMEDIATE NEURONES

Sensory neurones pass their signals to intermediate neurones. How intermediate neurones function individually was described in the previous section, and their bulk properties will be examined shortly. In the meantime a couple of points can be made.

The first concerns the quantity of intermediate neurones. A human brain contains 10^{11} of them. As a human has about 10^7 sensory neurones, it follows that there are approximately 10,000 intermediate neurones for each sensory neurone. This means that signals initiated by sensory neurones can be subjected to enormous modification by the intermediate neurones before being passed to motor neurones for transmission to muscle cells. As a result, human behaviour cannot be expected to bear any precise relationship to events detected by the sensory neurones. This expectation is of course fully in accord with observation.

The second point to make, which is related to the first, is that the actual, as opposed to the potential, number of intermediate neurones involved in passing any given sensory signal to motor neurones is highly variable. To understand why, it needs to be realized that the human neuromuscular system was not created in one go. It resembles by analogy a small bungalow whose successive owners have progressively enlarged it, until now it has become a gigantic palace. Additions to the building by each new owner have caused little demolition of existing structures. Somewhere in the palace is the original bungalow with its plumbing and wiring still intact. The processing of particular kinds of incoming signals by intermediate neurones may involve large parts of the palace or may hark back to earlier times when the building was smaller.

The simplest human behaviour, corresponding to the original bungalow, is the reflex. Reflexes are sensory-to-motor communications involving very few neurones and very little modification of the signal as it passes from sensory neurones to muscle cells. The famous knee-jerk reflex, which occurs when a person's knee is tapped with a hammer, is an example. The tap on the knee results in the stretching of an attached muscle. A sensory neurone imbedded in the muscle, whose role is to send a signal in response to increased muscle tension, is thereby caused to fire. Its signal is passed to a motor neurone which signals to the muscle cells which have stretched. When they receive the motor neurone's signal they change shape, contracting, causing the lower leg to extend.

There is a complication to even this simplest of reflexes. The sensory neurone, in addition to signalling to a motor neurone, also signals to an inhibitory intermediate neurone. This intermediate neurone signals to a different motor neurone, one which signals in turn to a muscle which would normally contract in such a way as to resist the extending of the lower leg. The complication to the simple reflex is thus aimed at preventing two opposing muscles fighting each other by contracting simultaneously.

The intermediate and motor neurones controlling this reflex are located in the spine, so the reflex happens before the brain gets to know about it.

Another reflex which is a little more elaborate — the bungalow plus some home improvements — occurs when someone's hand encounters an unexpectedly hot or sharp object. Even while pain signals are travelling up axons in the spine towards the brain, intermediate neurones in the spine are already signalling to motor neurones to contract muscles which will withdraw the hand towards the body and thereby away from the cause of the pain.

Because of their very simplicity, reflexes are not typical of human behaviour. Such activities as driving a car or playing football involve the use of a considerable part of the human neurological palace, with the participation of thousands of millions of intermediate neurones as against the handful in a reflex.

MOTOR NEURONES

Motor neurones closely resemble intermediate neurones. Their main distinguishing feature is that their axons do not form synapses with other neurones but with muscle cells instead. Normally a single motor neurone's axon will contact between three and several hundred muscle cells. Contact is via ordinary synapses using acetylcholine as the neurotransmitter. Motor neurones are atypical in that their axons may be up to a metre in length. This is a consequence of the cell bodies of most motor neurones being located in the spine.

MUSCLE CELLS

A muscle consists of a bundle of muscle cells. In muscles which are attached to bones (those responsible for movement around the environment, and which account, incidentally, for about 40% of the human body) each muscle cell stretches from one end of the muscle to the other and is about 10^{-2} mm in diameter. It is thus shaped like a long, very thin tube. A muscle consists of many such tubes.

Each muscle cell has its full complement of membrane proteins. This includes sodium pumps and potassium channels which give the cell a membrane voltage of about -90 mV. Receipt of a signal from a motor neurone by way of the muscle cell's membrane-imbedded acetylcholine receptor proteins reduces the voltage in the usual way (i.e. by opening gated sodium channels), invariably causing the membrane voltage to fall below -60 mV. Once that has happened, the muscle cell reacts much as a neurone does. A wave of voltage disturbance spreads over it for its entire length. The result of the voltage disturbance is different though. By a complicated process involving calcium ions, the influx of sodium ions causes proteins within the muscle cell to slide over each other in such a way that the cell shrinks in length. This pulls on the two bones to which it is attached (one bone at each end of the muscle cell tube) and produces movement. Following receipt of a single motor neurone signal, a muscle cell will contract for about 30 ms, the duration varying according to the type of muscle. Obviously, to keep a muscle in a contracted state the motor neurone making synapses with it must transmit a continual sequence of signals.

GLANDS

Finally there is glandular activity to consider. Glands are collections of cells which produce specific chemicals. When these chemicals are used for signalling within the body they are secreted into the blood and are called *hormones*. Hormones, depending on their chemical identity, can affect other glands, muscles, and even neurones in some parts of the brain. Hormones achieve their effects in a similar way to second-messenger neurotransmitters; that is, they react with receptor proteins in the membranes of the cells which the hormone is signalling to, and thereby bring about changes to the chemical reactions taking place inside those cells.

Some hormones have particular neurones amongst their targets, the hormones acting in these cases to modify the rate at which, or ease with which, the neurones transmit signals. Such target neurones are in this respect sensory by the definition of the term used in this book, and sense the internal hormone levels of the body. By doing this they provide the means whereby hormones can influence behaviour. For instance, neurones which detect hormones produced by the sex glands are able to modify the amount of sexual activity according to the quantity of sex hormones in the blood.

Certain neurones are able reciprocally to affect glands. The adrenal glands are a good example. When these two glands receive signals from their contacting neurones they pour considerable amounts of hormones called adrenalin and noradrenalin into the blood, which rapidly circulates them all over the body. The subjective effects of this are quite pronounced, as anyone who has suffered a sudden fright can testify, fright being very good at causing neurones which make synapses with the adrenal glands to transmit signals.

A noteworthy thing about the adrenal hormones is that they are also neurotransmitters. This serves to highlight the point that the body has two signalling systems, a nervous one and a chemical one. The chemical system, first mentioned in section 6.1 in the course of discussing cell coordination, is slower than the nervous system and is concerned with minutes, hours and even months rather than fractions of a second. It has nonetheless an important role to play in enhancing its possessor's chances of survival, and it cooperates closely with the nervous system to that end. How this cooperation is brought about — by a part of the brain called the hypothalamus — will be discussed shortly.

6.5 THE STRUCTURE OF THE BRAIN

<u>INPUTS TO THE BRAIN</u>

The paths taken by signals as they pass from one brain organ to another provide a convenient basis for a description. The obvious way to make a start is therefore to follow the course of input signals as they travel from peripheral sensory neurones into the brain's great system of intermediate neurones. Figure 6.1 shows the parts of the brain that will be referred to.

The most distant sensory neurones are associated with the body surface and with its interior. The signals they transmit are concerned with touch, pressure, vibration, itch, the position and movement of each part of the body with respect to other parts, surface temperature and pain. All these senses are known collectively as the *somatic senses*. Axons from somatic sensory neurones, other than those located in the head, carry their signals to the *spinal cord* where they make synapses with intermediate neurones. These neurones then pass the signals by means of their own axons up the spinal cord to the brain.

(A note for enthusiasts: somatic sensory neurones do not have the usual dendrites plus cell body plus axon form. The cell bodies of somatic sensory neurones are actually located in the spine alongside motor neurones. Consequently, instead of dendrites, they have a second axon. This second axon carries signals the wrong way, from where the touch, pressure, vibration, etc. detectors are located, *to* the cell body.)

There is no clear demarcation between spinal cord and brain. Indeed the spinal cord can fairly be considered to be an outgrowth of the brain (or vice versa). To emphasize this point, the spinal cord, as it enters the head, merges with the brain in an area called the *medulla*. This is where axons carrying somatic sensory signals from the head make synapses, so the medulla is to the head as the spinal cord is to the rest of the body. Axons carrying taste, balance and hearing signals from the mouth and ears also make synapses with intermediate neurones in the medulla.

Above the medulla is the *pons*, about which more will be said in a moment.

Behind the pons is the *cerebellum*, a sheet of ridged and infolded tissue (*tissue* in biology is any aggregate of cells with a common identity) which is divided into two halves, left and right. Each half consists of three surface layers of vast numbers of neurones, beneath which run an equally vast number of axons, which are myelinated (see section 6.3). Sensory signals coming directly from the spinal cord and medulla pass information about balance, body position and body movements to the cerebellum. This data about what the body is doing is supplemented by visual signals

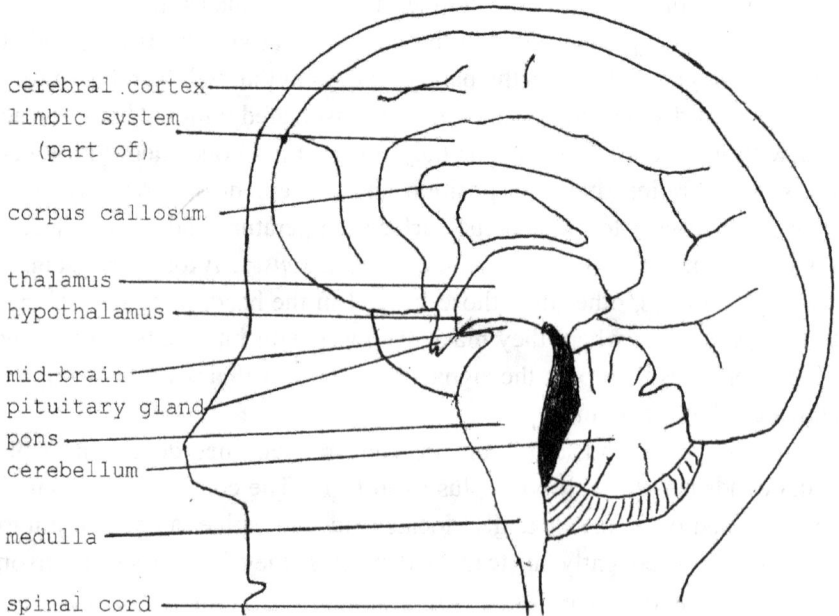

Figure 6.1 The location of some important brain structures. This view is of the right hand side of the brain, seen from the left; i.e. as if the left hand side has been removed.

informing the cerebellum of the body's orientation in space and of its immediate environment. The cerebellum is one of the end points for incoming signals.

Incoming signals not destined for the cerebellum pass through, or in some cases to, an area known as the *mid-brain*, which lies above the pons. One branch of visual signals joins the incoming signal route in this area. The mid-brain is primarily a relay station between lower and higher parts of the brain.

From the mid-brain axons pass to the *thalamus*, which lies above the mid-brain. Like the cerebellum, the thalamus is divided into left and right halves. Here a second branch of axons carrying visual signals forms synapses. By now, all the senses except smell have joined the incoming signal route. Thalamic neurones send their axons to the main end point for incoming signals, which is the largest organ in the brain, the *cerebral cortex*. This part of the brain, which will in future be referred to simply as 'the cortex', will be described in detail shortly.

Smell signals, it may be remarked, have what amounts to their own private route into the brain, arriving in a piece of cerebral tissue specifically devoted to evaluating them. They also travel to, and via, the thalamus but it does not stand astride their route into the cortex as completely as it does with the other senses.

All signals arriving in the cortex from the senses will be referred to as sensory signals even though they are almost always carried there by intermediate neurones. Sensory neurones originate sensory signals, but intermediate neurones carry those sensory signals to the cortex. Sensory signals are conveyed by both sensory and intermediate neurones.

OUTPUTS FROM THE BRAIN

Just as sensory signals are carried to the brain by intermediate neurones, so, in the opposite direction, motor signals too are carried from the brain by intermediate neurones. Motor signals are conveyed by intermediate neurones in the first instance and then passed to motor neurones.

Outputs from the brain concerning muscular activity begin their journey in the cortex. Axons from cortical neurones carrying motor signals pass through the thalamus with some synapse formation. They then travel through the mid-brain. In the mid-brain are the motor neurones which control movements of the eye muscles, so eye muscle motor signals leave the main output signal path here, passing directly to the eyes.

The remaining motor signals descend to the medulla where signals controlling the other muscles located in the head pass to their motor neurones and thence to their final muscular destination.

Motor signals for the rest of the body are conveyed down the spinal cord and to the motor neurones located there. In terms of outputs as well as inputs the medulla can thus be seen as being to the head what the spinal cord is to the rest of the body.

Not all motor signals pass to the muscles. Some are intercepted by the pons and relayed to the cerebellum. The cerebellum is in consequence in receipt of both sensory signals arriving from the spinal cord, medulla and eyes, and motor signals coming from the cortex. The sensory signals relate in particular to the position of the body as a whole (balance) and to the position of its parts in respect to one another and to its surroundings. The motor signals relate of course to what movements those parts will make next. The cerebellum is specialized to evaluate these two sets of signals simultaneously and to detect any discrepancies between them. This is quite a complicated computational process. All the positional signals in the brain are out of date because of the time taken for them to reach it. From distant parts of the body a fiftieth of a second may elapse between a sensory neurone initiating a signal and the brain receiving it. A limb engaged in vigorous movement can travel over a tenth of a metre in that time. So to compare its two sets of sensory and motor signals for discrepancies, the cerebellum must be able to predict where each limb will be and what each muscle will be doing when the motor signals it is being passed by the pons finally get to their target. The evaluating functions of the cortex do not give it a predictive ability of this kind, so its motor signals are inherently inaccurate, the more so the faster the movements concerned are. The cerebellum detects these inaccuracies and corrects them by sending signals via its own axons back up to the thalamus and cortex, and also to the pons and medulla. Thereby, by the usual means of neuronal excitation and inhibition, it modifies the motor signals being sent to the muscles so that they do what the cortex intended them to.

The predictive ability of the cerebellum may seem little short of magic. Here, three comments will be made that put it in perspective. The first comment is that modern electronic circuits can easily be designed which are capable of predicting the course of some mechanical event; there is no reason in principle why the brain should not contain circuitry which is functionally similar. The second comment is that cerebellar predictions can be regarded simply as output signals (from the cerebellum) which have the effect of stabilizing what would otherwise be a rather unstable system. The patterns of signals necessary to achieve this stability are learnt by the cerebellum during infancy, the learning taking the form of modifications to the synaptic connections amongst the cerebellar neurones. And the third comment is that the cerebellum, despite its comparatively small volume, contains roughly twice as many neurones, and thereby theoretically twice as much computing power, as the entire cerebral cortex. And most of this power is dedicated to fine-controlling motor output. Taking these various points into account, the abilities of the cerebellum do not appear quite so miraculous.

THE HYPOTHALAMUS AND LIMBIC SYSTEM

The signal paths considered so far have involved the external senses and the somatic senses on the input side and motor signals on the output side. But these are not the only signals entering and leaving the brain. Accompanying them are input signals concerned with the internal state of the body — for example, signals which reflect the body's internal temperature — and output signals which control its glands and the heart.

Signals concerned with sensing the body's internal state either originate in, or are passed to, the *hypothalamus*, a small piece of brain tissue located in front of, and a little below, the thalamus. The axons of hypothalamic neurones then carry the signals both downward to the mid-brain and medulla and upward to the thalamus. Hypothalamic signals relating to the internal state of the body are thus in a position to exert a considerable modifying influence on the external sensory and motor traffic heading to and from the cortex.

The hypothalamus is also concerned with monitoring external sensory inputs, and internal somatic sensory inputs, with a view to evaluating how harmful or beneficial they are to the body. This role it shares with a group of brain structures known as the *limbic system*. (This system is also often called the limbic lobe. The two names can be taken to be synonyms here.) The limbic system is located around, above, and largely adjacent to the thalamus and is so closely associated with the cerebral cortex that it is considered to be cerebral, though the layout of its component parts makes it distinct from the cortex proper. Signals emanating from the hypothalamus and limbic system concerned with evaluating harm and benefit are able, either directly or via the cortex, to produce behaviour characteristic of pain, fear, anger, pleasure, etc. This behavioural role of the hypothalamus and limbic system will be discussed further in section 7.2.

Another role for the hypothalamus is to coordinate the body's two signalling systems, nervous and hormonal. Regarding the latter system, the hypothalamus is both sensory, possessing receptors for various hormones, and motor, in the sense that it controls the pituitary gland, which in turn controls all the other glands by means of pituitary hormones. The pituitary gland lies at the base of the hypothalamus and is in direct physical contact with it, enabling the hypothalamus to exert its control in two easy ways. There is the normal way, which is by means of synapses of hypothalamic neurones with pituitary gland cells; and there is also what amounts to a glandular way. What happens is that some hypothalamic neurones secrete hormones — mostly formed of short amino acid chains — which affect pituitary cells. The hypothalamus is thus partly a gland as well as being an important component of the brain. Given it's in two-way communication with the glands, and is similarly in two-way communication with neighbouring parts of the brain, the hypothalamus gives the brain considerable influence over the body's glandular signalling system, and gives the body's glandular signalling system considerable influence over the brain.

Not content with control via the pituitary, the hypothalamus also sends signals by way of a chain of intermediate neurones directly to various other glands. Examples are the adrenal glands — see the previous section — and sweat glands. The heart and other rhythmically contracting muscles are also influenced by axons carrying output signals originating in the hypothalamus.

One other function performed by the limbic system that is worth mentioning concerns the sense of smell. The axons of smell-sensing neurones form synapses at a location just above the nasal cavity known as the olfactory bulbs: one on the left and one on the right. From there smell signals are conveyed by intermediate neurones straight to the limbic system as well as to the thalamus and cortex. The path followed by smell

286

signals is thus unique amongst the senses. The reason for the uniqueness is thought to be that when the first mammals appeared on the Earth smell was their most important sense. As a consequence of this it proved advantageous to have smell signals communicated directly to the limbic parts of the brain where harm and benefit are evaluated. Such an arrangement is irrelevant to present-day human sensory needs and it highlights one aspect of the analogy about the bungalow and the palace. The smell signal pathway is like a building extension put up by a long-departed owner. The present owners don't use it much and would not have designed it that way. But the building was not constructed for the present owners alone. It is a piecemeal creation put together over hundreds of millions of years of evolution. The smell pathway is an echo of long-dead owners whose sensory priorities were altogether different from those of modern humans.

SUMMARY OF THE PRINCIPAL BRAIN STRUCTURES

Quite a few names have been encountered in this section. It is unfortunately inevitable that this should be so, given that the object under investigation is one in which the complexity permitted by the laws of physics reaches its highest expression. Indeed, the complexity of the human brain has been greatly underplayed in this description, as attention has been confined to those structures that will be needed in the following sections and chapters.

To itemize these structures, here is a list of nine components, to each of which can be assigned a very broad-brush, simplified functional description. (Viewed close up, the brain's axonal wiring is far more complicated than is indicated here, meaning the function of each structure is also far more complicated.)

- The cortex is the largest single structure in the brain, where external sensory signals are mainly evaluated and where most motor output signals originate.
- The thalamus is the structure through which virtually all signals, both to and from the cortex, pass.
- The hypothalamus is a small but vital structure having to do with sensing and monitoring the body's internal state, and getting the

cortex to produce motor output signals relevant to the internal situation (e.g. hungry, thirsty, cold). It also greatly influences the body's hormonal signalling system.

- The limbic system is a group of cerebral structures concerned, along with the hypothalamus, in preventing self-harming behaviour and in encouraging self-benefitting behaviour. Individual parts of the limbic system have additional functions which are not of concern here.
- The mid-brain is a staging post for signals passing to and from the thalamus.
- The pons is another staging post, this time for motor signals passing to and from the cerebellum.
- The cerebellum is a large structure responsible for putting the finishing touches to motor output signals.
- The medulla is the structure at the base of the brain in which sensory neurones receiving signals from the head are located and in which motor neurones controlling the muscles of the head have their cell bodies.
- The spinal cord is an outgrowth of neural tissue extending most of the length of the spine, in which sensory neurones receiving signals from all parts of the body below the head are located and where motor neurones controlling the muscles of the neck and below have their cell bodies.

In the remainder of this chapter attention will be concentrated on the most masterful of these brain parts: the cortex.

6.6 THE CORTEX

STRUCTURE

The human cortex contains about half the mass of the nervous system and, cerebellum excluded, perhaps three quarters of its neurones. These neurones are segregated into six layers which, insofar as they are able to, run parallel to the skull. The thickness of each layer varies from place to place in the cortex but the combined thickness of all six layers is generally from two to four millimetres. Below the six layers run the axons that pass signals amongst the neurones of the cortex and to/from lower brain structures. The six layers are grey in colour — hence the term 'grey matter' — while the underlying axonal mass is white, a colour produced by the myelin insulation.

If the cortex could be removed from the skull and laid out flat, it would cover the surface of a square 350 mm across (14 inches). To fit such a large amount of tissue into the skull requires it to be rather buckled; the six layers of cortical neurones that make up the cortical surface take up a distorted shape comprising broad ridges and deep, narrow fissures, the ridges being against the skull. Figure 6.2 illustrates this.

The cortex is physically divided into two halves. Each half occupies one side of the head, the halves being known as the left and right *hemispheres*. Signals are passed between the hemispheres by several routes, the main one being a large bundle of axons located centrally and running both leftwards and rightwards. This inter-hemispheric connection is called the *corpus callosum*. Again see figure 6.2, and also figure 6.1.

Because the cortex is such a big piece of tissue, it is useful to break it down into smaller sub-units. A number of such mappings have been

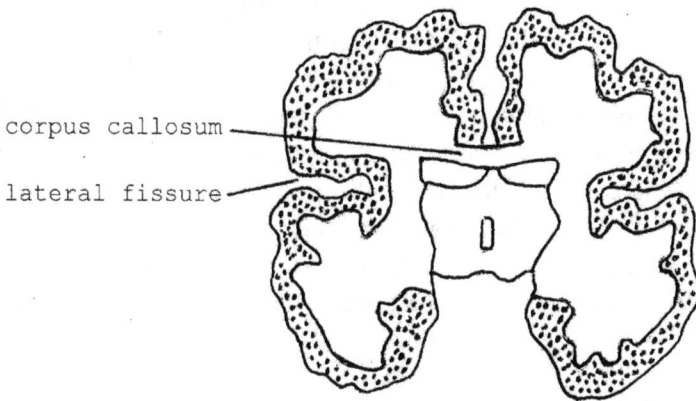

Figure 6.2 The two hemispheres of the cortex, showing the infolding of the cortical surface and the corpus callosum. This illustration is of a vertical slice of the brain running roughly between the two ears. The stippled area is the cortical surface.

made, based on criteria such as the comparative thickness of the six layers of neurones in different cortical locations. Here two kinds of mapping will be described, one drawing on the gross features of the cortical surface — its geography, so to speak — and the other drawing on the functions that particular cortical locations carry out.

CORTICAL GEOGRAPHY

Taking geography first, each hemisphere has a couple of prominent fissures. These fissures match up on both sides of the brain, the two hemispheres being more or less identical in appearance except that one is the mirror image of the other. Using the fissures as boundary markers makes it possible for four regions to be defined. The regions are called *lobes*.

The two fissures used to demark the lobe boundaries are called the central and lateral fissures, with imaginary lines added where necessary. The cortical surface in front of the central fissure is known as the *frontal lobe*; the surface behind the central fissure and above the lateral fissure is known as the *parietal lobe*; the surface below the lateral fissure is known as the *temporal lobe*; and the surface at the back of the brain (whose boundaries are more arbitrary) is known as the *occipital lobe*. Figure 6.3 illustrates this mapping of the cortex.

It should be kept in mind that these geographical subdivisions of the cortex do not represent any actual physical breaks in the cortical surface. The fissure-demarcated boundaries of the four lobes only mark places where the surface folds. This contrasts with the division between the hemispheres, which does involve an actual gap. Consequent on this, each lobe is of course really a pair of lobes, one in the left hemisphere and one in the right.

CORTICAL FUNCTION

Turning now to function, the deriving of a functional map requires the paths of input signals to be traced once more, starting this time at the point where they enter the cortex.

Axons carrying smell signals make synapses with cortical neurones in a part of the frontal lobes near the lateral fissure where the frontal lobes

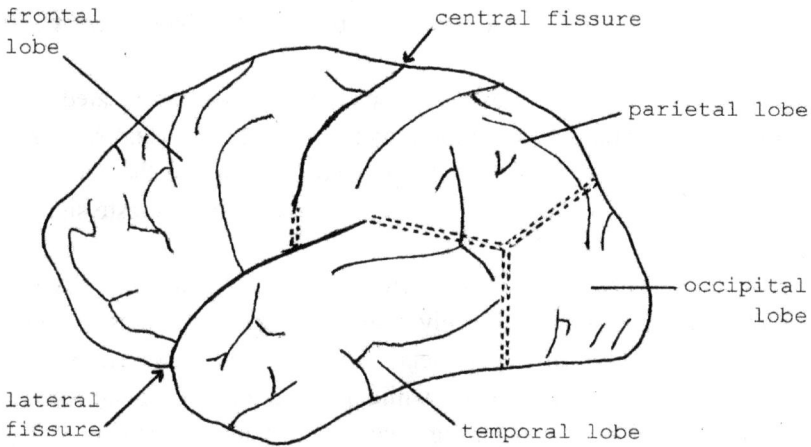

Figure 6.3 The left hemisphere of the cortex as it would appear if removed from the skull. The front of the head is on the left. The dashed lines, together with the central and lateral fissures, mark the borders between the lobes.

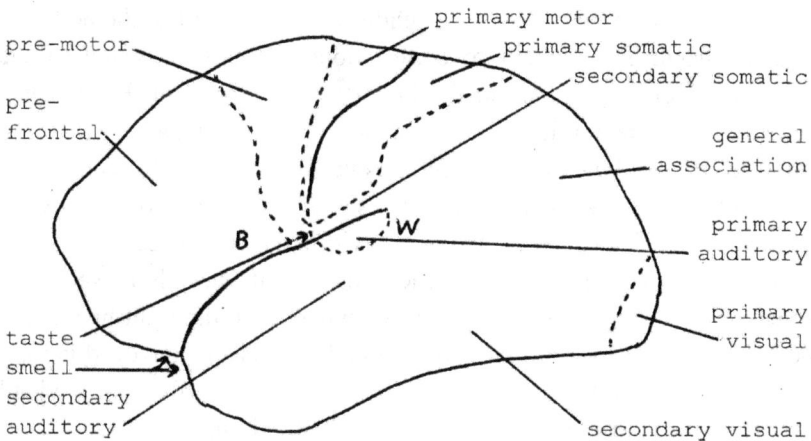

Figure 6.4 The cortex as in figure 6.3 with the function of each area indicated. Dotted lines mark the borders between functions. The taste and smell areas are found at roughly the locations shown but are not visible in this view. B = Broca's area, W = Wernicke's area. Note that the boundaries between secondary and association areas are poorly defined.

reach their lowest point. Synapses are also made with cortical neurones at the forward end of the temporal lobes. Figure 6.4 shows the sites of smell signal evaluation, as also the other functional subdivisions about to be described.

The sense of taste is conveyed from the taste sensors located in the tongue and other parts of the mouth and throat, via the usual subcortical structures, to the lowest part of the parietal lobes where the central and lateral fissures meet. The part of the cortex which receives taste signals is in fact inside the lateral fissure.

Somatic sensory signals make synapses with cortical neurones in an arch of tissue that lies immediately behind the central fissures; i.e. in the forward-most parts of the parietal lobes. Here, however, there is something of an oddity. Either while passing up the spinal cord, or alternatively while passing through the medulla, axons carrying somatic sensory signals from the left side of the body cross over to the right side, and axons carrying somatic sensory signals from the right side of the body cross over to the left. Because of this, each hemisphere of the cortex receives somatic sensory signals from the opposite side of the body. These crossed-over somatic sensory signals form synapses at different locations along the arch of parietal tissue, depending on which part of the body they originate from. The amount of cortex along the arch allocated to each body part is closely related to the importance of that part. For instance, face and hands have a lot of cortex evaluating their signals, whereas the abdomen, despite being much larger, has very little cortex dedicated to it.

Sound signals, like somatic sensory signals, also tend to be conveyed to the opposite side of the cortex from the side they are detected on, but in this case the crossing over is much less marked. It can only be said that a modest majority of left ear signals form synapses in the right hemisphere and vice versa. The location to which sound signals are conveyed is found in each temporal lobe bordering the lateral fissure, partly on the visible cortical surface there and partly buried inside the fissure.

Visual signals are also processed in an unexpected way. One of the properties of lenses such as the eyes possess is to invert the light that passes through them. As a result, the picture that falls on the back of the eyes is inverted top to bottom and left to right. The light sensing neurones located there transmit this picture to a central location inside the head where, without forming synapses, the axons carrying signals from the left half of each eye (which represent light from objects on the person's right, because of inversion by the lenses) pass to the left side of the brain. Similarly, axons from the right half of each eye pass to the right side. The signals, after being passed to thalamic neurones, are then carried to the back parts of the occipital lobes. The left occipital lobe therefore receives signals from the left side of both eyes about what is visible on the right side of the head, and the right occipital lobe receives signals from the right side of both eyes about what is visible on the left side of the head.

So far this functional mapping has covered only a small proportion of the cortex. A substantial part of the remainder can be accounted for when it is realized that somatic, auditory and visual signals require a great deal of processing in order to be fully evaluated. The parts of the cortex performing this processing lie adjacent to the areas of cortex where the somatic, auditory and visual signals are initially received. The initial receiving areas are known as primary areas, while the areas of further processing are known as secondary areas. Hence the brain possesses a primary visual (area of) cortex and a secondary visual cortex, a primary auditory cortex and a secondary auditory cortex, a primary somatic sensory cortex and a secondary somatic sensory cortex. The secondary visual and auditory areas surround their primary areas; the secondary somatic sensory cortex abuts the primary somatic sensory area in a location near the lateral fissure.

On the output side, activation of the muscles of the body is affected by the same crossing over of axons in the spinal cord and the medulla that occurs for incoming somatic sensory signals. The right hemisphere's output signals activate muscles on the left side of the body, and the left hemisphere's output signals activate muscles on the right side of the body. Muscular activity is initiated by the arch of cortex which lies immediately forwards of the central fissures in the frontal lobes. As with the somatic sensory area, the amount of this arch of cortex devoted to each body part is not proportional to the size of that part; it is instead proportional to the fineness of muscular movement required. Again the face and hands are best provided for. As the output signals are eventually destined for motor neurones, the arch of cortex from which they emanate is called the primary motor cortex.

Motor output signals may also need additional processing, though not in quite the same way that sensory signals do, and a secondary area in front of the primary motor cortex fulfils this role. It is called the pre-motor cortex.

Having covered those areas of the cortex which deal with sensory and motor signals, there still remain two large areas to account for. One, comprising the frontal parts of the frontal lobes and forward-most parts of the temporal lobes, is called the pre-frontal cortex. The other includes large parts of the temporal, parietal and occipital lobes. Here this second area will be called, for reasons which will become clear shortly, the general association cortex. The function of these two areas will be discussed in the next section.

6.7 DAMAGE TO THE CORTEX

CAUSES OF BRAIN DAMAGE

The fundamental principle of science, as has been made clear repeatedly, is that observation determines truth. But observation requires that there are observers; and where humans are concerned at any rate, to be an observer is to be conscious. An unconscious human makes no observations and knows no truths.

The scientific hypothesis concerning consciousness is that there is an identity between consciousness and the brain. The hypothesis states that it is the brain which observes, and it is the brain which is conscious and knows what is (apparently) true.

To be consistent with scientific method, validation of this hypothesis must itself rest on observation. Observers must observe themselves. In other words, observations need to be made of what the human brain does, especially when it malfunctions. If it can be shown that when neurological events cease occurring — when a brain is damaged — subjective experience is altered, this would provide strong support for the hypothesis that the brain is both necessary and sufficient for consciousness.

Brain damage occurs whenever the connections amongst neurones are severed, and usually this involves the destruction of the neurones concerned. As was stated in section 6.2, neurones lose the ability to replicate at about the time of birth. As a result, neurone destruction thereafter is permanent. Any knowledge or information stored by some means (to be discussed in section 8.1) in neurones which are destroyed, anything learnt by them, any behavioural ability conferred by them, will be lost with them and may be as permanently irreplaceable as they are.

To produce observable symptoms, neurone destruction must either be widespread or, if limited to a small part of the brain, must involve many of the neurones in that part. Loss of tiny numbers of neurones dotted here and there would pass unnoticed both subjectively and objectively. It has indeed been suggested that typical individuals lose ten thousand neurones every day through ordinary wear and tear. (This amounts to less than one per cent of the brain over the course of a long lifetime.)

Causes of major neurone destruction include strokes, severe blows to the skull, bullet wounds, drug abuse, persistent heavy drinking of alcohol, viral infection of brain tissue, inflammation of the brain or its surrounding tissues, poisoning by various chemicals, haematomas, tumours, and about fifty different types of dementia (still, as yet, mostly of unknown causation).

Any one of these numerous causes, plus any surgery carried out to treat or cure them, may result in non-fatal but significant amounts of neurone destruction. The behavioural consequences of such brain damage can be investigated by doctors specializing in that subject, and then the brain examined after the death of the patient to precisely locate the site of the damage. It is but a short step from there to deduce the functions that are associated with each piece of the brain.

Such a step is not, however, free from pitfalls. The greatest difficulty arises because damage rarely affects just one specific area, but crosses the boundaries between functionally distinct parts of the brain. Also, the damaged area no longer sends its signals to other healthy parts, which may themselves begin to work abnormally as a result. The situation can be further obscured by a certain amount of functional duplication between different brain locations, and by the ability of undamaged parts to take over or acquire the functions of parts that have been destroyed; while the brain damage is permanent, the behavioural damage resulting from it may not be. When relearning is possible, the person concerned has literally to reacquire the behaviour like a young child.

These days the functions of different parts of the cortex deduced from studies of brain damage can be corroborated by non-destructive methods. For example, electroencephalograms detect increased local activity in the cortex by measuring tiny voltage changes on the surface of the scalp. There are also computer-based scans of the brain which can identify the link between quite accurately localized cortical activity and the processes of sensing, thinking and behaving. These modern studies of intact brains produce results which match those from looking at brain damage.

So what then are the observational consequences of brain damage? In answering this question attention will be confined to the cortex, which is where the most sophisticated brain functions are carried out.

While reading the descriptions of the various clinical syndromes, it should be kept in mind that the brain is far more complicated than is being presented here. To do justice to the cortex would require a sizable textbook. Thus the roles assigned to each functional area are to varying degrees idealized, the intention being only to establish the broad principles governing how the cortex works.

THE PRIMARY VISUAL CORTEX

Destruction of the primary visual cortex leads to blindness. This is so even if the eyes themselves are in perfect working order. Blindness arises because visual signals are passed to the rest of the cortex from the primary visual area. If that area is put out of action, no signals from the eyes can get to the remainder of the cortex, which means that the cortex as a whole becomes incapable of reacting to visual input.

There is, however, a small let-out clause in this apparently non-negotiable state of affairs. Visual signals from the eyes make synapses in the thalamus as well as being merged there to some degree so that signals from the same location on the retina of each eye are brought together. This means that though the cortex is blind, the thalamus certainly isn't. Some of the visual signals received by the thalamus are sent to cortical locations other than the primary visual cortex; specifically, to parts of the temporal lobes. In time, a blind cortex can learn to interpret these signals. However, only the visual cortex is structured to process visual information, so the thalamic signals interpreted by the cortex are not experienced as vision, but rather as a kind of feeling. The cortex remains completely blind. People afflicted with damage involving the entire primary visual area are sometimes able, for example, to point to the location of lights flashed in their eyes, but deny seeing anything. If asked how they know where the light is, they reply that they just guessed. In fact they are interpreting thalamic signals and the feeling it gives them. The phenomenon is known as blind-sight.

Very often damage to the primary visual cortex is not total and parts of it remain functional. In these cases, affected people are aware of a blind area in their field of vision, the blind area covering the same part of the visual scene in both eyes. Sufferers are visually unaware of anything in that area or of what is happening there.

As a final comment on the primary visual cortex it can be remarked that direct electrical stimulation of this area during brain surgery leads, as might be expected, to the patient reporting seeing (in reality non-existent) flashes of light, coloured spots, and similar things, but nothing of a visually more complicated nature.

THE PRIMARY AUDITORY CORTEX

Because each hemisphere receives signals from both ears, damage to the primary auditory cortex on one side of the brain may pass completely unnoticed. Careful clinical experimentation reveals a raising of the threshold at which the ear opposite the damage can hear sounds, but that is commonly the only sign. Damage to the primary auditory cortex on both sides of the brain is very rare without being fatal because the two pieces of cortex are so far apart. (Contrast this with the primary visual cortices, which are side by side.) If both primary auditory areas were destroyed, deafness would result.

Direct electrical stimulation of the primary auditory cortex causes a patient to hear simple sounds.

THE PRIMARY SOMATIC SENSORY CORTEX

Damage to a part of the primary somatic sensory cortex results in a loss of feeling in the corresponding part of the body. Because of a degree of duplication of function between the primary and secondary somatic sensory areas, the cortex as a whole is not rendered totally unaware of the signals being sent to it by an affected body location. It continues to be able to feel what is happening there even when that part's primary somatic sensory cortex has been completely destroyed. Such feelings though are comparatively crude; the cortex can still sense pain and pressure and cold, for example, but may have only a rough idea of where the sensations are coming from in the affected body part.

When widespread damage to the primary somatic sensory cortex occurs, the position of each part of the body with respect to the rest ceases to be known. This leads to a less obvious consequence for behaviour. It will be noticed that the primary motor area lies adjacent to the primary somatic sensory area. The two areas merge into one another in the section of the cortex that lies at the base of the central fissure, and signals are able to pass directly between them. The motor cortex uses signals from the somatic sensory area relating to body position to keep it informed of where the limbs it is controlling are, and to tell it what state of tension the muscles under its control are in. If the motor cortex is deprived of these signals it 'loses' the part of the body concerned, and the activity of muscles becomes uncoordinated. The cerebellum is of little use in such circumstances because its task is to put the finishing touches to movements that the motor cortex has signalled it intends to make. In cases of somatic sensory deficiency, the motor cortex doesn't know what movements to make because it doesn't know where its limbs are. To get round the deficiency the motor cortex uses an indirect route: it draws on the eyes and the visual cortex to ascertain where its body parts are. But the delay involved in using such an indirect route disrupts the smoothness and precision of any movements the motor cortex wants to make.

It also seems that performance of reasonably slow activities such as picking up a cup or tying a shoe lace is controlled not by the motor cortex but by the somatic sensory cortex. This comes about because the somatic sensory cortex is able to learn what the activity feels like. Then, when the motor cortex performs the action, the somatic sensory cortex can tell if it is proceeding as intended and, if not, can signal to the motor cortex to make corrections. In these cases, the motor cortex pulls the strings under orders from behind the central fissure.

Electrical stimulation of the primary somatic sensory area causes the patient to feel some sensation or other in an appropriate part of the body.

THE PRIMARY MOTOR CORTEX

Damage to any part of the primary motor cortex leads to a degree of paralysis of the muscles on the receiving end of its signals. The paralysis is not total because some of the signals being sent to the muscles originate in the pre-motor cortex and travel via some neural structures at the base of the two hemispheres. Consequently, damage confined solely to the primary motor cortex leads to weakness and a loss of precision control rather than to complete paralysis.

Electrical stimulation of the primary motor cortex causes simple limb movements to be made by the body part receiving the resulting signals.

THE SECONDARY VISUAL CORTEX

Damage to the secondary visual cortex can lead, amongst other things, to a situation in which sufferers can see well enough but cannot interpret the picture. They cannot recognize what they are looking at. Their position is rather analogous to people who have never learnt the Russian alphabet being asked to read a Russian sentence out loud. They can see each letter clearly but do not recognize what sound each one should be turned into. Those with damage to the secondary visual cortex can see the world but cannot identify the splodges of colour, the outlines and the contrasts that their eyes and primary visual cortex detect.

One part of the secondary visual cortex with very specific symptoms of damage lies on the cortex's underside where it curves inward at the bottom of the temporal lobes and forward-most parts of the occipital lobes. Damage here, particularly if it affects the right hemisphere, leads to loss of the ability to recognize faces. An afflicted person may be unable to recognize by sight even close relatives. Recognition by means of the sound of the voice is unaffected.

It can be noted here that signals from the primary visual cortex do not pass solely to the secondary visual cortex. If the secondary visual cortex was the only area to receive signals from the primary visual area then destruction of the secondary area would render the cortex as a whole blind. People with such damage would be unable to describe their visual experiences. Yet afflicted individuals most definitely can do that. They are not blind. They simply cannot make sense of what they clearly see.

Electrical stimulation of the secondary visual cortex causes a patient to see complex visual objects, e.g. dogs, tables, etc. From this it can be inferred that the secondary visual area has the function of recognizing objects on the basis of the processed primary signals it receives. How it does this is poorly understood. The act of recognition will be discussed further in section 7.2.

THE SECONDARY AUDITORY CORTEX

The secondary auditory cortex, which extends deep inside the lateral fissure as well as occupying part of the outer surface of the temporal lobes, presents a new type of situation: an asymmetry of function. The function of the secondary auditory cortex in the left hemisphere is different from the function of the same piece of cortex in the right hemisphere.

In a large majority of people — different authorities give different figures, but 90% can be taken as a rough guide — the right hemisphere specializes in music and other non-speech sounds, while the left hemisphere specializes in language. In the rest of the population, either these roles are reversed or, more commonly, both hemispheres share both roles. It will be convenient in future to talk about the majority as if they include everyone. You should bear in mind that when this or that asymmetric function of the left hemisphere is discussed, the unmentioned qualification exists that 'for "left" read "right" or "left and right" in a small minority of the population'. Similarly vice versa for the right hemisphere.

Damage to the secondary auditory cortex on the right produces, as might be expected, a loss of ability to recognize musical tunes or to be able to reproduce them. The latter loss probably arises because in normal circumstances the right secondary auditory cortex monitors the sounds the primary motor cortex is producing, compares them with what it expects to hear, and issues correcting signals as necessary. Its function here is similar to that of the primary somatic sensory cortex in respect of slow motor activities. If the primary motor cortex is deprived of auditory signals telling it whether its singing is on the right track, it can be expected to lose its way very quickly.

While damage to the right secondary auditory cortex may pass unnoticed in an unmusical individual, damage to the left secondary auditory cortex is always noticed. The hearing of people with such damage is unimpaired but they find that they cannot recognize what they hear if the sounds in question are words. Severe damage to the left secondary auditory cortex can leave afflicted people incapable of understanding speech at all. They hear the words but they mean no more than if they were spoken in an unknown foreign language.

Another symptom of damage to this area is impairment of the ability to speak. Sufferers may know what they want to say but cannot monitor what they are actually saying because they no longer understand the sound of their own words. In the normal course of events the parts of the motor cortex controlling the speech muscles receive a constant flow of signals from the secondary auditory cortex. In their absence it is not surprising that speech is disrupted.

Electrical stimulation of the right secondary auditory cortex may cause the patient to report hearing musical sounds, or sounds such as that of running water. Electrical stimulation of the left secondary auditory cortex may cause the patient to report hearing a word or a phrase.

THE SECONDARY SOMATIC SENSORY CORTEX

Damage to the secondary somatic sensory cortex causes a person to be unable to recognize objects by touch. This has a follow-on effect on the signals output from the motor cortex, causing it to be unable to control precise movements. When control of the lips and tongue is affected, speech may become distorted. Again, disruption of motor output is due to the motor cortex not receiving signals informing it that what it is doing is on course.

Patients whose secondary somatic sensory cortex is electrically stimulated report experiencing sensations over wide areas of the body, or feeling that they are touching some specific object.

THE PRE-MOTOR CORTEX

It seems that the pre-motor cortex is responsible for large-scale sequences of muscular action involving the concerted activation of many different muscles. It also plays a part in the performance of fast activities where movement is too rapid for signals from the primary somatic sensory cortex to be sufficiently up-to-date to guide the process.

Electrical stimulation of the pre-motor cortex causes a patient to make complex body movements. This is consistent with the notion that while the primary motor cortex attends to the minutiae of movement, the pre-motor cortex deals with the broader strategy required for purposeful actions. The primary motor cortex contracts muscles; the pre-motor cortex signals to the primary motor cortex in a way that ensures that those muscle contractions in different parts of the body add up to achieving the desired behaviour. Its role is akin to that of a coordinator.

SPEECH

It is very common for cortical damage to affect speech. This is because speech is a many-faceted activity which involves large areas of the cortex acting in concert, especially those located in the left hemisphere.

In its broadest sense, speech includes reading and writing, and damage to areas of the brain involved in speech frequently affects these abilities too. For example, if the damage involves impairment of the ability to understand spoken words, the ability to understand written words is also impaired. This implies that when people read words they form an auditory image of the words in the word-recognizing part of the cortex. Subjectively it does indeed seem, when we read, that we think the words concerned as if we have heard them.

Three cortical areas take part in speaking (excluding the primary auditory area, which is concerned with sound rather than speech). None of these areas is exclusively devoted to speech, but certain parts of each area are assigned a speech task to perform. The first of the speech locations lies adjacent to the left secondary auditory cortex at the back of the temporal lobe and just below the parietal lobe. It is called *Wernicke's area* after the neurologist who first located it. The second speech location — which is called *Broca's area* for similarly appropriate reasons — is found in the left pre-frontal cortex above the lateral fissure and adjacent to the pre-motor cortex. Both these areas are indicated in figure 6.4. The third speech location is the primary motor cortex of *both* hemispheres and is that part controlling the muscles of the mouth, tongue and larynx. It lies inside the lateral fissure.

The speech task carried out by each location is as follows. Firstly, the thought to be spoken is constructed in Wernicke's area. It is then transmitted to Broca's area. Broca's area fills out the words by, for instance, adding appropriate word endings (-s, -ed, -ing, -'s, etc.) and tells the primary motor cortex what muscles to activate. The activation of the muscles of vocalization by the primary motor cortex is carried out in a sequence which produces the desired sounds.

Damage to Wernicke's and Broca's areas leads to fairly predictable consequences. In the former case speech comprehension is impaired and what the sufferer says, while grammatically correct, contains inappropriate words or even nonsense words. In the latter case speech comprehension remains intact but what the sufferer says becomes disjointed, hesitant, lacks connecting words such as 'or', 'while', 'which', and contains grammatical faults. When the primary motor cortex is damaged, speech comprehension is again unaffected and, in speaking, the sufferer knows full well what to say, but lacks control over the means of saying it. Speech becomes slurred as a result.

THE GENERAL ASSOCIATION CORTEX

It now remains to consider the effects of damage to the two large areas that have neither sensory nor motor functions. The larger of the two is the general association area. Cortical damage is unlikely to take in the whole of this part of the brain without being fatal, since the area constitutes a large fraction of the cortical surface. Partial destruction produces a range of different symptoms depending on the precise location of the damage. Nonetheless, varied as the symptoms are, they lead to a broad conclusion that the area carries out the integration of the many sensory signals the cortex receives. Associations are made between signals arriving from different senses as well as between different signals arriving from the same sense. The location of the general association cortex, bordering the visual, auditory and somatic sensory areas makes it ideally positioned for this role.

It would be impractical to describe all the possible consequences of damage to various parts of this area. Only a few of the most remarkable disabilities that can arise will be mentioned here.

One part with an easily understandable function lies between the secondary auditory and visual cortices in the left hemisphere. Damage there causes impairment of three functions: reading (called dyslexia), the naming of visually perceived objects, and the drawing of named objects. It may be deduced that this part of the general association cortex links visual input with speech. If such a link was broken, the area where words are stored, Wernicke's area, would not be able to know what the visual cortex was looking at. Hence it could not cause to be spoken the word corresponding either to the observed object or to the observed written word. In the case of drawing a named object, the output of the motor system in producing the drawing needs to be continuously monitored by the visual cortex and checked against the memory in Wernicke's area of the name of the object. Once again impairment arises out of a breakdown in communication between the area of the cortex where the object is visually represented and the area which hears the name of the object. To further emphasize this point, it can be remarked that a person with a brain damaged in this way has no difficulty in copying a drawn object. Only the link between vision and words is affected.

Damage to parts of the general association cortex in the left temporal lobe is often linked to a loss of ability to reproduce strings of words or short stories. When this is the case, the individuals concerned find it difficult to hold non-trivial conversations despite their comprehension of each separate word remaining normal.

It is of course not surprising that damage to the general association cortex of the left hemisphere leads to impairment of word-using abilities, since it is in the left hemisphere that words are stored. What is more surprising is that other brain functions having no apparent connection with language are also located only in the left hemisphere. People with left general association cortex damage, especially in the parietal lobe, may, for instance, find they've lost their sense of direction; they may be unable to demonstrate an understanding of what 'more' and 'less' of a quantity mean; they may be unable to manipulate objects in their head (i.e. to picture mentally such manipulation); they may confuse left and right; they may be unable to work out how to put their clothes on; and they may not comprehend the difference between such phrases as 'the master's dog' and 'the dog's master'.

All these kinds of symptoms, which have as their common denominator the need to relate different things together, arise much more commonly following left hemisphere damage than following right hemisphere damage. It would seem that the general association cortex of the left hemisphere is where thinking of the problem solving variety —

problem solving being taken in a very wide sense — is carried out.

What then does the right general association cortex do? The sort of abilities that are strongly linked to this side, and which are impaired following cortical damage, concern perception of the body, especially of its left side, and perception of the left hand side of the world in general. The parts of the general association cortex in the parietal lobe are particularly implicated in these kinds of deficits. When the impairment is severe, sufferers may simply not acknowledge that anything, including their own bodies, has a left side. If told to write on a sheet of paper, they will write only on the right hand half of it; if told to put on a pair of gloves, they will use both hands to put on the right glove and then insist they have completed the task. In the latter case, if attention is drawn to their ungloved left hand, they may claim that it belongs to somebody else.

It can be concluded from these various examples of the symptoms of damage to the general association cortex that in both hemispheres it is involved in performing the highest mental functions. It is in this area that a person's many sensory perceptions, both current and acquired in the past, are brought into association with each other. The nature of that association is two-fold. Firstly there is identification of objects of all sorts. This processing is carried out in the area's temporal lobe parts. Secondly there is spatial association; that is, locating where everything is. This processing is carried out in the area's parietal lobe parts. (The reason why I have called this the general association cortex should now be apparent.) The role of this area of cortex in consciousness will be speculated on in the next chapter.

THE PRE-FRONTAL CORTEX

To end this survey of localized cortical damage, there remains the pre-frontal cortex to investigate. There seems here to be little distinction — Broca's area aside — between left and right hemispheres. Damage produces relatively subtle effects, manifesting itself typically as an ease of distractibility. Afflicted individuals will flit from one thing to another according to what has currently caught their attention. As a consequence, their ability to carry out other than short duration tasks is compromised if the environment in which the tasks are being performed is in any way distracting. These behavioural deficits show that one of the pre-frontal area's functions is to ensure that sensory signals arriving in the cortex (and also in the thalamus) which are irrelevant to the task in hand are prevented from interfering with that task. In this role the pre-frontal area is acting in an inhibitory way.

The other main characteristic of pre-frontal damage is loss of behavioural inhibition. Sufferers say and do things without displaying any concern for the reactions of other people. This can also be understood in terms of neuronal inhibition, in this instance of the pre-motor cortex and speech areas. Since the degree to which people inhibit their own behaviour is one of the main components of what is called their personality, the pre-frontal area can be considered to be the main location where personality is stored; that is, it is an area which plays a major part in making the behaviour of each person unique to that person.

Beyond these two functions, generalizing on its role in behavioural inhibition and personality, the pre-frontal cortex plays a key part in deciding what to do and what not to do, and in issuing commands to put those decisions into effect and see them through. It is, in a sense, a pre-pre-motor area. To carry out this command role it obviously needs to be constantly receiving signals from the general association area. Its place in the flow of signals through the cortex, in a greatly over-simplified form, can be written:

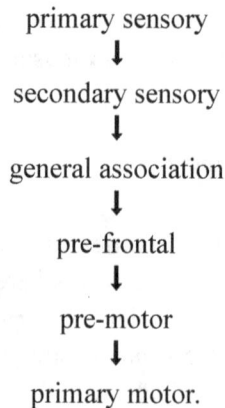

primary sensory
↓
secondary sensory
↓
general association
↓
pre-frontal
↓
pre-motor
↓
primary motor.

7

CONSCIOUSNESS

7.0 OBJECTIVE: To outline a physical basis for the subjective phenomenon of consciousness.

In the previous chapter, especially section 6.7, neurological events were shown to be related to various subjective experiences. This chapter will take the relationship much further and seek to account for the very existence of subjectivity as something which arises naturally from the way brains process signals.

In attempting an explanation of consciousness, we shall be concerned not so much with science as with philosophy constrained by science. As has been repeatedly stated before, science does not recognize the existence of anything that violates the laws of physics. This requirement is a non-negotiable scientific constraint imposed on philosophical speculation, and it applies to consciousness equally as much as to everything else. This is despite the fact that consciousness is a subjective phenomenon and hence cannot be objectively observed. There are, however, indirect observations. Specifically, consciousness manifests itself through purposive behaviour, and behaviour is objectively real and is required to be physically law-abiding. We shall accordingly consider one way in which it is possible to account for consciousness by drawing on behaviour and what it implies about signal processing within the brain. A number of topics will then be broached in the light of this account. This chapter will thus attempt to show how it may be possible for the subjective experience of reality to arise within a particular kind of objectively existing physical structure.

The discussion of consciousness given here amounts to one scientific view of the reality of the observer, seen from without and seeing from within. It is my personal view and you should not take it to be scientifically orthodox in the way that the previous chapters have been. It has as its basis some ideas due to Sommerhoff. (See Bibliography: Sommerhoff.)

7.1 WHAT IS CONSCIOUSNESS?

<u>SUBJECT AND OBJECT</u>

As was seen in section 6.7, damage to the brain can produce loss of specific components of consciousness — vision, for example, or language — as well as distortions to perceptions of what a person is conscious of. This evidence argues strongly for the hypothesis that consciousness resides in, and arises from, the physical brain. The challenge is to suggest how.

The search for a scientifically acceptable explanation of consciousness requires that it is first made clear what is meant by consciousness. The starting point for this task involves a return to section 2.1. There it was explained that there are two kinds of reality, subjective and objective. The relationship between them was expressed in the following way.

- Objective reality is the ultimate reality; the universe exists regardless of whether or not I do.
- Subjective reality lies within objective reality; I exist within the universe.
- An image of objective reality exists within subjective reality; I possess subjective images of the universe that I maintain correspond to how it truly is.

As will become clear, images are crucial to consciousness.

One of the things that can be said about most images of real objects is that they are given names. For example, the image of where a person lives is given the verbal label 'home'. But verbal labels are not limited to objects. There are also names for internal mental states such as happiness, anger, curiosity, and so on. These labels are subjective when used by people to describe how they feel. But what about a person using them to describe someone else? The labels are then used objectively. Do they mean the same thing in both cases? When I say 'I am happy,' I am happy. But when an observer remarks that I am happy, what is being referred to is how I am behaving and not how I feel. One word, two different meanings.

The point here is that consciousness itself is a word with two meanings, and to avoid confusion it is important from the outset to be clear which meaning is in use at any given moment. The subjective and objective meanings of consciousness are quite distinct.

When I say that I am conscious I am making a statement about subjective reality. The statement is an assertion that I know what is happening around me insofar as my various senses provide me with signals representative of those happenings. The signals from my sensory organs are perceived in relation to my own existence. (Without being perceived in this relational way I would not say that *I* perceive the happenings at all, though my senses and brain might.) This is what saying 'I am conscious' means subjectively.

When I say that someone else is conscious, on the other hand, I am making a statement about objective reality. In that case consciousness takes the form of observed behaviour; that is, only when sensory inputs to a person's brain affect motor output from it do I say that that person is conscious. Objectively, to be conscious is thus to be responsive to events in the world. This is clearly different from the subjective meaning of the word.

To illustrate the difference, imagine the following experiment. Take a volunteer and inject him with a substance that paralyses all his muscles. He is thereby rendered completely incapable of any movement. Next invite someone who is unaware of what has been done to come and determine whether the volunteer is conscious. The investigator will talk to the volunteer, may shake or pinch him, and will in general provide a number of sensory inputs for the volunteer to react to. Because the volunteer is paralysed he will not respond to any of them. His lack of responsiveness will lead the investigator into drawing the conclusion that the volunteer is unconscious. He is indeed objectively so. Yet when an antidote to the paralysing substance is administered, the volunteer will state when he recovers that he was fully conscious throughout the experiment.

The discrepancy between the investigator's and the volunteer's accounts of the experiment is due entirely to the fact that the meaning each attaches to the word consciousness is different. The investigator is using the word in its objective sense: 'The volunteer is not objectively conscious.' The volunteer is using the word subjectively: 'I am subjectively conscious.'

310

To avoid misunderstandings about which of the two meanings of the word is being used at any particular moment below, the convention will be adopted here that 'consciousness' refers to subjective consciousness only; 'responsiveness' will be used to mean objective consciousness. The assertion can now be made without apparent contradiction that the volunteer was conscious but unresponsive.

Defined in this way, consciousness is clearly a matter of outside events impinging on the conscious object, not of the conscious object impinging on events. It is a matter of inputs into the object, not of outputs from it. This contrasts with responsiveness, which consists of both inputs and outputs; i.e. an object is responsive if it behaves purposefully (as discussed in section 5.7) in the context of events external to itself.

The difference in meaning between responsiveness and consciousness raises the question of why the same word — consciousness — is normally used to mean both things. The answer lies in the relationship between them. Responsiveness is not identical to consciousness, does not require it nor even imply it, but nonetheless strongly suggests its presence. You and I know from experience that when we are conscious we are likely to behave in such a way that we appear responsive to an observer. Hence we conclude that responsiveness probably indicates consciousness. In normal circumstances the conclusion is very reliable, and that is why it is usually safe and unambiguous to use the same word for both subjective and objective forms of consciousness.

However, it is important to be aware that 'very reliable' doesn't mean 'certain'. Equating consciousness with responsiveness is ultimately an *assumption*. And furthermore it's an assumption which *cannot* be validated by objective means. Because consciousness is entirely subjective, there is no objective measurement that can establish whether or not an object is conscious. Only responsiveness can be objectively measured.

REQUIREMENTS FOR CONSCIOUSNESS

Having made clear what the word consciousness is being taken to mean, it is possible to list the properties which an object must have if it is to be capable of being conscious.

As a start, the point can be made that consciousness is not some sort of absolute state of being. It does not exist in isolation. The assertion 'I am conscious' contains within itself a hidden implication; namely: 'I am conscious of things.' Consciousness is necessarily consciousness *of*. The simplest things that I am conscious of are objects and events around me that my sensory organs react to. Hence I am conscious of what I am seeing, hearing, etc. This requires me to be able to process incoming sensory signals. That is, then, the first property.

- A conscious object must be a signal processing system.

But being conscious of objects and events is more than processing signals. It involves recognizing them; that is, storing images of them which can be recalled and compared. This is the second property.

- A conscious object, in the course of processing signals, must form internal images representing the sources of both present and past incoming signals.

This second property is demonstrably possessed by modern computers. While it is not possible to prove computers are not conscious, their responsiveness — that is, the way they respond — strongly suggests they are not. If this is correct, at least one additional property is required. That third property is self-knowledge: the ability to detect internally in the course of processing signals that there is a processor doing the processing.

- A conscious object must form an internal image of itself within the signal processing system; this image must not be of the system as an object, i.e. as just another source of incoming signals; instead, the image must be of the signal processing system as a possessor of other images which do represent sources of incoming signals.

Lastly, consciousness involves the continuous interacting of the images of external reality with the internal image of the self.

- A conscious object must be structured in such a way that interactions occur between the signal processing system's images of external objects and its image of itself; its knowledge of itself, by interacting with its knowledge of the world, gives it knowledge of its knowledge.

In summary, consciousness requires (1) a signal processing system which can (2) carry out internal processing and storage of input signals in such a way that it knows of the world, and (3) knows of itself, and (4) knows that it knows.

With the four properties required of a conscious system having been established, it is now possible to attempt a scientific speculation — the word 'speculation' needs to be stressed — about the neurological basis of consciousness. Although attention will be confined to the human brain, you should remain aware that the speculation may have a wider applicability. It should also be kept in mind that the developments proposed to occur as the brain matures are idealized: exactly what the developments are and how they take place can be modified in practice without compromising their value in principle. Finally, although some of the mental abilities leading to consciousness are attributed to children rather than babies, it would be wrong to take this as fixing an age of onset for consciousness. Exactly when a developing human passes from unconscious mental babyhood to conscious mental childhood is a contentious question. Discussion of this issue will be deferred until section 9.4.

7.2 THE PHYSICAL BASIS OF CONSCIOUSNESS

THALAMIC BEHAVIOUR

The newly formed (idealized) brain begins life with all its sensory neurones in place and functioning. 'Sensory' should as usual be taken to include not just senses external to the brain such as vision, body position etc., but also internal senses such as those concerned with, for example, monitoring the brain's blood supply. Sensory neurones thus include all neurones which begin signalling, or which change their rate of signalling, in response to anything other than signals from other neurones. 'Anything other' must be something external to the neurone itself reflecting some situation in the world, which includes the immediate environment of the brain within the body.

Motor neurones and muscles are also in place. A baby is born with a complete repertoire of reflexes and more complex behaviours produced by pre-determined neuronal wiring directly specified by DNA. These various at-birth behaviours do not involve the cortex. The behaviour of a baby born without a cortex — a rare defect — is initially quite normal to outward appearance. Only as the weeks pass and the baby fails to develop more sophisticated behaviour does its defect become apparent.

The absence of cortical involvement means that the baby's brain initially processes its neuronal signals in structures at the level of the thalamus and below. Except in an instinctive sense (see section 8.1) little if any of the brain's activities are learnt; i.e. arise as a result of previous experiences. The very structure of the neuronal connections, formed as part of the DNA plan, dictates the baby's behaviour. Its subcortical brain functions more or less like a system of interconnecting reflex chains, where the baby's response to what it senses is automatic. The baby's brain simply does what it is wired up to do. There is no sign here of consciousness as it has been defined above.

Despite its reflexive nature, the subcortical brain is still required to make decisions. It will be in receipt at any one time of many sensory inputs, all of which will be capable of causing output behaviour. This is because every sensory input signal must ultimately lead to one or more motor neurones. (Input into a closed loop of neurones which pass the signals nowhere but round and round is theoretically possible but biologically absurd.) If all sensory signals are capable of producing output behaviour, it is clearly necessary to decide which ones will actually do so. Without such decisions two or more separate unrelated but simultaneous sensory signals would each be able to initiate behaviour. Attempting to perform two or more motor actions at the same time practically guarantees failure to do either. A choice must be made as to which sensory signal will be permitted to produce motor output.

In addition to the thalamus there are two other parts of the brain which play a role in making the choice. They are the hypothalamus and the limbic system. These two structures are particularly concerned with sensory inputs relating to the inside of the body. The sensory input signals they process are generally known as need inputs. As an example of this kind of input, there are neurones which begin signalling, or which change their rate of signalling, when the amounts of sugar or amino acids in the blood fall to low levels. The need in this case is of course for food. Other need input neurones signal the need for water, the need to increase or reduce body temperature, the need to breathe, and so on. The three structures, thalamus, hypothalamus and limbic system possess neuronal wiring arranged in such a way that the various need signals can be evaluated and the most urgent attended to first.

At the start of life, before the vast bulk of the cortex has been brought into play, need signals always take priority over non-need signals from the external senses. Hence, when signals are being received, the choice of which ones will produce motor output is easily made. DNA specified neuronal wiring settles the issue. The motor output that results is always uncomplicated and often includes producing the general purpose need noise: crying.

The newly formed brain is thus an organ which can perform simple evaluations and make simple choices. But it is, for all that, no more than a piece of automatic machinery. It may have the potential for consciousness but that potential has yet to be realized.

CORTICAL LEARNING

Above all this activity in the thalamus, hypothalamus and limbic system sits the mighty but so far largely empty cortex. Signals from both the internal and external senses pour into it via the subcortical brain structures.

While signals from the internal senses are quite straightforward, those from the external senses are inevitably extremely complicated. Somehow, in ways as yet only poorly understood, the cortex is able to make sense of them. And thereby it begins to play its part in the behavioural development of the baby.

Consider, for instance, one of the first problems the baby confronts: how to recognize its mother. For many months, all (or almost all) its needs are supplied by her. When she appears in its vicinity its motor output may need to be changed to respond to her presence. Now, the picture of the mother that falls on the baby's retinas varies from moment to moment. Consequently so must the signals transmitted by the retinal neurones into the brain. The signals will be affected by how close the mother is, whether she is smiling or frowning, whether she is flushed with warmth or blue with cold, whether she is facing the baby or looking to one side, and so on. Change amongst these factors will obviously produce variations in the input signals received by the baby's cortex. And just to add to its troubles, the picture the cortex receives from one eye will differ slightly from the picture it receives from the other eye. Out of all these varying features the cortex must somehow extract the one thing they have in common, namely that they come from its mother.

Studies of the primary visual cortex (of various mammals) have given some indications of how the cortex goes about the task of recognition. The first step is to break the input signals down into a collection of simple signals. Neurones have been found in the primary visual cortex which fire in response to a straight line at a particular angle lying anywhere on the retina. Others fire when lines or spots move across the retina in particular directions, and yet others fire in response to particular colour contrasts. Every set of visual input signals received by the primary visual cortex is broken down into a collection of lines, angles, colour contrasts, movements, etc. The first step is thus one of analysis.

The second step, which takes place mainly in the secondary visual cortex, is synthesis: the recombining of all the simple signals coming out of the primary visual cortex. This must happen in such a way that the signals emerging from the synthesis — the signals which are passed to the rest of the cortex — are always the same in response to the same object regardless of that object's variable features. When the baby recognizes its mother, the signals that emerge from the secondary visual cortex will be a set of signals which only arise when the mother is being looked at. The production of that unique set of signals in the cortex is indeed exactly what recognition is. The baby can only be said to recognize its mother when the sight of her produces unique motor output on its part. Unique motor output clearly depends on unique input into the motor cortex. Hence recognition is equivalent to the passage between the sensory and motor cortices of a set of signals unique to what is being recognized.

The synthesizing powers of the various secondary cortices are something that can be inferred from objective observations, but how the secondary areas achieve this marvel is not yet known. In the case of mother recognition, despite its sophistication, the feat is one a baby is able to accomplish within three or four months of birth. Although she looks much like any other woman, the baby is able to distinguish from the analysis its cortex makes of her variable appearance, and from the subsequent synthesis it makes of the results of the analysis, whatever it is about her that is unique.

Armed with the ability to recognize, the cortex is also in a position to be able to match need inputs to recognizable objects in its environment. When it receives a need input signal, the cortex can recognize what it is that assuages the need. The cortex comes to know not only *that* it wants, but also *what* it wants. (As a word of caution, 'know' is not to be taken here in any sense subjectively.)

There is one object that the cortex can learn to recognize which stands in a special relationship to it. That object is the body in which the cortex is housed. It will be recalled that the primary somatic sensory cortex is directly connected to the primary motor cortex. Because of this, incoming somatic sensory signals can be expected to get passed to the motor cortex and produce motor output even before the somatic sensory cortex is in a position to do much recognizing. As it learns to analyse and synthesize its sensory inputs, the somatic sensory cortex will inevitably find that some of its inputs are closely related to earlier outputs. This is because signals it outputs to the primary motor cortex produce limb movements. And limb movements cause somatic sensory neurones to send signals to the somatic sensory cortex. A signalling loop is thus established around the somatic and motor cortices via movements of the limbs and other parts of the body.

This signalling loop becomes important when the baby has learnt to recognize what it wants. The baby will find, perhaps by chance, that particular output signals from the somatic sensory cortex can bring to it, by way of the motor cortex, the object of its need. In the same way, output signals can also keep the baby away from any object it needs to avoid. The cortex is then in a position to add to its knowledge of what it wants the knowledge of how to go about getting it.

CONTROLLING THE MOTOR OUTPUT SYSTEM

As already remarked, in the early stages of behavioural development need inputs always predominate over external sensory signals. The various needs are evaluated subcortically and the most urgent attended to first. In effect, the winner in the urgency contest takes control of the motor output system and excludes all its lesser rivals by means of inhibitory signalling.

In some ways, the production of motor output to assuage needs is the reverse of the recognition process. Sensory inputs start as a variable signal which is converted into a unique signal; conversely, motor outputs begin as a unique signal — a signal to grasp a particular object for instance — which has to be converted into a variable signal. This is because the final output signal must take into account such things as where precisely the object is and where the limbs involved are positioned at the start of the movement. Unique signals entering the pre-motor cortex are turned into variable motor signals by the time they leave the primary motor cortex. The sensory cortices are essential to this process since they alone can fill in the details of exactly where everything is, including of course where the parts of the body are.

Thus, specific motor output — as opposed to nonspecific output such as crying — cannot proceed without guidance from the external senses. This means that need inputs cannot directly take control of the motor system. What they do instead is to dictate to the cortex which particular sensory signals it should pass to the pre-motor area. In effect need inputs establish a behavioural goal. It is the cortex's job to use sensory signals to convert these goal-establishing need inputs into motor output. For example, if the need is for water, the pre-motor cortex cannot initiate drinking unless it knows where the water is in relation to where its body is. This knowledge is provided by the external sensory inputs.

Not all need signals, it should be noted, are concerned with what the body lacks or with avoiding what it has too much of. Muscle fatigue signals, for example, are also able to operate in a need input role. And of course pain signals are the most powerful need signals of them all.

While needs are present, the predominant one will determine motor output. But what about when the cortex is in receipt of no need inputs? When that situation arises, external sensory inputs continue to be received by the cortex, but it now lacks the need criterion by which it normally selects which of its many concurrent sensory inputs to pay attention to. The choice must now be made within the cortex itself instead of being determined from below. It appears that the cortex is able to make such choices innately. The baby concentrates its attention on the most novel feature of its environment. All irrelevant sensory inputs are excluded. The signals relating to the object of attention are then able to take over the motor system.

It seems plausible that when an external sensory signal acquires control over the motor output system it uses the same neural machinery as need inputs do to sustain its control; that is, it uses the same inhibitory wiring to exclude other irrelevant sensory inputs. The wiring can be expected to involve the subcortical brain because that is where need inputs come from. In this way the cortex can create its own substitute need input, a need arising entirely within the cortex. To successfully accomplish the motor activity initiated in response to a novel sensory input itself becomes a need.

The situation that the idealized young brain finds itself in at this stage in the discussion would be worth summarizing.

- Specific motor outputs are always related to external sensory inputs.
- Where an internal need is present, the selection of which external sensory input signals to pass to the motor cortex is determined by the need. For this to be possible, the cortex must learn to match needs to sensory signals.
- Where no internal need is present, the cortex itself selects the sensory signals to pass to the motor cortex on the basis of their novelty. It excludes all irrelevant sensory signals by taking over the need input parts of the brain to produce a purely cortical need.

With these abilities the baby's brain can recognize objects in its environment and knows what to do with them in response to its needs. It knows how to output motor signals to achieve particular goals. It also knows of the existence of its body, though that body is still just an object, albeit one with special features. None of this knowledge argues compellingly for the presence of consciousness in its possessor. It forms a necessary step on the road to consciousness but it would not appear to be sufficient to bring it about.

IMAGES

The analytic and synthetic properties of the sensory cortices enable the signals emanating from those areas to correspond to particular aspects of the world. Any given pattern of signalling neurones matches uniquely a specific (and recognized) perception. Such patterns may be spatial or temporal. The neurone signalling pattern that corresponds to the mother is presumably spatial; that is, a collection of neurones signalling more or less simultaneously. The neurone signalling pattern that corresponds to a particular tune is presumably temporal; that is, a collection of neurones signalling over a period of time as the tune progresses.

Since a particular neurone signalling pattern corresponds to a unique sensory perception, the signalling pattern can be considered to be an *image* of what it corresponds to. The signalling pattern which is activated when and only when some particular object is seen or heard or otherwise sensed represents that object inside the cortex. Functionally it is an image of the object.

(At the risk of triviality it should be stated that this neurone-based image is not an image in the way that, say, a photograph is an image. The neuronal image need not, and almost certainly does not, bear any spatial or temporal resemblance to what it is an image of.)

In the course of time, many images will be formed in the cortex. Each one will be a unique set of signalling neurones and each one will correspond to a specific perception. The neurones which constitute any given image will be caused to signal whenever the thing they are an image of is perceived. The thing, be it visual, auditory, tactile, or whatever, can then be said to have been recognized.

At any given moment many different things may be present in the cortex's perceptions. Each will have its corresponding neuronal image. Each will be perceived in relation to all the others. As indicated in section 6.7, integration of perceptions takes place mainly in the general association area. This part of the cortex can thus be thought of as containing a complex image of the whole world as it appears at any instant.

The neurones whose signalling corresponds to some perceived thing are not restricted to signalling only when that thing is actually present; they may also signal in the absence of sensory input. This can be inferred to be the case from the ability of people to describe a familiar thing despite its being entirely absent from their current sensory inputs.

Image formation is a crucial step on the path to consciousness. According to the requirements listed in section 7.1, a signal processing system which is conscious is required to possess images of the world and an image of itself as a perceiver of those images. The first of these two requirements has now been met. The cortex contains images of objective reality which it can relate to one another. These can be images of what it is currently perceiving, and can also be images of past perceptions which it can recollect.

But the other of the two requirements is still absent. The image which comes the nearest to being an image of the signal processing system itself is an image in the cortex of the body that houses it. This is an essential ingredient of its present image of the world and a frequently needed ingredient of recollected world images. The image is, however, of an object. The image of the self being sought is not this one. There is a way yet to go.

SUBCORTICAL EVALUATION

The ability of the cortex to generate needs on its own account is one of the great strengths of the human brain. It makes it possible for images of the objective world to be refined by means of exploration. The cortex is thus enabled to learn not only what it wants but what best satisfies its wants, and to learn not only how to go about getting things but to discover better ways to go about getting them.

But as it stands this is a rather dangerous development. Suppose, by way of an illustration, that a baby's visual cortex detects a novel object, a flame. Suppose further that the baby's brain is in receipt of no need inputs. The cortex is thus free to activate a standard set of input signals into the pre-motor area that it often uses for novel objects. The signals are an instruction to grasp. The motor cortex responds accordingly and the baby's hand closes on the object. Immediately, damage sensors in the hand initiate the withdrawal reflex (see section 6.4) and send strong pain signals to the brain. The subcortical brain receives these signals and duly passes them to the cortex. As need inputs go, none are more commanding than pain signals, and their sudden arrival forces the cortex to produce the motor output that corresponds to pain need inputs — crying. Eventually, however, the pain subsides and the cortex once again finds itself with no need inputs to attend to. And the flame hasn't gone away. It remains a novel object — even more novel given the unanticipated reflex response it was able to produce — so the grasping motor action is reinstated....

The danger should now be clear. A cortex which can pursue its own goals when there are no need inputs to attend to can act to the detriment of its body as well as to its advantage. What is required to overcome the danger is a means of ensuring that the cortex takes full account of need inputs that arise during the course of its behaviour.

When a cortical action goes towards meeting a need, the need sensing parts of the brain are able to reward the cortex by sending it excitatory signals connected to the need in question. These signals act to initiate or maintain the activity or to cause it to be repeated. Conversely, when a cortical activity goes against meeting a need, or indeed creates a need as in the flame example, the need sensing parts of the brain are able to punish the cortex by sending it inhibitory signals connected to the offending activity. These signals act to stop the activity or to cause it not to be repeated. Pain signals, as everyone knows, are especially effective in this latter respect. Thus the cortex not only learns how to achieve goals, but also what goals should and should not be attempted.

The danger of possessing a freely acting cortex is thus largely avoided by making sure it learns under the supervision of subcortical brain structures. Every action the cortex ever carries out is vetted by these structures in terms of whether it should be encouraged or discouraged. The cortex is never truly free of the tyranny of the subcortical brain.

Suppose the cortex carries out an action which produces an encouraging or discouraging response from the subcortex. The subcortical response will become associated with the sensory image — the unique sensory signal — which led to the action. Next time the object is encountered, one of the neuronal paths that will begin transmitting signals alongside the sensory image of the object will be a path signifying what happened last time. Was the action on previous occasions encouraged or discouraged? Once the baby has put its hand in a flame, then in future when signals emerge from the visual cortex that correspond to a flame, they will cause the associated what-happened-last-time neurones to transmit signals too — inhibitory ones in this case. The baby will recognize the flame and, additionally, will associate the flame with the pain-need response which resulted from grasping it. The what-happened-last-time neurones triggered into firing by the image of the flame will stop the grasping motor action before it begins. The cortex has acquired the ability to predict.

The predictions the cortex makes are an addition to the external sensory signals it receives. The neuronal pathway that signifies 'this is what happened last time' lies wholly within the cortex. It is not sensed from without but *added from within.* It constitutes a facilitating or discouraging addition to the unique set of signals that the sensory areas send to the rest of the cortex. And the importance of this to the development of consciousness is that it shows that the cortex isn't simply a processor of external sensory inputs. It can also identify, recognize and remember *internal* evaluations. Signalling pathways that serve these functions do not represent objects; they represent the brain's internal responses.

It can be noted, incidentally, that because the facilitating/discouraging additional signals lie within the cortex and do not involve the subcortical structures, the *memory* of what happened last time will produce qualitatively different behaviour from the *experience* of what happened. In the flame example above, for instance, the sight of a flame is not itself painful, so the baby won't cry in response to seeing it; but its memory of the pain will cause it to respond to the flame by avoiding it.

SCENARIOS

Think about what takes place (by inference) when a child — 'child' is now more appropriate than 'baby' — describes a scene that he (or she, naturally) can remember but which he is not presently experiencing. The neuronal signalling patterns representing the scene and all its components, in other words the images which make up the scene, are caused to signal. The cause may be someone asking questions, some other sensory input which excites the pattern, or some less obvious cause not necessarily associated with sensory input at all (subject of course to conservation of mass-energy).

The child may be asked to imagine moving about in the scene. He may for instance be asked to imagine walking down a familiar street describing what he sees as he does so. Anyone with a mature brain, child or adult, can do this sort of thing. It can be inferred that as the scene changes, so the neuronal signalling patterns that comprise the scene must also change. Recollected scenes need not be static.

To a certain extent, just as the scene was learnt and thereby became an image in the cortex, so the changing scene has also been learnt and become a changing image. However, this is not entirely adequate as an account of what is taking place. It is possible for the child to freeze the image at some location where he has never stopped in actual life. It is possible for him to inject other things into the changing scene which he has never seen as part of it. He is able to picture his street blocked by a fallen tree even if he has never known it to be blocked by one. And so on. The ability to manipulate images inside the cortex cannot be completely accounted for in terms of previous experiences with whole scenes.

Image manipulation can only be carried out if each feature of a scene can be called upon separately from any others. The neuronal signalling pattern that is the image of a fallen tree, for example, can be caused to signal whenever the image of any scene is being manipulated which requires a fallen tree. And that is so even if that particular scene has never contained a tree before. The general association cortex must be able to combine quite separate images together to achieve this. By such means it can not only recollect images of the world; it is also able to create imaginary images of the world by means of novel combinations of component images. The only constraint on cortical image manipulation is that each of the component images in its most basic form must have been previously experienced; the complete world image need not have been.

The manipulation of images within the cortex I shall hereafter refer to as *scenario enactment*. 'Scenario enactment' will always and only mean manipulating images entirely within the cortex. Similarly, the word *scenario* will always and only be used to refer to scenes — collections of images — within the cortex.

Much of the time, the child enacts scenarios as a part of deciding how to go about assuaging some need he is either experiencing or else expects to experience. In this situation a scenario will have an outcome. It will be judged successful if the need, as featured within the scenario, is satisfied when the scenario enactment has run its course; it will be judged satisfactory if no do-not-do-this inhibitory consequences arise within the scenario enactment. In scenarios of this type, which are presumably the first ones a child acquires the ability to enact, the enactor of the scenario must almost inevitably be a participant in it. These scenarios can therefore be expected to contain image components representing objects in the environment, components representing need inputs, and components representing the behaviour of the enactor.

A scenario, enacted wholly within the cortex, may produce

unsatisfactory consequences to the enactor within the scenario or else may simply fail to achieve the desired result. When this happens, other scenarios can be tried using different image components. This process of evaluating several scenarios — choosing a course of action — may take seconds or even hours. When a scenario is found that works, it may be physically put into operation. The real outcome of actual activity can then be stored as a new image in the cortex and itself used as an image component to make future scenarios more realistic and hence enhance their predictive power.

During scenario enactment no physical activity takes place. Yet the scenario enacting parts of the cortex are not separate from those parts which hold the actual real scene. This lack of separation is suggested by subjective experiences. For example, people find that visualizing a scene is easier with their eyes closed; closing their eyes cuts out real interference with the imagined scene. The difficulty people have in recalling one tune whilst hearing another is also a good example. In addition, brain scans of people seeing an object, and of people imagining seeing the same object, show that the parts of the cortex which are active in these two situations are largely identical.

If the scenario enacting parts of the cortex are also those parts where real scenes are processed, it would be expected that scenario enactment should feed signals into the motor cortex and produce behaviour just as real events do. But this is not what happens. It is possible to enact a scenario within which one moves while nevertheless outwardly remaining completely motionless. It can be inferred from this observation that during scenario enactment any signals passing from the enacting area (e.g. the general association cortex) to the motor cortex are inhibited strongly enough to prevent movement actually occurring. The source of this inhibition is probably the pre-frontal area. Damage to that area produces symptoms (see section 6.7) which do not conflict with such a hypothesis.

There are thus two ways in which scenario enactment differs from real activity. One is the absence of sensory input relating to the images in the scenario; and the other is the absence of somatic sensory signals reflecting the consequences of motor output. Just as it was remarked above that memory is qualitatively different from experience, so scenario enactment is qualitatively different from real activity.

Now, the two differences that distinguish scenario enactment from real activity must be manifested by neuronal signalling. There must be neurones whose signalling makes the activity in the scenario enacting parts of the cortex into a scenario rather than a representation of current objective reality; i.e. there must be neurones that signal only when a scenario is being enacted. These neurones will provide a substitute for the absent sensory input by feeding their signals into the scenario. They will also produce the absence of motor output by inhibitory signalling into the motor cortex. Both input-substitution and output-inhibition are functions that are required for scenario enactment, and the neurones that carry these functions out can be considered to correspond to the scenario as an entity. They are an image of the scenario: a scenario-image.

What are the attributes of the scenario-image? The image is not of an observable object. True, scenarios have an objective existence in that a neurosurgeon could theoretically watch neurones signalling when scenario enactment is in progress. But the scenario-image is not formed as a result of analysis of sensory input. It cannot therefore be equivalent to the perception of a sensed object; that is, to something perceived as being part of the objective world. It cannot be assigned any attributes of the kind that the cortex would give to objects. Accordingly it must be an image of something entirely distinct from the objective world, and that distinction will be its prime attribute.

THE I-IMAGE

To take the final step on the road to consciousness, the question must be addressed of whereabouts in the cortex the scenario-image neurones can expect to be found. The cortex has a number of functions to fulfil, and it has been determined observationally that these functions are not carried out equally by all parts of it; different areas of the cortex are specialized to carry out particular tasks. Do all of these parts need a scenario-image?

As a starting point it can be stated that the pre-frontal area (with its output inhibiting powers) and the general association area (where most scenarios are enacted) will certainly contain a scenario-image. They both need to be able to distinguish scenario enactment from real activity. But limiting scenario-images to these two areas is probably too restrictive. It seems reasonable that any part of the cortex which can form images — that is, any part of the *synthetic cortex* — will possess scenario-image neurones. Such neurones would enable the neurone signalling pathways that are the images in these areas to signal in the absence of real sensory input. They would also enable these same pathways to signal without producing motor output. On this reckoning the only areas in which scenario-image neurones would not then expect to be found are the primary sensory areas and the primary motor area. These are analytic rather than synthetic in function. Apart from them, it is possible that scenario-image neurones may be located in much of, or all of, the rest of the cortex.

Indeed there is even a possibility that the scenario-image extends outside the cortex to the cerebellum. Computerized brain scans show that the cerebellum participates in scenario enactment and so may well have within it neurones whose signalling distinguishes scenarios from actual behaviour. In the remainder of this chapter it will be taken that scenario-images are confined to the cortex, but it should be kept in mind that this may be doing the cerebellum an injustice.

The next question to ask is: how many scenario-images are there? Will there be one image for each functionally distinct part of synthetic cortex? The key here lies in the ability of each cortical area to produce neuronal signalling elsewhere. For instance, when I see a familiar object I not only recognize it (secondary visual area), I know what it is called (Wernicke's area), know what it feels like to touch (secondary somatic sensory area), know what it is used for (general association area), and so on. This ability clearly requires that the participating areas of cortex are extensively interconnected.

The extensive nature of intra-cortical interconnections will apply to scenario-image neurones as much as to neurones representing images of externally sensed objects. Given that scenario-image neurones permeate much of the cortex it can be proposed that, whatever part of the cortex is engaged in enacting some particular scenario, not only will the scenario-image neurones in that area be signalling; those in the rest of the cortex will be signalling too. This seems plausible because the scenario-image neurones carry out the same functions applicable to all scenarios. It makes no difference whether the scenarios are visual or auditory or more general in nature. Scenario-image neurones will always be firing so as to inhibit unrelated scenarios from starting up in other areas, and to prevent motor output taking place.

These considerations lead to the conclusion that when any one part of the synthetic cortex is engaged in scenario enactment, all the other parts will be in receipt of signals, not to do with what the scenario is about, but representing the part of the cortex carrying out the enactment — a sort of 'scenario in progress' signal. For those cortical parts not involved in the scenario, the 'scenario in progress' signal is tantamount to an image of the scenario-enacting part of the cortex. Scenario enacting area X's scenario-image becomes scenario enacting area Y's representation of area X. It is an image of another part of the synthetic cortex.

But as has just been remarked, when one area's scenario-image neurones fire, every area's scenario-image neurones fire. Hence scenario enacting area Y's representation of area X is functionally inseparable from Y's representation of all the scenario enacting parts of the cortex. This integration of the scenario-images across the entire synthetic cortex into a single unitary whole means the integration similarly of what the images are an image of. The synthetic cortex as an integral signal processing system thus comes to contain neurones whose signalling is an image of itself. *The scenario-image is an image of the synthetic cortex.*

This line of reasoning can be summarized as follows.

- Each functionally distinct part of the synthetic cortex needs a scenario-image in order to distinguish scenarios from real activity.
- When area X enacts a scenario, other cortical areas will interpret its scenario-in-progress signals as an image of area X.
- The huge amount of interconnectivity throughout the cortex means that scenario-images of the various functionally distinct areas integrate into a single image of the synthetic cortex as a whole.

As explained before, the scenario-image is not an image of a physical object. It is not therefore an image of the objectively existing physical cortex. It is instead an image of the cortex which has been sensed entirely *from within*. It cannot be described in objective terms and has no properties save that it exists. The verbal labels that accompany this image are 'I', 'me' and 'myself'. The possession of the image — I will henceforward call it an *I-image* — confers on the synthetic cortex the property of viewing its scenarios and images, not only in relation to each other, but in relation to itself as the possessor of those images. It enables the cortex to know that it knows, to perceive that it perceives.

And that is consciousness.

7.3 LEVELS OF CORTICAL ACTIVITY

AMPLIFICATION AND ATTENUATION

Consciousness is a subjective property of the synthetic cortex. It is not, however, an unvarying property. There are times when consciousness is absent or is experienced oddly. What is going on in these situations? In order to answer this question in neurological terms, the stability of cortical activity needs to be considered.

Picture a small block of cortical neurones. The block will be in receipt of incoming axons from neurones outside it. There will also be axons emerging from the block going to other parts of the cortex. Assume that the block of neurones has the same number of axons entering it as leaving it.

Imagine a signal arrives by way of the incoming axons for the cortical block to process. The overall signal may well involve a large proportion of the incoming axons emitting their neurotransmitters more or less simultaneously. The neurones in the cortical block will process the signal in the way outlined in section 6.3, some being excited enough to trigger off a pulse of voltage disturbance, some not. Very soon the axons emerging from the block will begin having pulses of voltage disturbance passing down them, thereby carrying the signal to the next piece of cortex for further processing and eventual output to motor neurones. For every piece of cortical tissue the picture is one of signals entering it, being processed inside it, and then leaving it.

This simple picture establishes no requirement that the proportion of outgoing axons carrying the signal is the same as the proportion of incoming axons carrying it. The number could be much higher or much lower. If the number is higher it means that that particular block of neurones has amplified the incoming signal, making it stronger. Other blocks of neurones downstream from the block under consideration will receive a magnified signal. If they magnify the signal in turn, then by the time processing is complete the signal to the muscles is likely to be positively convulsive. Conversely if the number of outgoing axons carrying the signal is lower, the signal will be attenuated. Other cortical blocks downstream will receive a weakened signal. If they in turn produce a similar attenuation, the signal may die out altogether during its processing and never reach the muscles at all.

While a degree of amplification or attenuation may be beneficial in certain situations, the cortex needs some means of ensuring that signals emerging from any part of it do so in the main with roughly the same strength as they enter with. The sensitivity of the cortex to input signals must be neither too great (amplification) nor too little (attenuation). The overall effect of inputs to the cortex — it must be stressed here that what is being talked about are overall effects — should be to excite the cortical tissue just enough to produce motor output.

THE MECHANISM OF CORTICAL EXCITATION

Signals input to surgically isolated blocks of cortex are always severely attenuated. On a larger scale, if the entire cortex is separated from the mid-brain and below, sensory signals entering it don't get very far before dying away. The cortex is rendered completely unresponsive to sensory inputs and completely incapable of initiating motor output. In effect it ceases to function. It can be concluded from these observations that the cortex, in part and in whole, needs to receive excitatory signals from the mid-brain in order to maintain cortical sensitivity at its proper level.

The mid-brain neurones responsible for exciting the cortex by the requisite degree form what amounts to a network of interconnections. The functional neural unit that they constitute is known as the *reticular formation*. The reticular formation is found not only in the mid-brain but also extends downward into the pons and medulla.

The reticular formation has at least four functions: it influences the strength of spinal reflexes; it helps to regulate breathing and the rate at which the heart beats; it can strengthen or weaken pain signals on their way to the cortex; and it controls cerebral excitation. It is the last of these functions which is of interest here.

How might reticular formation signals excite the cortex? The answer lies in the way neurones are caused to fire. Straightforwardly enough, the nearer a neurone's cell-body membrane is to the critical $-55\,\text{mV}$

threshold, the more likely it is to respond to an incoming excitatory signal. A neurone with a high membrane voltage will ignore incoming signals unless a lot of excitatory ones arrive at once; a neurone with a comparatively low membrane voltage will need to receive far fewer excitatory signals before a pulse of voltage disturbance is triggered off and the neurone fires.

In the absence of reticular formation input, experiment shows that cortical neurones have high membrane voltages. Signals entering the cortex are attenuated because the neurones ignore most of the signals they receive, passing very few on. The signals soon die out. Conversely, in the presence of reticular formation input, cortical neurones on average have membrane voltages close to about -60 mV. It can be concluded that the reticular formation excites the cortex by sending signals more or less continuously at just the right rate to lower the voltage across neurones' cell membranes from -70 mV to -60 mV. Sensory input signals to the cortex will then no longer be attenuated.

It needs to be understood here that even such a conceptually simple idea as excitation of the cortex is not nearly as straightforward as it sounds. In fact, the reticular formation doesn't excite all the cortex's neurones uniformly. Quite the opposite. While the average excitation increases, the majority of neurones are actually more inhibited. The way to think of this is to picture a blurred, dimly-lit scene with almost no contrasts — an analogy for an unexcited cortex. Then picture the same scene in sharp focus, with stronger illumination and much better contrasts, the brightest parts being brighter (excited) and the dark parts being darker (inhibited) — this is the cortex being excited by the reticular formation. Thus, cortical excitation is as much about highlighting the differences amongst signalling neurones as about making them more responsive to sensory inputs.

Given that the reticular formation excites the cortex, a good question to ask is: what excites the reticular formation? The answer is quite predictable. Since the reticular formation's role in cortical excitation is to facilitate processing of sensory input signals, it is not surprising that it is these same sensory input signals, making synapses in the reticular formation on their way to the cortex, which excite it so that it can excite the cortex in turn.

Interestingly, the wiring of sensory inputs to the reticular neurones looks decidedly casual. Axons carrying signals from all the different senses get mixed together, frequently forming synapses on the same reticular neurone, so that any one reticular neurone might receive signals from several senses. This means that when reticular neurones transmit signals all they are saying is that something is happening. It is impossible for them to convey information about what that something is because the signals which cause them to fire are so disorganized. Any neurone on the receiving end of a reticular neurone's signals can get excited, but it can't tell what it is getting excited about.

From the mid-brain, reticular formation signals travel upward to the thalamus where one of two things happens to them. Either they pass straight on to the cortex, or they make synapses in the thalamus where they are processed to some degree. The majority take the former of the two paths and end up forming synapses with cortical neurones distributed widely throughout the cortex. They show no preference for one cortical area over another and, because of their diffuse nature, produce a widespread reduction in cortical neurone membrane voltages. They do this without actually causing membrane voltages generally to cross the $-55\,\mathrm{mV}$ signalling threshold. This is exactly what is required if the cortex as a whole is to be excited.

Reticular signals which make synapses in the thalamus come to be associated thereby with sensory signals passing through the thalamus on their way to the cortex. These reticular signals tend to be passed to specific areas of sensory cortex rather than to the whole cortex. Nonetheless they retain their generally diffuse character and merely add to the excitation of the neurones in that specific area.

The mechanism of cortical excitation can thus be summarized as follows: sensory input signals excite the reticular formation to the extent that its neurones begin sending signals to the cortex. These signals are widespread and diffuse and serve to make cortical neurones more readily able to transmit signals in response to sensory inputs.

One important qualification has to be made to this account of cortical excitation. It is not strictly sensory signals that trigger off the reticular formation to excite the cortex but *changes* in sensory signals. At a cellular level this is not surprising, because it is a characteristic of most neurones (damage sensors excepted) to respond to changes rather than to things which are static. Behaviourally it is not surprising either. People don't want to be roused — have their cortex excited — because their skin is being touched by their clothes, or because their alarm clock is ticking. They only want to be roused if something changes. Only a changing situation is likely to require a reaction on their part.

This characteristic of the reticular formation ties in with one particular subjective experience. Because the reticular formation excites the cortex when sensory input signals change, it readily does so when something stops happening as well as when something starts. Say a clock in someone's room stops ticking. Cortical arousal is duly triggered off, causing the person concerned to suddenly become more alert, with a feeling that something has happened. However, because the reticular formation only tells the cortex something has changed without giving it a clue as to what, when a cessation of sensory input produces excitation the aroused person will not find the cause immediately obvious. Most people have this experience occasionally. They are suddenly alert, but are unaware why.

FEEDBACK

The input side of cortical excitation is only half the story, for there is also an output side. Excitatory input signals from the reticular formation must be of just the right strength to ensure that other signals passing through the cortex are neither excessively amplified nor excessively attenuated. Signals are output from the cortex back to the reticular formation to enable the cortex to ensure this is so.

These output signals give the excited cortex control over the excitatory signals it receives — a kind of control called feedback. What happens is that if the cortex begins to be over-excited it feeds back inhibitory signals to the reticular formation, so damping the incoming excitation down. Conversely, if the cortex begins to be under-excited it feeds back excitatory signals to the reticular formation, so stimulating the latter to increase its excitatory output. The process of cortical excitation consequently involves a two-way flow of signals passing between the cortex and reticular formation: an ascending input set (mid-brain → thalamus → cortex); and a descending output set (cortex → thalamus → mid-brain). The descending set enables the cortex to keep its excitation at the desired level.

One instance of the value of feedback control arises when there are few changing sensory input signals being received. There would then appear to be little to excite the reticular formation. However, feedback signals passing from cortex to reticular formation can take on a similar role, as far as excitation is concerned, to that of sensory input. Hence they are able on their own account to produce excitatory signals from the reticular formation to the cortex. These in turn produce feedback signals from the cortex to the reticular formation which in turn etc. etc. A cycle is thus established, and it is this cycle that enables the cortex to remain excited even when nothing much is changing in the environment. Signals from the cortex substitute for externally originated input.

In the absence of changing external sensory signals, the feedback nature of cortical excitation has as a consequence that if the signalling between cortex and reticular formation drops below a critical value, the collapse of the feedback loop follows rapidly. If too few signals enter the cortex, too few signals come back down, with the result that even fewer signals go back up next time around the cycle. Once the feedback cycle begins to collapse, it does so very quickly.

The converse is also true. If too many signals pass to the cortex, too many pass back down, and an even greater excess of ascending signals results. Normally the cortex is able to damp down any excess excitation before it gets out of hand. It does this by means of inhibitory signalling as already explained. But in some individuals this ability is compromised by structural or biochemical defects in the brain. When excessive cortical excitation occurs, the excitatory feedback loop quickly magnifies it to such an extent that the great bulk of cortical neurones are so excited they begin signalling all at once. An attack of epilepsy is the result. If it wasn't for the eventual exhaustion of the neurones involved in the attack, and perhaps an increase in inhibitory signalling within the reticular formation, the attack would last indefinitely. Fortunately the vicious circle is broken by these two means, usually in a couple of minutes.

UNCONSCIOUSNESS

Armed with an understanding of cortical excitation, it is now possible to answer the question asked at the start of this section about how consciousness can be variable.

In its most extreme form, variable consciousness is no consciousness at all. What happens when a brain is unconscious?

There are no structural changes accompanying unconsciousness. All the neurones remain in place and functioning. All the synaptic connections between neurones remain in being. What does change is, not surprisingly, the input to the cortex from the reticular formation, which is greatly reduced in quantity. Signals travelling through the cortex are thereby heavily attenuated and die away very rapidly.

The linking of unconsciousness with signal attenuation accords well with an objective basis for consciousness. Consciousness, if the definition of it given in section 7.1 is correct, must depend on neuronal signalling; neurones whose signals represent the internally sensed cortex must signal in conjunction with neurones whose signals represent some aspect of objective reality, thereby enabling the two representations to interact. When neurone signalling is sufficiently attenuated, cortex-representing I-image neurones will cease to signal and the interactions between them and other neurones will stop occurring. One of the requirements for consciousness will no longer be met. Unconsciousness results.

In summary, subjective consciousness ceases when there is an objective severe reduction in reticular formation input into the cortex. The absence of this input leads to strong neurone signal attenuation and a concomitant inactivation of the I-image neurones.

SLEEP

From the comments on feedback above it is clear that the reticular formation must also play the key role in sleep. Sleep is initiated by neurones located in the mid-brain which are intimately connected synaptically to the reticular formation. What happens is that these neurones begin to transmit signals which inhibit the reticular neurones, causing the feedback loop with the cortex to break down. As already remarked, this happens rapidly and accounts for the subjective and objective observation that the transition from wakefulness to sleep takes place very quickly.

As sleep begins, blood pressure drops, breathing decreases and the muscles relax, the last of these things due to the lack of, and also active inhibition of, motor signals from the cortex. During this time fairly strong sensory inputs are needed to excite the reticular formation because it continues to receive inhibitory signals from the neurones which initiated the onset of sleep.

After a variable amount of time, but usually within a couple of hours, a change takes place. Other groups of neurones located principally at the base of the mid-brain begin to strongly inhibit any signals that might otherwise be sent to the muscles, causing muscle relaxation to become very pronounced. At the same time the inhibition of the reticular formation becomes much reduced so that, from the point of view of excitation, the cortex wakes up; that is, it becomes excited. The subjective experience that accompanies this activity is dreaming.

An episode of dream sleep can last anything from a few minutes to more than an hour. Typically the longer episodes occur in the second half of a person's sleep period. When dream sleep ends, the strong inhibition of signals to the muscles (which prevents dreams from being physically enacted) declines to pre-dream sleep levels, the inhibition of the reticular formation returns and the cortex goes back to what may be called inactive sleep. In mature adults, the several periods of dreaming which occur during the seven or so hours of sleep each night normally take up about a quarter of the time.

There are a number of hypotheses about why the human brain needs to sleep. One possibility is derived from the reasonable suggestion that during wakefulness some cortical neurones will do a lot of signalling while others will do very little. Those neurones in constant use will gradually begin to adjust to repeated signalling by becoming less easy to excite and so less willing to pass on the signals they receive. (This is not to be confused with learning; see section 8.1.) The effect on the cortex of possessing numbers of increasingly recalcitrant neurones will be to produce a gradual deterioration in the appropriateness of the brain's responses to events in the environment. Such deterioration can certainly be observed in people deprived of sleep. Sleep may thus be a way of giving the heavily used neurones a rest.

Whatever the reasons why people sleep, when the objectives have been reached, the neurones inhibiting the reticular formation cease signalling, and some other neurones begin transmitting excitatory signals in their stead. Reticular formation signalling picks up, and feedback with the cortex ensures a rapid transition to wakefulness.

It may be remarked in passing that this daily cycle of wakefulness and sleeping indicates that the brain has at least one internal cellular clock. Not surprisingly, given the hypothalamus's wide range of regulatory functions, it is in the hypothalamus where this clock is to be found.

DREAMS

Unconsciousness and sleep occur when reticular formation signalling to the cortex is absent or at a low level. An increase in the signalling above a threshold value starts up the feedback loop between the reticular formation and the cortex and thereby brings about the activation of the cortex-representing I-image neurones and the onset of consciousness. This can have two distinct subjective effects. One is ordinary wakefulness; the other, which occurs during sleep as explained above, is that oddly experienced form of consciousness known as dreaming.

What happens objectively when dreaming takes place? An important observation is that sensory signals are strongly inhibited. This has the effect of releasing cortical signalling activity from the wakeful imperative of reflecting objective happenings detected by the senses. With no sensory input to impose order on the paths taken by the neuronal signals, the electrical activity in the 'awake' cortex is able to pass relatively freely amongst the cortical neurones. Random or semi-random patterns of neurones representing images of different components of objective reality can be caused to signal quite easily and, in combining together, can create complicated scenarios that have no counterpart in objective reality.

Accompanying this undisciplined signalling in the neuronal pathways which correspond to images of objects is signalling by I-image neurones. The I-image takes part in dreams and its involvement enables them to be experienced; that is, it makes them subjectively real. I am conscious in my dreams just as I am conscious when awake (though what I am conscious *of* differs radically in the two cases).

Thus it would appear that dreaming is normal cortical activity divorced from sensory inputs. But there is more to dreaming than that. Wakeful scenario enactment is also cortical activity divorced from sensory inputs but, needless to say, scenario enactment is significantly dissimilar to dreaming. The following differences spring readily to mind. When I enact a scenario I know that that is what I am doing, whereas when I dream I do

337

not know I am dreaming. When I enact a scenario I can remember afterwards having done so, whereas when I dream I usually cannot afterwards remember what I was dreaming about. Finally, when I enact a scenario I choose the images to be included in it according to particular criteria, whereas when I dream, the images that arise are at best only loosely related and are not ones I choose.

Dreams are thus like scenario enactments but with some of the ingredients of the latter missing. The neurological equivalents of these subjectively missing ingredients are largely not obvious, though in one or two instances there are clues. At a neurological level, brain scans reveal that cortical excitation is different in dreaming than in wakefulness. For instance, in dreaming the limbic system is more active, whereas some parts of the frontal lobes are less active. And failure to remember dreams is probably due to a reduction in signalling by neurones using the adrenalin family of neurotransmitters, and may be associated with the hippocampus, one of the component parts of the limbic system.

But these clues don't amount to an explanation. None of the theories concerning the function of sleep explain satisfactorily why dreaming occurs. It cannot even be said for certain whether dreaming is necessary of itself or is merely a harmless side effect of some other necessary process. Dreams are a puzzling phenomenon.

7.4 VOLITION

VOLUNTARY VERSUS INVOLUNTARY

Volition may be regarded as the behavioural counterpart to thinking. It is the doing that accompanies or follows scenario enactment. Subjectively, the activity which takes place — a person's behaviour — is described as being voluntary. But there is also involuntary behaviour. What is there to say about the neurological equivalents of these things?

Behaviour, both objectively and subjectively, can be defined as any change in the position of a part of, or all of, the body by means of its internal musculature. The subjective epithet of voluntary or involuntary which is added to this definition is chosen according to how the behaviour is perceived by the behaver. If a person perceives behaviour as having been self-caused, then it is voluntary. If the behaviour is perceived as just happening, it is involuntary.

In terms of images, voluntary behaviour involves the participation of I-image neurones in the signals being sent to the motor cortex from relevant parts of the synthetic cortex. Such behaviour is experienced in a context which includes the world and the self, as represented by I-image neurones. It is behaviour 'I' bring about. Involuntary behaviour, on the other hand, does not involve I-image neurones, nor generally speaking the cortex at all. It cannot therefore be experienced as something 'I' bring about; people can experience the effect of involuntary behaviour but they do not experience the cause of it because they — their I-image neurones — are not involved in the cause. All reflexes are examples of involuntary behaviour, as are the actions, for instance, of the muscles controlling the size of the pupils of the eyes and some of the muscles involved in the sex act.

It is significant that both involuntary and voluntary behaviour can sometimes be caused during brain surgery. If the surgeon electrically stimulates the primary motor cortex, muscle movement ensues. Patients in this situation may raise an arm, for example. But although they are aware of the movement, these patients do not consider it to have been brought about by them, and so regard it as involuntary. ("You made me do it," they will say to the surgeon.) This accords well with the view expressed in section 7.2 that the motor cortex should not be expected to contain I-image neurones. Without the participation of these neurones, any cortical signalling will be seen in a context which excludes the I-image and cannot as a result be experienced as a voluntary act. On the other hand, electrical stimulation of the secondary somatic sensory cortex may lead patients to report that they want to perform some action, and any movement they then make in consequence of their desire will be perceived by them as voluntary. And electrical stimulation of some parts of the pre-motor cortex may cause patients to carry out some movement and to claim the action resulted from a decision they made. These observations again accord with the proposed distribution of I-image neurones.

It is interesting to note that certain kinds of brain damage can convert what was previously voluntary into involuntary behaviour. One instance concerns damage to the parietal part of the general association cortex of the right hemisphere. Damage to this area may result in affected patients losing some or all of the left side of their body, 'losing' being in the sense of no longer considering their left side belongs to them. The behavioural effects of this were mentioned in section 6.7. It seems that patients continue to be aware of their left side but only as an object. Presumably when it obeys their wishes they regard it in much the same way that they regard involuntary behaviour; left side behaviour is something they perceive as just happening. It can be concluded that the general association area is where the image of the body and the I-image interact with one another and that in the absence of that interaction there is no consciousness of the body.

(It can be noted in passing that left hemisphere damage does not usually produce loss of the right hand side of the body in the way that right hemisphere damage produces loss of the left. This suggests that the image of the right hand side of the body is present in *both* hemispheres, while the image of the left side is found only in the right hemisphere. It is not clear why this should be so.)

It might be argued that to some extent involuntary behaviour is consciously controllable. People may be able to make their pupils dilate, for instance, by thinking appropriate thoughts. Neurologically what happens is that signals from the synthetic cortex pass to the subcortical structures where control of involuntary behaviour resides. There they form synapses with subcortical neurones, causing them to emit signals which duly bring about the desired event. But the point here is that the event is brought about by the exerting of influence rather than control. Dilated pupils are perceived as an involuntary by-product of thinking the thoughts and not as a voluntary consequence of doing so.

It can be concluded that the difference between voluntary and involuntary behaviour depends on which parts of the brain directly initiate motor activity. Initiation by the synthetic cortex, with the participation of its I-image neurones, is subjectively voluntary; initiation by any other part of the brain is subjectively involuntary.

SPONTANEITY

Subjectively the thoughts which lead someone to initiate a piece of voluntary behaviour may sometimes apparently come from nowhere, arising spontaneously. Objectively it would not be expected that spontaneous events occur in the brain. The only instances of spontaneity in physics are quantum events and these take place on scales many orders of magnitude smaller than those of neurones and synapses. What then can the neurological equivalent of subjectively spontaneous behaviour be?

Once again it is a question of involvement of the synthetic cortex. As long as neurone signals avoid synthetic areas of the brain, they will not activate the I-image and so will not be experienced. When they do enter the synthetic cortex, it is where they come from which determines how they are perceived. Signals entering the synthetic cortex from the primary sensory areas and the need sensing limbic system are connected with what has been sensed, so these signals give rise to thoughts and behaviour which are not considered spontaneous. Signals entering the synthetic cortex from non-sensory parts of the brain will also give rise to thoughts and behaviour, but in this case the thoughts and behaviour will subjectively seem to be spontaneous; that is, they will not be connectable to any sensed cause and may not relate to any immediately previous thoughts. Subjective spontaneity is thereby a consequence of the restriction of I-image neurones to the synthetic cortex.

It is fascinating to find that subjective spontaneity can be detected objectively by using an encephalogram. The classic experiment involves volunteers being asked to perform some simple motor activity, such as pressing a button, but not to do it until they feel like it. Their conscious decision to initiate the requested motor activity is taken about a quarter of a second *after* an electroencephalogram detects changed electrical voltages in the frontal lobes. In other words, an experimenter watching the encephalogram knows when a volunteer is going to take action before the volunteer makes the conscious decision. In this particular experiment, the input triggering off the spontaneous thought in the volunteer's mind starts in an area of the frontal lobes known as the supplementary motor area. This is a piece of pre-motor cortex which lies mainly on the cortical surface between the two hemispheres. Input that initiates activity in the supplementary motor area presumably originates in a location which contains no I-image neurones. Hence, to the volunteers, the moment 'when they feel like it' is experienced as being spontaneous. Spontaneous thoughts are not a matter of volition; it is what happens to those thoughts after they have arisen that is a matter of volition.

341

SPLIT MINDS

'Schizophrenia' is a near literal translation into Greek of the words 'split' and 'mind'. The symptoms of this disease are varied but they frequently include two that have a bearing on volition. One is a feeling that thoughts are being controlled by some outside agency; that they go beyond mere spontaneity and seem to compel behaviour without the sufferer's consent. The other is the experiencing of hallucinations, especially hearing voices which are not under voluntary control.

At its most basic level, schizophrenia is probably due to biochemical faults in neurotransmission or signalling, but could it be that the overall effect of these faults is to prevent the proper integration of the I-image across different cortical areas? If a part of the synthetic cortex passes signals to motor areas of cortex without the I-image in non-involved parts of the synthetic cortex being aware of it having done so, the resulting behaviour would be experienced in the uninvolved cortical parts as taking place without their consent. And to have a fully enacted scenario come into one's mind without being aware that 'I' enacted it would conceivably be experienced as a hallucination, quite probably one with a pronounced voice component. The whole subjective nature of volition would be compromised by a fragmented I-image.

Fragmentation of the I-image can also occur in a different way. In this second case, the fragmentation takes the form of a precise disconnection between the two cerebral hemispheres. This two-fragment split is the result of either a defect — a very rare one — in brain development which results in the corpus callosum not forming, or a surgical operation called commissurotomy which severs the corpus callosum. (The corpus callosum, it will be recalled, is the main connection between the left and right halves of the cortex; see section 6.6.)

The corpus callosum contains at least a hundred million axons passing from the left to the right hemisphere and a matching number passing in the opposite direction. It is the main route by which the two hemispheres 'talk' to each other. When these axons are absent or cut, very little direct inter-hemisphere communication is possible.

Not surprisingly, clinical tests on patients whose brains are split in this way show that both hemispheres are responsive largely independently of

342

each other. That implies, as would be expected if I-image neurones on each side are unable to initiate signalling on the other side, that each hemisphere is independently conscious; i.e. has its own I-image. And that in turn implies, since the I-image plays the key role in volition, that each hemisphere also experiences volition separately. Which means the left hemisphere regards behaviour it initiates as voluntary, and behaviour the right hemisphere initiates as involuntary; and the right hemisphere has the same view identically vice versa.

However, it is best not to get too carried away with the idea that there can be two separate consciousnesses inside the one head. Both sides of the cortex continue to share possession of the various subcortical parts of the brain, including especially the reticular formation. They also inhabit the same immediate environment and receive the same internal sensory signals. And they can communicate indirectly by means of producing motor output for each other to detect. So the degree to which consciousness and volition can diverge between the two sides of the cortex when they are out of direct contact with one another is necessarily limited.

MIND OVER MATTER

It is conceptually possible to go beyond volition as the initiator of behaviour and speculate that it may extend to other things than muscles. This notion can be encapsulated by the phrase mind-over-matter. ('Mind' here can be taken to be synonymous with consciousness.) The idea behind mind-over-matter can be expressed as an assertion; namely, that it is possible to bring about objective events independently of muscular movement by doing no more than thinking strongly about them. For some people this possibility is part of their subjective reality, or else the subjective reality they ascribe to certain other people. Can it be accommodated neurologically?

The idea of having conscious control over objective events by non-muscular means probably arises from the way the body reacts to thoughts. If people so decide, their body does what they want it to do without the need normally for assistance from other objects. Objectively, neuronal signals to the muscles, by a chain of biochemical causes and effects, move the body. Subjectively 'I' do it. Hence I have control over the voluntary muscles of my body — a straightforward equivalence between experience and neurology. Mind-over-matter takes this experience and extends it in either or both of two ways: to events internal to the body and to events external to it.

Taking the more restrictive of the two ways first, the suggestion is that it is possible to bring about non-muscular events in the body by thinking about them; that is, by enacting scenarios in which the desired events occur.

It is not hard to find a potential neurological basis for this proposal. The biochemical processes taking place in the body are greatly influenced by chemical emissions from the brain, particularly those arising from hypothalamic control of the pituitary gland. If these emissions can be influenced by the synthetic cortex in the same way that it can influence involuntary behaviour, then experiences of things such as psychosomatic illnesses and faith healing can be accounted for neurologically. With caution, observation suggests the synthetic cortex does have this power of influence, though the biochemistry involved is mostly unknown. People therefore probably have more influence over events in their bodies than purely muscular considerations would suggest.

It remains to extend mind-over-matter to events external to the body. Here are found such things as levitation, telekinesis, people bending pieces of metal without the use of physical force, and so on. It is extremely difficult to conceive of any neurological connection to these kinds of events which is compatible with the view of subjectivity elaborated in this chapter.

In the absence of a connection, the objective reality of the supposed events must be severely doubted. Observationally there is no rigorously obtained evidence which makes mind-over-external-matter plausible. Furthermore such events almost inevitably violate some law of physics, usually conservation of mass-energy. The objective brain cannot be expected to violate physical laws. Therefore, according to the view taken here of the basis of subjective reality, neither can 'I'. Given that objective reality and a person's subjective image of it may not correspond, it can be concluded that this external sort of mind-over-matter rests on self-deception or fraud. It is not objectively real.

7.5 OBJECTIVITY AND SUBJECTIVE EXPERIENCES

It cannot be denied that between the objective neurological equivalent of some subjective experience and the experience itself there is an enormous gulf. The dichotomy between the objective account and the subjective experience is qualitatively unbridgeable. Its sheer size highlights the unsatisfactory nature of any account of consciousness based on objective events (as a scientific account must be). In defence of science it can only be pointed out that unsatisfactory does not, of course, imply wrong.

As examples, here are two types of experiences — those concerned with sensations and those concerned with emotions — matched with the objective neural activity that must produce those experiences.

<u>Sensations</u>. The issue here is that I experience sensations differently from one another. Take visual experiences as an example. The neurones of the visual cortex are little different from the neurones anywhere else in the cortex; their axons are the same, their synapses are the same, their signals are transmitted by the same means. And yet signals in the visual cortex are experienced in a way that is utterly different from the way that signals in, say, the auditory cortex are experienced.

The only plausible way to account for these subjective differences is to speculate that they are due, not to differences in the neurones themselves, but to the way neurones process their signals. The pattern of interconnections amongst the neurones of the visual cortex must be unique to the visual cortex and determines both how visual data is processed and how it is experienced.

Currently little is known, in theory or from observation, about the nature of the different kinds of interconnectivity utilized by different parts of the brain. The only exception to this state of ignorance is the cerebellum, the principles behind its wiring being at least partially understood. But ironically, the cerebellum is an organ whose involvement in subjective experience is an open question.

The different experiencing of signals in different parts of the synthetic cortex is only one of the problems in this area. The dichotomy between subjectivity and its neurological equivalents goes further than that. Within one kind of experience there are many different sensations. Taking vision as an example again, there is a subjective difference between looking at a green light and looking at a red light: the two colours are perceived to be different. What possible distinction can there be between neurones which signal when something green is looked at and neurones which signal when something red is looked at? It seems most unlikely the difference is down to the patterning of neuronal interconnections. Perhaps it is simply the destinations of the axons concerned. Axons signalling 'red' make synapses in slightly different places to those signalling 'green'.

And there is also the matter of *what* sensations are experienced. It has been pointed out that neuronal signals must enter the synthetic cortex to be experienced. But that cannot be the sole criterion. Many signals entering the synthetic cortex — the phenomenon mentioned in section 6.7 of thalamic signals giving rise to blind-sight is an example — are not subjectively experienced. Why not? Again, is it the pattern of the neuronal signals that is significant? Is pattern an additional criterion of whether an input signal is experienced or not? Just as it has been proposed that signals patterned in a particular way are experienced as vision, and that others are experienced, for instance, as sound, are still others patterned in a way that isn't experienced at all?

Emotions. Emotions are a range of experiences which accompany a disposition (or which are a disposition?) to act in particular ways. If I wish to cry — whether I actually do so or not — the associated emotion is sadness; if I wish to run and hide, the associated emotion is fear; if I wish to smash something, the associated emotion is frustration; and so on. Some emotions accompany more wide-ranging behaviours: love, greed, ambitiousness and envy, for example.

Emotion is something sensed entirely subjectively and that makes it quite unlike things which also have an objective existence. If I sense something is red, it is possible for someone else to make an objective measurement of electromagnetic waves of the appropriate wavelength. But if I sense I am angry, there is nothing external to the brain that can be measured. Someone can tell me I am definitely looking at something red, but no one can tell me I am definitely angry. They may measure tell-tale physiological signs of anger such as raised blood pressure, but only I can say whether or not I am angry. Emotions are things which exist entirely internally within the brain.

Because of this, it is in non-sensory inputs to the cortex that a neurological equivalent of each emotion must be sought. There must be some objective happening which takes place inside the brain when, and only when, any given emotion is experienced. There are several aspects of neural science that are relevant here.

Firstly there is the role of the subcortical brain. The deliberate damaging of subcortical structures in various mammals can affect their global behaviour, for example changing an aggressive animal into a docile one. Extrapolating to human beings, such a change in behaviour would indicate an alteration in the subjective emotional state of the person. In a similar vein, electrical stimulation of subcortical structures, including particularly parts of the limbic system, can produce behaviour in mammals which would, if occurring in humans, indicate a range of emotional states.

Secondly there is the role of the general association cortex of the right hemisphere. One of the symptoms damage to this area can produce is unconcern about, or even a failure to notice, other symptoms of the damage (such as those described in section 6.7). This contrasts with other types of brain damage which cause the sufferer to experience the emotion known as depression. The lack of emotion expressed in some cases of damage to the right hemisphere's general association area clearly links that area with the experiencing of emotion.

Thirdly there is the role of the pre-frontal cortex. This area seems to be responsible for adding the emotional content to pain signals. The evidence for this is that severance of the neuronal connections between the thalamus and the pre-frontal cortex (an operation sometimes performed on patients with chronic pain arising from fatal illness) leads a patient to be aware of pain but not troubled by it. Pain is still experienced but it no longer hurts.

Fourthly there is the role of neurotransmitters. Certain drugs which affect emotional states are known to interfere with the mechanism of neurotransmission in one way or another.

Somehow these four kinds of observations contain within them the neurological equivalents of emotion. Quite possibly, different emotions are associated with different objective mechanisms. The emotional aspect of pain, for instance, is not surprisingly connected with signals from the pre-frontal cortex. That area, as was seen in section 6.7, is primarily inhibitory. Pain, too, is primarily inhibitory, bringing about a marked decrease in a person's ability to think and do. Other emotions, especially of an extreme nature, such as fury, terror and joy, are clearly associated with the limbic system. Emotions such as depression and elation, on the other hand, are more likely to be due to disturbances from equilibrium of neurotransmitters in the cortex. But the point about every one of these various objective mechanisms is that none of them comes remotely close to conveying what it is like to experience emotions. Subject and object are far apart.

These two wide-ranging forms of experience — what it feels like to discriminate between one sensation and another, and what it feels like to have emotions — should have made you aware how subjectively unsatisfying the objective account of consciousness is. This is not so much in terms of consciousness itself as of the experiences attendant on consciousness. And the issues raised above are not the only ones. More will be encountered in section 9.4. It can only be asserted, not very helpfully, that if, as an observer, you have ever wondered what it must be like to be a signal processing system of a hundred billion extremely interconnected neurones, well, you know the answer. Subjective experience is just what being a human brain feels like.

In view of this inadequate state of affairs, it would be useful to list three statements which must be true for any scientifically acceptable account of consciousness.

- Every subjective experience matches some objective event in the brain, e.g. a pattern of neurone signals, a flux of neurotransmitters, etc.
- Every event, without exception, which occurs in the brain conforms to the laws of physics.
- Consciousness arises because self-knowledge is a natural consequence of the structure of the brain.

The explanation of how consciousness comes about that I have advanced in this chapter is one I find fully compatible with my own subjective experiences. It is not a complete explanation. Nor may it be called a scientific theory; it lacks predictive power and precise enough detail for that. It is a speculation constrained by scientific knowledge. The accompanying speculations on the objective basis of various components of subjectivity I also find compatible with my subjective reality, though they admittedly and inevitably leave much to be desired. I have in the course of speculating made many assumptions, some of which you may disagree with. You must decide for yourself whether the overall tone of the speculation is reasonable in the light of your own subjective experiences. It is in the nature of subjectivity that this must be the case.

8

IDEOLOGY

8.0 OBJECTIVE: To discuss the nature of ideas, their constitution, their classification and their behaviour.

Now that we have put forward a scientifically based view of the observer, our next task is to look more closely at the images that make up the observer's view of reality. Those images, combined in particular ways, form ideas. Ideas are the units out of which any view of reality is built.

There is little in the way of a science of ideas — a science which I shall call ideology. ('The science of ideas' is the original meaning of the word 'ideology' before it started being used to describe particular sets of beliefs, e.g. communist ideology, Christian ideology. In this book, the word will always mean what it meant originally.) The reason for this lack of scientific data is probably that ideas are dependent on the existence of people and cannot be studied in isolation from them. It is consequently virtually impossible to perform experiments on ideas or, come to that, to make measurements of them. Even collecting systematic data about them is difficult. Ideas would thus seem to be more a matter for philosophy than for science. It would certainly be fair to describe section 2 onwards of this chapter as primarily philosophical. But I contend that a complete account of reality *must* include an account of ideas, since ideas are the units out of which that very description of reality is built. This chapter accordingly seeks to put forward a basic (and largely speculative) account of ideology, drawing on objective observations of ideas as manifested in human behaviour.

As with consciousness, the views expressed in this chapter, other than in section 8.1, are my own. It is for you to decide if they are reasonable.

8.1 LEARNING AND MEMORY

<u>DEFINITIONS OF SOME TERMS</u>

Much of the truth of chapters 7, 8 and 9 rests on the ability of events external to the brain to alter neuronal signalling pathways. When an image is formed in the cortex, a neuronal signalling pathway must be established which did not previously exist. That pathway constitutes the image. So far nothing has been said about how these paths are created. What biochemical processes take place to make a signalling pathway? Before delving into ideas, it would be as well to look at what is known of how neuronal signalling pathways come to exist. Ideas are, after all, in a sense made of these pathways.

The first step to take is to define some terms. An obvious word to begin with is *learning*. In this book, learning will be taken to mean the storing of information in such a way that future behaviour may be modified by it.

This definition calls in turn for a definition of *information*. In the context of ideas, information can be defined as some arrangement of matter that is related in a non-fortuitous way to an arrangement of matter at some earlier time; the later arrangement (which includes neuronal signalling pathways) is information about the earlier arrangement (which includes events external to the cortex). The creation of a signalling pathway brought about by some preceding external event is equivalent to learning (about), or storing information about, that external event.

Another word to define is *symbolize*. When one arrangement of matter is related to another, there is no necessity for the relationship to involve resemblance. Where there is no resemblance between the information and what it is information of, the former is said to symbolize the latter. For example, the markings used on maps to denote various geographical features — churches, forests, etc. — are symbols; they relate to, but do not resemble, the things they symbolize. The neuronal signalling pathways of the cortex are similarly symbolic.

A symbol may be converted into the thing symbolized and vice versa. Also, a symbol may be converted from one form to another. Conversions of both these kinds will be referred to as *translations*. One example that has already been encountered is the translation of DNA base sequences into proteins; DNA base sequences symbolize proteins. Another example is the translation by a DVD player of digital symbols stored on disc into images on a screen; the arrangement of the laser-read pattern on the disc symbolizes the images. The human brain is also able to translate symbols, as any cook working from a recipe, or any secretary converting the symbolic sounds of speech into writing can testify.

It is important to note that for symbol translation to take place there must be a translator; i.e. a piece of machinery capable of bringing the translation about. The human brain is just such a translator.

A final word to define at this time is *memory*. While learning is the storing of information, memory will be taken to mean the information so stored. Learning is, if you like, putting the writing on the wall; memory is what has been written there.

To tie all these definitions together, it can be stated that:

- when an event occurs which causes a neuronal signalling pathway to be created or modified, that created or modified pathway is information about the event
- because it does not resemble the event, the information is in symbolic form
- since every neuronal signalling pathway is able to affect motor cortical output — due to the interconnectivity of cortical neurones — every piece of stored information is able to affect behaviour; the act of storing the information is therefore an act of learning
- the stored information is a memory of the event it symbolizes
- when the memory in the cortex is converted into another form, written down for instance, or used to physically recreate the event it is a memory of, it is said to be translated.

It should be noted that while these definitions have largely been couched in human terms, there is no inherent justification for this restriction. Any living thing or machine which is capable of internally driven movement, and whose movements can be modified by internally stored information about past external events, can be said to learn and possess memories. This will be discussed further in section 8.6. In the meantime, it will be convenient to continue concentrating on the human brain in most of what follows.

INSTINCTS

To investigate how images are formed in the brain — i.e. how learning comes about — a useful step to take is to classify the different kinds of learning. Learning can be divided into two distinct types. In one, what has been learnt is stored in the DNA base sequence and can be passed to offspring so that they do not themselves have to go through the learning process. In the other, the learning is not communicable in this way and must be acquired afresh by each new generation in turn.

The first of these two types of learning produces memories which are known as *instincts*. The way in which they are formed is fairly simple in principle if not so well understood at a neuronal level of detail. What happens is that some chance modification in the DNA acquired by an organism causes some of its neurones to produce a protein which affects neuronal wiring. The resultant change in neurone connectivity alters some specific piece of behaviour, triggered usually by a specific set of sensory inputs. A sequence is thereby established: a given external event → appropriate sensory inputs → signalling path determined by DNA → motor output matching the external event. Thus the DNA-specified neuronal wiring symbolizes the external event and functions as information about it. In effect the event is recognized; the brain contains an instinctive memory of it. As long as the behaviour produced by the wiring is beneficial to the organism, the DNA concerned is likely to spread throughout the population of the species. The species as a whole will then have acquired the piece of information and the equivalent instinctive memory.

Instinct is the first kind of memory to have appeared on the Earth. As such it can be seen in very primitive organisms. Jellyfishes have learnt to swim by this means. It is also displayed in far more complex creatures. The cuckoo chick knows what to do with other eggs and hatchlings in its nest because of neuronal wiring in its brain which is defined, not by previous inputs from the environment of the individual cuckoo, but by the DNA which it inherited. Humans too possess a considerable amount of instinctive memory. The behaviour of a newly born baby, for example, is entirely governed by instinct. And as will be seen in section 8.5, instinct plays a part in getting the second type of learning under way.

NON-INSTINCTIVE MEMORIES

The second type of learning is that due to individual experience. It can be classified in several ways. Here three forms will be identified, distinguished by the extent to which consciousness is involved. They are short-term memory, permanent recallable memory and permanent skill memory.

In *short-term memory*, the information is learnt for up to about three minutes. After that time it is forgotten unless it is refreshed by being mentally rehearsed. The information in short-term memory at any given moment is very limited in quantity, and as time passes and events occur it is replaced by newly sensed information easily. While information is in short-term memory its possessor is conscious of it.

The neuronal mechanism producing short-term memory probably involves the setting up of feedback loops between the places in the cortex where the information is being held and subcortical brain structures. The capacity of the brain to support these signalling loops is limited to a very few at a time. The feedback loops ensure that the neurones whose signalling represents the information are continuously active, while participation of I-image neurones enables the person to be conscious of that information.

The next type of memory, *permanent recallable memory*, is also called declarative memory. Memories of this kind are not held in consciousness but can be brought into consciousness in response to sensory input or other information being consciously processed. Declarative memories can be expressed verbally or by some other deliberate motor activity (e.g. playing a tune on a musical instrument).

The key brain component associated with declarative memory is one of the main parts of the limbic system: a piece of cerebral tissue called the hippocampus (mentioned before in section 7.3 in connection with remembering dreams). Its absence from both sides of the brain stops declarative memories from forming. People with extensive hippocampal damage can recall their lives up to around the date when the damage occurred but can remember nothing after that. Their experience of life thereafter is like continuously waking from a swiftly forgotten dream.

Clearly the hippocampus is essential for conscious learning (and for the initial formation of the I-image?). It is not known how the hippocampus achieves memory consolidation, but it can be inferred that its signals bring about changes to the neurones whose signalling symbolizes the information in short-term memory. These changes enable the neurones concerned to cease their activity without losing the ability to signal in an identical way at later times. In this way, information can be lost from consciousness as it slips into the past, but can be recalled in the future should it be needed again.

The final type of memory, *permanent skill memory*, is also referred to as procedural memory. Memories of this kind are not directly accessible to consciousness and mostly relate to motor outputs; i.e. how to do things. They are the sort of actions which you do automatically or semi-automatically, without having to think about them — the sort of actions which become disrupted if you are interrupted half way through and have to think about how to resume them. These skill memories are mainly stored in the primary motor cortex. Their unavailability to consciousness is then readily understandable given the proposed absence of I-image neurones in that location. Procedural memories about motor skills can form normally, or nearly normally, despite hippocampal damage, and so must use a different method of consolidation to that used for declarative memory.

There is a particularly intriguing instance of procedural memory, one which involves learning how to make correct choices in laboratory test situations. Patients with hippocampal damage, who do not even recall entering the laboratory from one day to the next, let alone taking the particular test, can nonetheless learn to make the correct choice in a test if it is repeated often enough. However, lacking any awareness of previous tests, they do not know how they get the answer right.

THE NEUROBIOLOGY OF MEMORY

The cellular mechanism by which *short-term* memories are stored must involve reversible changes to the neurones whose signalling symbolizes those memories. The kind of processes that are probably at work here are those of habituation and sensitization.

In *habituation* a neurone which is repeatedly signalling emits less neurotransmitter as the signalling continues. This reduction is brought about by calcium ions which enter an axon terminal from the outside when the pulse of voltage disturbance arrives. (Recall the role of calcium ions in the release of neurotransmitter; see section 6.3.) Continual arrival of pulses reduces the openness of the calcium channels, thereby decreasing the amount of neurotransmitter released.

Habituation may also involve a degree of depletion of the supply of neurotransmitter. This comes about because neurotransmitter molecules in the axon terminal are kept in tiny bubbles. To be released the bubbles must move to the release site within the terminal — a process called mobilization. The rate at which mobilization can take place is limited, so when signalling is continual the amount of neurotransmitter available for synaptic signalling may decrease.

Sensitization, conversely, is the opposite of habituation: as signalling continues, the amount of neurotransmitter released increases. Clearly, sensitization has to overcome the habituation mechanisms just described. How this comes about is by way of a particular kind of synapse, where one axon forms a synapse with another axon. Call them axons A and B respectively. Axon B is the one sending its signal to the next neurone down the line, neurone C. Axon A forms synapses with axon B close to where the latter forms synapses with neurone C, in effect enabling axon A to reinforce the signal that has travelled down axon B. Signal reinforcement by A is brought about using a second-messenger neuro-transmitter (see section 6.3). The second-messenger neurotransmitter causes chemicals to be produced inside axon B which close potassium channels. The effect of this is to make the duration of each individual release of neurotransmitter at the axon B/neurone C synapse last longer, and this in turn increases the amount of neurotransmitter released.

Since short-term memory requires feedback loops, and these must involve a lot of neurones, it is plausible that some neurones in the loop will use the habituation mechanism while others will be sensitized.

What then of permanent memories, where the changes do not depend on feedback loops and can last indefinitely?

As cells develop in a foetus, they differentiate (see section 6.1). The differentiating process involves switching on or off specific parts of the DNA sequence, thus causing particular proteins to be made or causing particular proteins to stop being made. Once the switching on or off has occurred it is permanent. A cell which has become a kidney cell, for example, stays a kidney cell. Differentiation applies to neurones too, of course, and it is a form of differentiation which holds the key to permanent memory.

When a group of neurones stores a memory, they must collectively be different in some way after the memory has been stored. Since the only thing neurones do is signal to one another, the difference must involve signalling.

There are a range of mechanisms which could alter a neurone's signals (see below). Whatever form the mechanism takes, it can only be brought about by proteins inside the neurone, and that implies new proteins must be made which were not made before. Memory storage thus utilizes the standard way that cells differentiate, harnessing it to store information.

What happens in many neurones involves calcium ions — the same ubiquitous ions which are essential to muscle contraction and neurotransmitter release. The calcium ions gain entry to the neurone receiving a signal, doing so by way of a novel membrane protein. This protein is a special kind of glutamate receptor. (Recall glutamate is a classical neurotransmitter.) When the glutamate emitted by the sending axon reacts with this particular receptor, the ion gate in the receptor stays closed. Unless, that is, the receiving neurone, adding up all its incoming inhibitory and excitatory inputs, decides to fire. When that happens, the resulting voltage drop opens the glutamate receptor's ion gate, permitting calcium ions to enter the neurone. The calcium ions react with internal proteins which use the second-messenger system. In due course some of the second-messenger chemicals attach themselves to a section of DNA and, in so doing, cause that section to be switched on. New proteins are thereby added permanently to those found in the neurone.

How might the new proteins modify the neurone? Potential ways include catalysing the production of second-messenger chemicals directly without the need for an incoming signal, thereby changing the excitability of that neurone; increasing the amount of neurotransmitter released when a neurone fires; adding new synapses or deleting them; and identifying — somehow — the precise synapse where the calcium ion influx occurred, in order to make permanent modifications to that specific synapse.

In fact, bearing in mind the many different types of neurone and the many different neurotransmitters, it seems likely that a variety of ways will eventually be found by which neurones and synapses are permanently modified by external events. The formation of neuronal signalling pathways in the cortex probably rests on a number of mechanisms, of which those mentioned here are but a few.

As a final comment, it is interesting that the mechanism just outlined for long-term memory formation may also provide a glimpse into instinct. Suppose, instead of second-messengers in particular neurones binding to DNA and switching on new protein production, the protein production in those neurones was initiated as part of cell differentiation during foetal development. In that case the memory corresponding to activity in the involved neurones would be one the baby is born with. It would be a memory inherited from its parents: an instinct.

8.2 CONCEPTS AND IDEAS

WHAT IS AN IDEA?

Having seen how learning and memory can be accounted for in terms of biochemical and neurological processes, attention can now be turned from how memories are formed to the nature of memories as objects in their own right.

As a first step in this direction, the word 'memory' will be supplanted by the word 'idea'. Memories are always of events that have happened in the past. The past event might be something that has occurred, or it might be a past thought of something that has yet to happen, or it might be a part of a dream, etc. The memory — a past event — can thus relate to past or future, to possibility or impossibility. In other words, the *content* of the memory is different from the existence of the memory. The content of a memory is what will here be called an *idea*.

In terms of observation, ideas are more important than memories, for it is more correct to say that ideas determine behaviour than that memories do. To illustrate this and to clarify the distinction between memories and their contents — that is, between memories and ideas — consider an example. The example is: 'I heard that sodium cyanide is poisonous.' The memory is one of hearing that sodium cyanide is poisonous. The idea is 'sodium cyanide is poisonous'. On the next occasion I encounter sodium cyanide, it would be more correct to say my response to it will be determined by the idea it is poisonous than by the memory of hearing about it. Indeed, I may well have forgotten the circumstances in which the idea was communicated to me. I can possess an idea without retaining the memory of where the idea came from.

To avoid any confusion over what an idea is, it would be useful to give it a precise definition. Subjectively an idea is any experience (i.e. a thought) that someone is, or can be, conscious of and that is independent of present sensory input, relating to some actual or potential or supposed or imagined aspect of subjective or objective reality. An equivalent objective definition is: an idea is any information held in a signal processing system containing an internally sensed image of itself, which relates to some actual or potential or supposed or imagined aspect of internally or externally sensed objective reality.

It can be noted here that scenarios (see section 7.2) are ideas or groups of ideas. Also, to emphasize one part of the definition, there is no requirement that any idea about objective reality actually matches up with what is really out there. All that can be said is that possessors of an idea about objective reality implicitly *suppose* it corresponds to the truth. Using the formalization of section 2.1, an $S(x,t)$ statement — an idea about objective reality — is assumed to correspond to the truth, $O(x,t)$, but the assumption may be unwarranted.

CONCEPTS

To have an idea about something inevitably involves relating two or more things together. The idea that the moon is made of cheese relates the moon and cheese. An idea of the solution to a problem relates the parts of the problem together in a suitable way. Simply to think of a tree is not of itself to have an idea about the tree; but if I think 'the tree is in my garden', then I have an idea about the tree, an idea which relates tree and garden. Every idea must consist of a relationship between two or more things. These things are the components of the idea. The components of ideas I shall call *concepts*. Concepts are to ideas as words are to sentences.

Concepts, like the ideas they constitute, can refer to any actual or potential or supposed or imagined aspect of subjective or objective reality.

To acquire a concept in the first place requires that inputs to the synthetic cortex create a unique neuronal signalling pathway correspond-ing to that concept. As an example of concept acquisition, consider hardness. Sensory input from various hard objects must be analysed by the primary somatic sensory cortex in such a way that the result of the analysis includes one neuronal signalling pathway which always and only signals when hard objects are touched. The synthetic cortex will perceive this particular signal as a symbol representing hardness. The same signal being activated in the absence of sensory input will be perceived by the synthetic cortex as the concept of hardness.

It can be remarked in view of this that concepts and the images which take part in scenarios are one and the same. The combining of concepts to make ideas is equivalent to the combining of images to make scenarios.

In general, most concepts are accompanied by a verbal label. For example, the concept of resistant-to-deformation has the verbal label 'hard'; the concept of covering-to-a-building has the label 'roof'; the concept of travel-rapidly-on-foot has the label 'run'; the concept of together-with-and-contrary-to has the label 'but'. And so on. It may be proposed that every word barring interjections corresponds to a concept, and that for most concepts there will be corresponding words.

Concepts can be divided into two classes: basic concepts and complex concepts. Taking two concepts with verbal labels as examples, 'small' and 'river' are basic concepts which can be combined to form a complex concept, 'small river'. It will be noted that combining concepts in this way does *not* produce an idea. It will also be noted that 'small river' is equivalent to the single word 'stream'. This means that there is no clear-cut correspondence between basic concepts and single words on the one hand and between complex concepts and phrases on the other. As a rule, a single word can be expressed as several other words combined together, as any dictionary bears witness to. Below, 'complex concept' will be taken to mean any concept which cannot be expressed as a single word.

A final point to make about concepts is that the first ones to be acquired must be realistic, insofar as they will correspond to analysed features of sensory input. But as the brain matures it becomes quite capable of forming concepts, albeit rooted in realism, which are nonetheless totally unrealistic. The concept of spirit, for instance, falls into this category (see section 10.4). The concept of the square root of -1, which mathematicians have many uses for, and which crops up a lot in quantum theory, is another case of an unrealistic concept.

THE CONSTITUTION OF IDEAS

It will be apparent from the foregoing that the combining of concepts can lead to new concepts of increasing complexity. But this is not the only possible outcome of combining them. Concepts can also combine to make ideas. The difference between the two outcomes is this: if concepts are combined in such a way that they *define* something, they form a complex concept; if they combine so as to *describe* something, they form an idea.

Consider a linguistic idea: 'the king, who conquered Savoy while a young man, had three sons'. Each word of the sentence represents a concept. Some of the concepts are combined into complex concepts: 'a', 'young', and 'man' make a complex concept as do 'three' and 'sons'. Furthermore, 'who conquered Savoy while a young man', rather than describing the king, defines which king is being talked about. The words from 'the' to 'man' thus form one large complex concept. The words 'had three sons', on the other hand, do describe the king, and so join with the definitive concepts (the king...young man) to form the idea.

Ideas made from concepts for which there are verbal labels can invariably be expressed as complete sentences. All complete sentences are ideas. Not all ideas though can be, or need to be, expressed verbally. Here are two non-verbal examples. The first concerns the manipulation of objects. Given a round peg, a square peg and a round hole, I may form the idea that the round peg will fit in the hole and that the square one won't. I do not need to express the idea in words in order to be conscious of it. The second concerns music. The concepts of music are pitch, intensity of sound, duration of sound, etc. These are the images analysed from musical auditory input. The complex concepts formed from them are chords and rhythms. The ideas of music are tunes and chord progressions. (How, or whether, a musical idea is descriptive is not a matter I will pursue here.)

Concepts, then, combined together definitively, make complex concepts. Concepts, from simple to very complex, combined together descriptively, make ideas.

- defining: basic concepts → complex concepts
- describing: basic/complex concepts → ideas

The practical limitation on the number of concepts that can be combined to define complex concepts or to describe ideas is set by the ability of the cortex to sustain the requisite number of neuronal signalling pathways in short-term memory.

FOUR ASPECTS TO AN IDEA

The word 'idea' has four slightly different meanings. (The same also applies to concepts.) Each meaning refers to an idea in one of its four distinct forms. These are the four forms:

- the subjective idea — the conscious thought
- the objective idea — the neuronal signalling pathway that corresponds to the conscious thought
- the encoded idea — the neuronal signalling pathway converted into a form in which it can be communicated from one time or place to another; examples are diagrams, written sentences, musical symbols on staves, patterns on computer discs and sound waves in air
- the realized idea — the physical object or event that the idea is of; not all ideas are physically realizable, some being unrealistic.

Of these four forms of ideas, only the third, the encoded idea, and the fourth, the realized idea, are directly observable. And of these two, the realized idea is of little use in terms of observation since most ideas are never realized. This is either because they are unrealistic or, much more likely, because to realize them would be impractical. Imagine trying to express the ideas in this book in their realized rather than encoded form! Observations must therefore be largely of encoded ideas, and it is with ideas in that form that this chapter will be primarily concerned.

8.3 THE PROPERTIES OF IDEAS

COMMUNICATION INTERACTIONS

In order to observe encoded ideas — e.g. words on paper — it is first necessary to find a way to identify them. This is not as easy as it might seem. What for instance is the difference between an idea encoded in the form of a typed sentence and a random sequence of characters?

The key to this is that encoded ideas interact in specific ways with what may be called idea translation systems or, more briefly, idea translators. (The comment on the role of translators made at the start of section 8.1 may be recalled here.) The human brain is an idea translator, and it is the one to which attention will be confined (though there is no reason in principle why other idea translators might not also exist).

The identification of encoded ideas by their association with idea translators may appear to be circular reasoning, for idea translators are identified reciprocally by means of their interactions with encoded ideas. Encoded ideas and idea translators identify each other. But what this means is that observations are not of ideas as such but are of a particular kind of interaction — an interaction which uniquely identifies both encoded ideas and idea translators.

The interaction — which it would be unwise to associate too closely with the interactions of chapters 1, 3 and 4 — is one that can best be described as a communication. A *communication interaction* takes place when the following sequence occurs.

- An idea translator generates movement which produces an encoded idea — an arrangement of matter — that can be either spatial or temporal in nature.
- The arrangement of matter is detected by the sensory apparatus of another idea translator.
- This second idea translator stores the arrangement of matter within itself in some symbolic form.
- The internally stored symbols give the second idea translator the potential to generate movement of its own which will produce an equivalent (but not necessarily identical) arrangement of matter to the one produced at the start of the sequence.
- The sequence can (but does not have to) be repeated, with the idea stored inside each idea-translator being encoded, passing to another idea translator, being re-encoded, passing to another idea translator, being re-encoded, and so on indefinitely.

To this definition of a communication interaction the requirement might be added that an idea translator only counts as such if it possesses an internally sensed image of itself (an I-image). This restriction derives from the definition of an idea in section 8.2, where it was explicitly stated that ideas exist subjectively as well as objectively; i.e. they are contained within a structure possessing an I-image. The need for I-image involvement in ideas has yet to be justified. This is an issue which will be addressed in section 8.6. For the time being, since ideas are being considered here largely in connection with human brains, and since human brains are conscious, the link between consciousness and ideas will be taken for granted.

It can be seen from the communication interaction sequence above that encoded ideas provide the means by which subjective/objective ideas are communicated from one human — from one idea translator — to another. They are in fact essential for communication. That's because ideas in their subjective form are entirely private to the person experiencing them, and their equivalent objective neuronal firing patterns are also effectively unobservable. Ideas *must* be encoded if they are to be communicated.

(It can be noted here, incidentally, that from now on I will use the expression 'cerebral idea' as a convenient abbreviation for 'subjective idea or its equivalent objective neuronal firing pattern'.)

LAWS OF IDEOLOGY

Encoded ideas can be recognized by their being communicated. This makes them observable. They are also, in general, symbolic, and as such they will possess a material form which is distinct from any realization of the idea.

The arrangement of matter in the material containing an encoded idea will correspond in some way with the arrangement of material in the realized idea; or if the idea is unrealistic, with the neuronal signalling pathways that are equivalent to the subjective idea. Encoded ideas thus take the form of particular arrangements of the material of some object. It is not possible to observe an encoded idea in isolation from the material which contains it. A material that contains an encoded idea is called a *medium*. Air while it is conveying speech, DVDs and ink-on-paper in books are examples of media.

A law can be formulated which reflects these observations of ideas. It is a law of ideology. Specifically: communication interactions (as defined above) always take place between brains and media or vice versa. The brain contains a cerebral idea which is encoded and placed in a medium; a medium contains an encoded idea which becomes a cerebral idea in a brain. Alternatively, the communication interaction can involve the idea in its realized form instead of its encoded form, but in this form too it must be made of matter functioning as a medium.

The next thing to notice is that communication interactions can often be broken down into several steps. For example, the interaction between the encoded ideas on a piece of paper and the human brain involves the illumination of the paper, thereby converting the symbols on the paper into patterns of electromagnetic waves. It is in this second encoded form that the idea is sensed and thereby becomes a cerebral idea in the receiving brain.

Not all interactions take place in stages though. The production of encoded ideas in the form of sound waves — spoken words — is a single stage process, as is the interaction of those waves with the sensory apparatus of a brain which detects them.

One point about communication interactions to arise from this multi-stage observation is that they take place over time. The human brain processes information sequentially and so can only interact with ideas if they are presented to it one concept at a time; that is, temporally. Contrast this with encoded ideas. These are generally found in spatial media; that is, they are arrangements in space. Writing is the most obvious example here. A written idea consists of marks which vary from place to place on paper.

Ideas in a spatial medium are of no use to a brain since brains require ideas to interact with them temporally. When encoded ideas are in a spatial medium they cannot interact with a brain unless there is an intermediate stage in which they are converted to a temporal form. For instance, the light waves passing between book and human eye are detected sequentially. Alternatively there needs to be a means by which the spatial medium can be sensed directly in a temporal way as, for instance, braille writing can be.

This rule, that encoded ideas only interact with brains temporally, can be considered a second law of ideology. It is a law which applies in both directions. When a cerebral idea is converted into its encoded form, the conversion always involves muscular movement. This is something that happens temporally. Thus when encoded ideas are produced, they are always produced over time.

IDEAS AND GENES

In looking at communication interactions thus far, a highly significant characteristic has been overlooked; namely, that when a communication interaction occurs, the idea concerned is *replicated*. Furthermore, communication interactions involving any particular idea can take place over and over again, increasing almost without limit the number of places where the idea may be found. This ability of ideas to be replicated indefinitely is indeed one of their most prominent characteristics. It enables them to spread from one brain to another in such a way that the number of brains possessing them increases. An idea which begins its existence as a cerebral idea in only one brain can, by way of communication interactions, soon come to be stored in many places, both in other brains and in encoded media. This is a readily observable property of ideas. Think, for example, of popular songs — a kind of abstract idea — or the teachings of Jesus to see how ideas can become disseminated.

Replication is a property previously associated with DNA. Finding something else which also replicates invites comparison. Is the replication of ideas in any way comparable to the replication of DNA?

To make such a comparison it is necessary first to identify how ideas are organized. It has already been established that ideas are made of concepts combined descriptively. But ideas do not generally exist in isolation from one another; they occur in groups, where the members of a group are related through being about the same subject matter. These groups of related ideas will be referred to as *belief-sets*. Each chapter of this book could be considered a belief-set. Belief-sets in their turn combine together; that is, within any one person's consciousness, there will be found a number of belief-sets. The total of all the belief-sets a person has will be called the *world-view* of that person. Lastly there is what I will call the *idea-environment*. The idea-environment is the physical environment in which cerebral ideas exist; that is, the aggregate of all human brains.

How do these ideological entities match up with DNA's biological ones? The answer to this question can be set out as follows.

BIOLOGY	IDEOLOGY
A DNA 3-base sequence, specifying an amino acid, is the irreducible 'atom' of biological replication.	A concept is the irreducible 'atom' of ideological replication.
A chain of DNA bases grouped together and specifying a protein is a *gene*. (This definition of a gene is good enough for comparing biology with ideology. More generally it has the usual if-and-but reservations.)	A collection of concepts grouped together descriptively is an idea.
Groups of genes, expressed as functionally interacting proteins, cooperate to perform biological functions (e.g. controlling and harnessing energy; see section 5.3).	Groups of related, mutually supporting ideas constitute belief-sets.
A complete set of genes constitutes an organism.	A complete set of ideas constitutes the world-view of a human.
DNA, genes and organisms exist in the biosphere, the physical environment consisting of the water, land and air of the Earth.	Concepts, ideas and world-views exist in the idea-environment, consisting of all human brains.

Since it is complete proteins which are the realized form of the DNA base sequence, it is actually individual genes rather than DNA as such that are the units of biological replication. Similarly, it is ideas rather than concepts that are the units of ideological replication. Hence the following list of characteristics compares ideas with genes. You need to be aware that pointing out similarities and differences between the two is highly speculative and that no great importance should necessarily be attached to it. It is rather suggestive, though, that in a number of respects ideas and genes behave in similar ways.

Characteristic no. 1: genes are inseparable from matter but exist independently of it. A gene is information on how to make a protein. This information is stored in DNA but it need not be so. Theoretically the information could be symbolized in other molecules than DNA. So long as a translation mechanism exists for them, genes may reside on any suitable form of matter. In other words, it is genes which are replicated; the matter being used to achieve replication is incidental. Ideas are also inseparable from matter but exist independently of it. An idea may be symbolized in many different ways both in brains (or their equivalent) and in a wide range of media. So long as a translation mechanism exists for them, ideas too may reside on any suitable form of matter. As with genes, it is ideas which are the prime replicator, not the matter being used. [✔]

Characteristic no. 2: genes exist in the biosphere and cannot replicate if removed from it. This is because genes don't interact directly with other parts of the world; they need to be translated into proteins before they can do so. Consequently, a gene loses its significance — ceases to be a gene — if it never meets up with a translator. Encoded ideas similarly exist in the idea-environment and cannot replicate if removed from that. Encoded ideas don't interact directly with other parts of the world unless they are converted into motor output. In their encoded form they too lose their significance — cease to be ideas — if they never meet up with a brain to translate them.   [✔]

Characteristic no. 3: every gene is translatable into a protein. A protein is a realized gene. This is a point of dissimilarity between genes and ideas for not all ideas are translatable into a realized form. The significance of genes lies not in their translation as such but in their realization. The significance of ideas, on the other hand, cannot lie solely in realization. For unrealizable ideas, significance rests on translation alone — translation from cerebral idea into encoded form and back again.   [x]

Characteristic no. 4: genes do not exist singly but occur in functional groups, and a biologically healthy organism's functional groups all support each other in building the organism. Ideas are similar. They largely occur in belief-sets, and in a psychologically healthy person all the belief-sets (plus non-belief-set ideas) support each other in comprising that person's world-view.   [✔]

Characteristic no. 5: genes replicate. When placed in a suitable environment a gene will be copied. As was remarked above by way of introducing this comparison of genes and ideas, ideas also replicate. In the case of genes the mechanism is understood and is physically inevitable. In the case of ideas replication is more mysterious. It can be observed to happen, but why it does so is not readily apparent.   [✔]

Characteristic no. 6: gene replication is purposive. Viewed from the biological level of complexity (see section 5.7), the behaviour that genes, via proteins, produce in the organisms which contain them has replication as one of its prime aims. Idea replication also appears purposive. The communication of ideas does not take place passively. The possessors of ideas actively seek, to varying degrees, other brains to pass their ideas to. It seems that the behaviour produced by ideas in the brains that contain them has idea-replication as one of its prime aims.   [✔]

Characteristic no. 7: genes compete with one another. The replicative ability of organisms is so great that in a finite environment overcrowding is inevitable. The result is that organisms have to compete for the available space and resources. Some organisms, and the genes they contain, must lose out in the competition and as a result cease to exist — genes can become extinct. Ideas similarly replicate in a finite environment; that is, the idea-environment. And they too replicate rapidly enough to run up against the limitations imposed by finite numbers. Ideas must compete for space in people's brains. Mutually exclusive ideas, especially, must strive against each other since only one of them can be accommodated in any one brain. Ideas which lose out in the competition can cease to exist, becoming extinct. However, because they are able to lie dormant in encoded form for long periods of time — media containing encoded ideas can be thought of as idea stores — extinction for ideas need not be as final as it is for genes. Dormant ideas can be reactivated if they encounter brains whose existing ideas are compatible with the dormant ones.   [✔]

Characteristic no. 8: the replication of genes is very accurate; technically it is said that they have a high replicative fidelity. Errors occur during gene replication only rarely, so that a gene is likely to remain unaltered across many generations. This is not a marked characteristic of ideas. Ideas have a low replicative fidelity. While it is true that in encoded form they may be replicated many times without error, their interactions with brains and subsequent re-encoding seem to produce errors almost as a matter of course. Brains reinterpret ideas they have received before they re-encode them. In part this may be because the meaning ascribed to concepts is likely to differ subtly between one brain and another; that is, the translation machinery in each brain is slightly different.   [x]

Characteristic no. 9: genes evolve. When replicative errors do occur, a slightly altered resultant gene may occasionally be better at replicating for some reason than its parent. It may then come to replace its parent as the two replicate subsequently in competition with each other (see character-istic no. 7 above). By this means, genes with a particular function, and organisms of a particular species, are able to change — to evolve — as time passes. Ideas, with their low replicative fidelity, are able to evolve in a similar way but very much faster. Indeed, so rapid is idea evolution that it can be impossible to tell whether an idea has become extinct or merely changed beyond recognition. This makes idea evolution significantly different from genetic evolution.   [~]

Characteristic no. 10: replication enables a gene to achieve a kind of immortality. The lifespan of every DNA chain is limited, but a given base sequence can persist indefinitely so long as it replicates. Ideas too can outlive the individual brains and media that contain them. Given a genetically replicating population of brains, an idea can, through being communicated, become immortal.   [✔]

Characteristic no. 11: all an organism's genes replicate together. The environment always contains whole organisms, never isolated genes. In this respect, ideas are completely different. Idea replication can take place one idea at a time. Consequently, while an organism's genes all come from one place (or two in sexual replication) and are acquired in full at the moment of conception, a person's ideas can, and do, come from many places and are acquired piecemeal.   [x]

The above list of characteristics, most of which are common to both genes and ideas, gives some grounds for thinking that their behaviours are governed by related sets of laws. Why this is so is currently an unanswered question. The consequences of this similarity, on the other hand, are readily observable. Some of these consequences will be described in section 8.5.

8.4 BELIEFS, DOUBTS AND LIES

<u>THE CLASSIFICATION OF IDEAS</u>

An idea has been defined as relating to an actual or potential or supposed or imagined aspect of subjective or objective reality. Any thought (or its objective equivalent, or its encoded equivalent, or where possible its realized form) comes within this definition of idea, whether it is linguistic, diagrammatic, artistic, or whatever. Such an all-embracing collection of items cries out to be classified.

The most significant characteristic of ideas to use as a basis for classification is the subjective evaluation that all ideas are given in terms of truth and falsity. The objective equivalent of this evaluation will be suggested in connection with signals to the motor cortex, discussed below. Four categories of ideas can be discerned using this criterion.

- Neither-true-nor-false ideas. These ideas can be considered to be abstract. They are mostly artistic and are held to be independent of any correspondence to truth. For this reason they have little bearing on scientific views of reality and will not be considered further.
- True ideas. These ideas will be referred to as *beliefs*. A belief is an idea that is held by its possessor to correspond to reality.
- Either-true-or-false ideas. These ideas will be referred to as *doubts*. A doubt is an idea that its possessor holds may, but only may, correspond to reality.
- False ideas. These ideas will be referred to as *lies*. A lie is an idea that is held by its possessor not to correspond to reality.

With the exception of neither-true-nor-false ideas, a further subdivision is possible according to whether the reality concerned is subjective or objective. For example, the ideas 'I dislike getting wet' and 'primroses are yellow' are about subjective and objective reality respectively. In general, ideas about subjective reality tend to be simple and their correspondence to that reality absolute. Ideas about objective reality, on the other hand, are frequently complex, and their correspondence to reality is never certain. As ideas about the latter reality are scientifically the more accessible and the more controversial, it is with this type of ideas that the rest of this chapter will primarily be concerned.

SECTION 2.1 REVISITED

In section 2.1 the relationship between the subjective and objective aspects of events was established. How does this relationship appear when put in terms of beliefs, doubts and lies?

At any given place (x) and time (t) the situation can be expressed in a statement $O(x,t)$. $O(x,t)$ is the truth. The situation makes an impression on an observer's senses, whereby it is experienced. The experience of the situation at x and t can be learnt and then becomes an image of the situation. The image is likely to be complex and so will consist of a collection of more basic images. Each of these images is a concept. The concepts can be combined and incorporated into a statement, $S(x,t)$, which is an idea. In statement-cum-idea form, the idea in $S(x,t)$ is held to be equivalent to $O(x,t)$.

(A possible confusion may arise here in that the statement $S(x,t)$ has previously been described as an image of objective reality rather than an idea of it. Strictly speaking it should always have been called an idea. The image in $S(x,t)$ is related to the idea $S(x,t)$ quite simply. Say, for example, I experience seeing a tree in my garden. The image is 'a tree in my garden'; the idea is 'there is a tree in my garden'. Subjective reality contains images of objective reality. When those images are put into statement form, $S(x,t)$, they become ideas. Since only now are restricted meanings being given to the terms 'image' and 'idea', the earlier slightly inaccurate use of the more evocative 'image' can hopefully be excused.)

There are important assumptions underlying the relating of $S(x,t)$ and $O(x,t)$. It is assumed that human senses accurately reflect $O(x,t)$, and that human brains analyse and synthesize incoming sensory signals without error. The subjective experience or memory that matches the signals passing through the synthetic cortex can then be held to correspond to the true situation at x and t. $S(x,t)$ equals $O(x,t)$.

Now, it has been frequently remarked that it is not possible to be sure that no errors occur in building images of objective reality. The equality between $S(x,t)$ and $O(x,t)$ must always be regarded as questionable.

However, just for now, take it that $S(x,t)$ really does equal $O(x,t)$. What can then be said about beliefs, doubts and lies in terms of $S(x,t)$ and $O(x,t)$?

Beliefs, doubts and lies are all $S(x,t)$ statements. The following equations can be written:

- for a belief, $S(x,t) = O(x,t)$
- for a doubt, $S(x,t) = O(x,t)?$
- for a lie, $S(x,t) \neq O(x,t)$.

Each of these equations is itself a belief. For instance, in the case of $S(x,t) \neq O(x,t)$ it can be stated as a belief that $S(x,t)$ is a lie. To make such a statement is to turn the concepts in the $S(x,t)$ statement into a definition (as opposed to a description). This makes the particular $S(x,t)$ into a complex concept. The relating of the complex concept $S(x,t)$ to 'lie' is clearly descriptive and therefore an idea. As the new idea is implicitly true it is also a belief.

The three equations can be modified to show that each is actually a belief. When this is done, they become respectively:

- $\{S(x,t) = O(x,t)\}$ $= S(\text{belief})$
- $\{S(x,t) = O(x,t)?\} = S(\text{belief})$
- $\{S(x,t) \neq O(x,t)\}$ $= S(\text{belief})$

where $S(\text{belief})$ is the statement 'I believe it is true that $\{....\}$'.

Putting these equations in terms of a concrete example, consider the next three $S(x,t)$ statements. Neptune orbits the sun. Neptune is made of ice. Neptune is a star. Symbolize these ideas as $S_1(x,t)$, $S_2(x,t)$, $S_3(x,t)$, respectively. Then, referring back to section 4.1:

- $\{S_1(x,t) = O_1(x,t)\}$ $= S(\text{belief})$
- $\{S_2(x,t) = O_2(x,t)?\}$ $= S(\text{belief})$
- $\{S_3(x,t) \neq O_3(x,t)\}$ $= S(\text{belief})$.

These three equations, expressed in English, are:

- (I believe) it is true that the idea 'Neptune orbits the sun' corresponds to objective reality
- (I believe) it is true that the idea 'Neptune is made of ice' doubtfully corresponds to objective reality
- (I believe) it is true that the idea 'Neptune is a star' does not correspond to objective reality.

The conclusion to be drawn here is that each person's total of ideas of objective reality consists of a set of beliefs: beliefs about what is true, about what is doubtful, and about what is false.

The fact that all ideas of objective reality are ultimately expressible as beliefs is a consequence of the ability of every idea, by way of its identification with neuronal signalling pathways, to produce behaviour. Behaviour is not something which can be true, doubtful, or false. It either happens or it doesn't happen. The motor cortex cannot evaluate whether the signals reaching it are thought to be true or not. So the synthetic cortex must ensure that all signals sent to the primary motor cortex reflect truth rather than lies. The primary motor cortex will then produce behaviour appropriate to what is thought to be objectively real.

Subjectively speaking, this means that it must be possible to express all ideas, to the motor cortex, to the world, and consequently to oneself, as beliefs. In this way people can act on what they believe to be true, avoid acting on what they believe to be doubtful, and act against what they believe to be false (though see the comments on lies below).

THE DIFFERENCE BETWEEN BELIEFS AND DOUBTS

What about the questionable assumption mentioned above that $S(x,t)$ equals $O(x,t)$? If it is agreed that the assumption is unreliable — that human brains can make mistakes — it follows there must be doubt associated with all beliefs. So what then distinguishes beliefs and doubts? Is it simply a matter of degree?

To answer these questions look again at the doubtful statement made above about Neptune. If it is a belief of mine that Neptune is made of ice then I cannot also believe it is made of rock, since the two beliefs are mutually exclusive. However, if I doubt that Neptune is made of ice then I can also consider it doubtful *at the same time* that Neptune is made of rock. Mutually exclusive doubts can exist side by side without difficulty or trace of irrationality. Hence while all my current ideas about Neptune must be considered doubtful as a matter of scientific principle, they must be classified as beliefs as long as I am unable to accept as possibly true other ideas about Neptune which are incompatible with what I currently believe. The distinction between a belief and a doubt is thus that, given two ideas which are mutually exclusive, if I hold one idea as a belief then I will hold the other as a lie, whereas if I hold one idea as a doubt, then I may hold the other as a doubt also.

STRENGTH OF BELIEF

It is a matter of experience that a person's beliefs are hard to change. When people communicate beliefs that are contrary to someone else's, the someone else may at best treat what is being said skeptically, and at worst refuse to listen. The more strongly held a belief is, the more likely a person is to resist countenancing a contradictory belief.

In objective terms it is clear why beliefs are resistant to change. As was stated earlier, every idea in the synthetic cortex is connected to the motor cortex so as to be able to produce behaviour when appropriate. The neuronal wiring to achieve this must signal to the motor cortex in ways that reflect whether the idea is held to be true or not. Now, the moment an idea is stored in the cortex which is contrary to a belief already held, then either the new idea must be stored as a lie, or else the existing belief must undergo a change of status. 'I believe $S(x,t)$ is true' must become 'I believe $S(x,t)$ is doubtful' or even 'I believe $S(x,t)$ is false'. This change in status will be manifested by a change in behaviour and must accordingly be brought about by alterations to neuronal signalling pathways. Neuronal signalling pathways are not easy to alter once laid down because, as was described in section 8.1, the changes which accompany their formation are permanent. Presumably, to convert a belief into a doubt or a lie additional pathways have to be established which inhibit the existing ones. Whatever the precise case, the resistance of neuronal pathways to being changed accounts for the difficulty people have in changing their beliefs.

There is more to a belief's resistance to change than this, however. There is also the question of degree: some beliefs are more negotiable than others. Every person is aware of having beliefs which would be easy to change, while the changing of others would be seriously traumatic. It is reasonable to assert that the resistance of a belief to change in its status is an indication of the strength of the belief. A strongly held belief is strongly resistant to being changed into a doubt or a lie.

At first sight this may seem somewhat lacking in profundity. However, on further thought it becomes apparent that with the possible exception of instinctive beliefs the strength of a belief is a *consequence* of its resistance to change rather than merely a restatement of it. So what makes a belief resistant to change?

The answer lies in the fact that almost no belief exists in isolation. Practically every belief is associated with others which support it, and to which it in turn gives support. The resistance of a belief to change is determined by the consequences for the supporting beliefs of altering its status. Some of them would have to alter their status as well, becoming doubts or lies. The greater the number of these additional supporting beliefs, the more resistant to change the belief in question will be. For example, a belief I have about the number of bacterial cells inhabiting the human body (an estimated 10^{13}) would not, if it became a lie, make much difference to any of my other beliefs. It supports them, and they support it, only weakly. It is therefore easy to change and is in consequence a weakly held belief. On the other hand, if my belief that matter is atomic (as discussed in section 1.3) became a lie, huge numbers of associated beliefs — most of chapters 3, 4, 5 and 6! — would have to undergo drastic alteration. My belief about the existence of atoms is strongly supported by, and gives support to, many other beliefs. It is consequently highly resistant to change: a strongly held belief.

SOME REMARKS ABOUT LIES

Until now this discussion has primarily been about beliefs and, to a lesser extent, their weaker relatives, doubts. To conclude this section a few words on lies are in order.

Lies are statements about objective reality that are held to be false. What makes them interesting is that they can be, and more important usually are, encoded for communication to others as if they are beliefs. The encoder of the lie believes it is a lie but expects anyone receiving the encoded idea to take it to be true.

Of course, lies can be communicated quite legitimately in the form 'I believe that S(x,t) is a lie'. In this form they are stated as part of a genuine belief and are incorporated into a true statement. But they can also be communicated in the form 'I believe that S(x,t) is true' when in fact I believe it is false. Additionally I can turn a belief into a lie by stating 'I believe that S(x,t) is a lie' when in fact I believe it is true. In every case, whether genuinely (as in the first instance) or falsely, the lie is always communicated in such a way that it is presented to the audience as a belief or as part of a belief.

The communication of lies as fraudulent beliefs is an example of acting on a lie. Normally, as remarked above, people can be expected to act on what they believe to be true, because ideas in their objective neuronal form will produce behaviour appropriate to what is thought true, but not to what is thought false. Encoding a lie or, even worse, going so far as to perform actions on the pretence that a lie is actually a belief, would seem to go completely against this reasoning.

To account objectively for the telling of lies and behaving as if lies are true, it is necessary to invoke additional signals in the cortex capable of overriding inhibitory signals with excitatory ones. More often than not, the source of these putative signals is external to the brain and takes the form of either a potential punishment or a potential reward. People tell or act on a lie if they believe that to do otherwise will be to their disadvantage. But sometimes it seems that the source of the signals leading a person to lie comes entirely from within their own brain. One curious example is that a lie may be told if it makes the communication of a person's genuine beliefs easier. For example, someone might say: "I can confirm so-and-so is true [genuine belief] because I've seen it for myself [facilitating lie]." And occasionally lies are told for no apparent reason at all. Instances of this last kind, if habitual, cannot be considered within the bounds of normality.

It will be noticed that even when discussing lies the statements being made are beliefs. Apart from abstract ideas, ideas in encoded form are stated as beliefs. Every statement made is implicitly taken to be true. Only if an indication is given to the contrary is this rule breached, and even then the indication of contrariness is taken as true and to be believed.

Because observations of ideas are always by way of communication interactions, and because communication interactions are always concerned with beliefs, any observations made of ideas will actually be of beliefs. So what do observations of beliefs reveal about their properties? It will be found that a novel light is cast on human behaviour by such observations.

8.5 BELIEFS AND BELIEVERS

THE ACQUISITION OF BELIEFS

Consider a newly-formed brain. It will be devoid of all beliefs except instincts. Subsequently it will acquire many beliefs in the course of interacting with its environment. These non-instinctive beliefs are learnt.

The acquisition of learnt beliefs takes place in one of two ways: unintentionally, and by design.

Unintentional beliefs are acquired mainly passively. They arise out of the general vagaries of an individual's experience, out of sought-after experiences and communication interactions which turn out unexpectedly, and out of unsought communication interactions, especially from contemporaries, parents and teachers. In this last case, the recipient of the belief is passive but the communicator of the belief is, more often than not, active in its communication. In this way the idea-environment — all the other brains in the world — plays a strongly positive role in the acquisition of beliefs by its members.

Beliefs-by-design in contrast are deliberately acquired; the behaviour undertaken by the individual has as its purpose the acquisition of information. Once obtained, the information becomes part of the individual's world-view.

As a person acquires more and more beliefs, they increasingly control belief-by-design acquisition. This is because the beliefs people hold determine what beliefs they seek to acquire. For examples of this, you will be able to draw on personal experience. A topic that interests you, whether it is archaeology or dressmaking or whatever, is one you will communicate interactively about. You will read about it and talk about it and seek to expand your knowledge of it, steadily acquiring a growing stock of beliefs as you do so. A topic that does not arouse your interest is conversely one you are unlikely to acquire beliefs about, since you will not seek to acquire them. What this means is that the acquisition of beliefs-by-design is not simply controlled by beliefs already held, but totally depends on them.

Furthermore, the influence of existing beliefs extends beyond beliefs-by-design to unintentional beliefs. Existing beliefs influence behaviour generally in a way that results in the opportunity to acquire beliefs unintentionally being somewhat circumscribed. For example, a person who believes all party political broadcasts warrant turning the off switch on the radio or television is unlikely to acquire, even unintentionally, beliefs being communicated by politicians. Existing beliefs influence the acquisition of unintentional beliefs as well as controlling the acquisition of beliefs-by-design.

Now, a newly-formed brain is born with no learnt beliefs at all, so the amassing of beliefs-by-design can be traced back to when a very first belief of this kind was acquired. The acquisition of that first intentional belief must depend in turn on beliefs already held which are either instincts or previously acquired unintentional beliefs. Hence *all* beliefs, including those which have been intentionally sought, owe their acquisition ultimately to instinct and/or unintentional beliefs, the latter consisting, as already remarked, of chance interactions with the environment and unsought incoming communications from other people. The unsettling conclusion follows that people's world-views are not acquired by choice.

You may feel this conclusion to be subjectively questionable. Surely we do choose what we believe. But do we truly? What you choose to believe depends on what you *already* believe: about the most dramatic experiences in your past; about the people who have influenced you; about what is most important in life; about what you instinctively believe to be 'right'; about the validity of reason and superstition. And how negotiable really is this platform of your underlying formative beliefs?

The sources of a person's beliefs can be summarized as follows.

- Some beliefs are instinctive. They are inherited as DNA base sequences which specify proteins that cause particular neuronal wiring to be set up.
- Some beliefs are unintentional. They arise
  (a) through chance interactions with the environment

(b) when deliberately sought beliefs produce unexpected results

(c) as a result of incoming communication interactions from other people.

- Some beliefs are deliberately sought. Beliefs-by-design are sought on the strength of existing beliefs.

BELIEF CONFLICTS IN INDIVIDUALS; CREATIVITY

In the course of acquiring beliefs, people will more or less inevitably come across instances where different beliefs are incompatible with one another. When an incompatibility is encountered, they cannot incorporate both incompatibles into their subjective image of reality. A choice must be made. Not surprisingly the choice is determined by beliefs a person already holds. People choose whichever of the two incompatible beliefs best fits in with their existing beliefs. If neither fits in, they may of course choose to regard both incompatible beliefs as lies.

This simple resolution of the problem of conflicting beliefs is sometimes greatly hampered by the presence of supporting beliefs. What happens is that a person assimilates two distinct belief-sets without any awareness of a conflict between them. Each belief-set thus comes to contain a lot of strongly held beliefs. Only after that has happened is a conflict discovered.

Since the incompatibility now involves not just two rival beliefs but their supporting belief structures too, it becomes a conflict between two belief-sets. An example of this might be of a person who acquires the belief that supernatural events are real, and also the belief that physics describes reality. Initially there may be no apparent contradiction, and the person may acquire a large set of beliefs about the supernatural and another large set about physics. As the two belief-sets are in fact incompatible (see section 2.5), it is likely that eventually this state of peaceful coexistence will be shattered by the discovery of some new idea to add to one of the belief-sets which is overtly ruled untrue by beliefs in the other.

To settle the conflict, the believer has three basic options. One is to convert the weaker of the belief-sets into a set of lies. When this happens it can be subjectively very distressing, a distress which arises from the need for large amounts of changes to neuronal signalling pathways.

Because of the difficulties attendant on it, it is not a much used option. The second option is to convert both belief-sets into sets of doubts. This is unsatisfactory as a final solution to the problem but may be used as a temporary measure until something better can be thought of.

The third option is to go ahead and think of 'something better', and this is the option most often taken up. What happens is that the conflict is treated as a perfectly ordinary problem. There is the existing situation — the conflict — and there is the desired situation — restoration of peaceful coexistence. It is no different in principle from problems of the 'how to get hold of an apple three metres above the ground' kind. In both cases a solution can expect to be found by manipulating concepts within scenarios. And in both cases the solution is a belief about reality created within the general association cortex.

In the case of conflicting beliefs, the created belief usually acts as a kind of insulator which separates the rival belief-sets and so prevents them contradicting one another. In the example given above, it is the supernatural and physics belief-sets which need to be mutually insulated and rendered non-conflicting. So how about creating a new belief that the presence of a systematic observer or of measuring equipment somehow prevents supernatural events occurring? The conflict is resolved! So painless is this third option that it can readily be seen why it is the preferred one.

It may be noted that the third option is an instance of human creativity. The act of idea creation is in essence the manipulation of concepts into novel combinations until one is arrived at which solves the problem under consideration. Problems involving conflicting beliefs are of course only one application of creative thinking. There are several other reasons why ideas are created. Some are thoroughly practical (e.g. inventions), and some are completely abstract (e.g. musical compositions). Self-creation is a fourth source for the beliefs a person acquires, and can be added to the three (instincts, unintentional, and by-design) already listed above. Like beliefs-by-design, created beliefs are based on beliefs already held.

BELIEF CONFLICTS IN THE IDEA-ENVIRONMENT

The acquisition of beliefs and belief-sets by each brain has consequences for the idea-environment. The combined effect of all the individual acquisitions produces a cumulative global belief behaviour. Although some aspects of this behaviour have already been mentioned in the course of comparing ideas and genes, the topic is worth further thought.

Consider a single belief in the idea-environment. Sooner or later every one of the brains containing it will die. When that has happened the idea will cease to exist. Unless, that is, it is communicated. Then, by continually taking root in young brains it evades the deaths of its possessors. (The botanical analogy of taking root seems appropriate.)

Being communicated can result in a belief becoming widespread in the idea-environment. This is all to its advantage because the more people who possess the belief, the more people will be communicating it, and the more certain will be its persistence. Without anything to limit its spread, a belief may end up communicated to virtually everybody. Some beliefs have indeed reached this position — the belief that the Earth is spherical in shape is an example. But a great many beliefs do not become so widely disseminated and secure. Their spread is subject to a troublesome limitation.

The source of the limitation is conflicting beliefs. These other beliefs are rivals for possession of each brain in the idea-environment. When two rival beliefs have each taken root in a proportion of the brains of the populace, they clearly prevent each other from achieving universal dissemination (as beliefs). How this situation comes about in the first place, and how it may develop, can perhaps best be described with reference to a specific case. A good example would be beliefs about the origin of the universe.

There are many sets of beliefs about this, almost all of them exclusive of all the others. Here two will be considered: the Biblical account and the scientific account. The Biblical account is a belief-set usually known as Creationism. The scientific account is the belief-set described in sections 4.6, 5.5 and 5.6.

To describe Creationism in a paragraph, the belief-set contains the principal beliefs that the universe was brought into being in a series of discrete steps; that each step consisted of events which did not follow inevitably from the laws of physics, but which had an extra-universal cause; that the extra-universal cause — called God — is a conscious entity having absolute control over the universe; that the period of time taken by God to complete the Creation was six days; that at the end of that time the universe was as it is now except for the number of people on the Earth; and that the Creation occurred about six thousand years ago.

Creationism is obviously grossly at variance with the scientific account of the same events. As the two belief-sets stand, they are incompatible and hence will come into conflict with each other when taking root in new young brains. The conflict is made more complicated because both belief-sets are but small parts of vast structures of supporting beliefs — Judaeo-Christian religion on the one hand and science on the other. How then has this conflict, as a typical case, progressed?

Take as the starting point for study the time several hundred years ago when there was no scientific belief-set dealing with the creation of the universe. For simplicity, attention will also be confined to those parts of the world where the Christian religious belief-set was pre-eminent. The conflict between the Biblical and scientific accounts of the Creation can thus be taken to begin with the Biblical ideas almost universally disseminated and held to correspond to objective reality.

The undermining of the Creationist supremacy started with the use of observation in a novel way — systematically — to discover new unexpected truths about local reality. The scientific beliefs resulting from this did not at first conflict with any significant part of the Christian belief-set — at least not amongst reasonable non-fundamentalist people — and so they were able to be communicated with little opposition. The manifest success of science as a tool of understanding provided the impetus for the spread of scientific beliefs and for the making of ever more observations leading to yet more scientific beliefs. The scientific belief-set grew.

In due course, this expanding observational base gave rise to scientific theories which contradicted Creationism, particularly with regard to the date of the Creation. A conflict was discovered. By that time the scientific belief-set was sufficiently disseminated, and in individual terms held strongly enough, that its extinction was not a viable option.

The conflict was resolved by a few individuals who invented compatibility restoring beliefs. Since dates were the problem, the invented beliefs involved dates in one way or another. For instance, one belief to resolve the conflict is that the universe was created at the Biblically appointed time, but in such a way that it looks old; the universe is like a piece of modern furniture made to imitate an antique. And with God as its maker, the imitation is bound to be perfect. Science and the Bible were reconciled by the belief that while the scientific observations were correct, the conclusions drawn from them rested on a false assumption.

(To digress briefly, taking this line of thinking to its extreme, it can be proposed that the Creation happened literally yesterday. If everything was created then, including all our memories of — non-existent — earlier times, there would be no observable difference between this one-day old universe and a universe which began 10^{10} years ago. However, although such proposals are perfectly defensible, they — and ideas of imitation antiquity generally — are not suited to the present-day idea-environment.)

Another example of a belief to resolve the conflict between Creationism and science is that the Bible's six days are not terrestrial days at all but some vastly longer period of time; the creation date then recedes into the distant past, thereby removing the source of conflict.

These two conflict resolving beliefs, and others like them, are themselves mutually exclusive rivals. They will take root and spread in the idea-environment according to which is the most compatible with the existing beliefs of individuals. In the event, the 6 Creation-days = 10^{10} years belief has proved quite successful in terms of proliferation.

But this development in the Creationism versus science conflict is not the end of the matter. The trouble with science from the viewpoint of the Creationist belief-set is that its beliefs are continually increasing in number and expanding to include more and more aspects of reality. Almost as problematic are its successes by way of realized ideas (inventions), which facilitate the spread of its beliefs by making them more believable. As old brains have died and young ones have taken their place, the idea-environment has become increasingly full of scientific beliefs. This allows young brains to solve the conflict between the Creationist and scientific accounts by simply opting not to believe the Biblical account at all. The fact that they can do this while their forebears couldn't indicates that the idea-environment is an evolving one. A successful (i.e. widespread) belief at one moment in history can become a failure at another.

And so to the present state of the battle. The dynamically changing idea-environment now contains belief-sets about the beginning of the universe which come into three broad categories. People can be similarly categorized according to which belief-set they believe. In one category are the die-hard Creationists who regard the scientific account as lies. Their beliefs concerning the rest of science, which supports the scientific account, and on which they depend for much of their well-being, rest on denial and ignorance. Then there are die-hard scientists who regard the Biblical account as lies. Their beliefs concerning the Bible, if not Christianity in general, are dismissive.

Finally there are those who believe both accounts. As the quantity of scientific beliefs has increased, conflicts with Creationism have grown in number. This has had the result that people in the 'science *and* Creationism' camp have been forced to invent more and more conflict resolving beliefs. Usually these invented beliefs rely on imputing imprecision to the Biblical account. For instance, when the Bible says that God began by creating heaven and a 'form-less' Earth, what 'Earth' actually means in this context is matter and energy. The next event, 'let there be light', then falls more or less into line with the Big Bang theory.

These three belief-sets, science, Creationism and hybrid, are each incompatible as a whole with the other two. The consequent competition amongst them is a straightforward one of each spreading as widely as possible in the idea-environment at the expense of the others.

How each can achieve the widest distribution is perceived by the communicators of the respective belief-sets as yet another problem to be solved. The problem comes down to how to enhance the likelihood of the communicated beliefs being accepted as the truth by would-be recipients. The solution is even more created beliefs.

One example of this final twist in the saga of competing belief-sets will suffice. It concerns Creationism and one of the more recent beliefs to be invented by those who wish to enhance Creationism's replicative success. For the communicators of Creationism, who have lost much ground over the centuries, the nub of the problem is that everyone now acquires scientific beliefs. The Creationist belief-set is less likely to be believed if it denies science, and yet what choice does it have? Now Creationists, having themselves rejected science, know very little of its beliefs. What they do know is that a belief which is described as scientific is more likely to be accepted as true by uncommitted brains than one that isn't. So an obvious move for them to make is to invent a belief that Creationism is a scientific theory. This belief does not correspond to objective reality, since the 'scientific theory' of Creationism makes no predictions and explains none of the details of the universe (the H to He ratio for instance). But as has been repeatedly stated, there is no necessity for a belief to correspond to reality. The belief that Creationism is a scientific theory has entered the idea-environment solely as part of an attempt by Creationists to solve the problem of how to enhance Creationism's replicative success. The question of realism is not relevant.

Two additional noteworthy points come out of this Creationism-versus-science example. The first is that when a new combination of beliefs — such as the Creationist belief-set plus the belief that Creationism is scientific — enters the idea-environment, it faces the world like a new or mutated gene. The state of the contemporary idea-environment alone determines its success or failure.

The second point concerns the extent to which beliefs are communicated individually rather than only in complete belief-sets. The belief that scientific ideas are likely to be true is an example of this. A statement made by someone who is perceived to be a scientist is more likely to take root in a person's brain as a belief than is a statement made by a complete stranger or a politician. And yet that same person will often have little idea of what science actually is, or why its pronouncements are believable. A related example is that most people believe the Earth orbits the sun. But how many of them also possess the supporting beliefs about gravity and the laws of motion? Beliefs are quite capable of spreading separately from their supporting belief-sets.

The example of Creationism gives a representative account of how belief-sets rise and fall and replicate and compete. The fate of a belief-set is determined by the state of the idea-environment, which is reciprocally determined by the belief-sets it contains. It is a dynamic, changing and constantly interacting system. Its similarity to the dynamic, changing and constantly interacting gene-based system of living things on the Earth is remarkable and obvious. Which leads to an altogether more sinister side to beliefs and believers. What happens when the behaviour produced by beliefs comes into conflict with that produced by genes?

THE SUBVERSION OF BIOLOGY

The brain evolved because it enhanced its owner's chances of producing offspring. Yet the beliefs that take root in people's brains are capable of causing them to devote much of their time to activities which have no bearing on this end. Individuals who spend all day disputing the truth of Creationism could have spent their time gathering food or protecting their children or doing countless other biologically productive things. Instead they advocate their ideas and argue with others about them, thereby wasting time on things irrelevant to survival and biological replication.

And belief-centred behaviour can go much further than mere time-wasting (from a genetic point of view). Some believers' determination to communicate cherished beliefs can lead them to override the genetically beneficial instinct to avoid aggression and potential injury. In practice, advancing a belief forcefully or even violently makes little difference to its acceptance by uncommitted brains, but such tactics are still sometimes resorted to, and the accompanying risks run, both by individuals and by organizations of a religious or governmental nature.

It is plain that where behaviour produced by beliefs risks aggression and injury it can create unnecessary dangers to the wellbeing of believers (not to mention their target audience) and can have a markedly negative impact on biological replicative success. Until the present time, judging by the number of humans on the planet, the drawbacks inherent in this situation from a gene replication point of view have been outweighed by the advantages; survival assisting beliefs have made a greater contribution to human behaviour than survival hindering ones. What is rather surprising is that evolution has not favoured brains more strongly resistant to the latter type.

In individual cases it is possible for survival hindering beliefs to outweigh survival assisting beliefs so greatly that a person may choose to die rather than act in contravention of them. Biologically speaking such behaviour is insane. But ideologically it may not be. The ability of a belief-set to spread can actually be enhanced if its possessors are prepared to die rather than say (or act as if) it is untrue. It is a somewhat macabre paradox that a belief-set which is able to cause its possessors to sacrifice themselves in its defence may become very widespread and successful. Early Christianity and modern Islam are excellent examples of this.

Not that a life-and-death conflict between beliefs and genes has a certain outcome: whether to die for one's beliefs is very much a choice. Brains built by genes have been around for hundreds of millions of years, far longer than the idea-environment, and for all that time have been programmed by DNA to resist death with the utmost vigour. The way this works in humans is that DNA specified neuronal wiring produces instincts, amongst which one of the most potent is the belief that self-death is something to be frightened of. It is a highly effective belief; DNA strongly biases people to choose survival over death.

A believer-sacrificing belief-set has to overcome this instinctive fear of death, and it does this typically by incorporating within itself beliefs which contradict instinct. There is, for example, a belief that death is like waking from a dream and finding oneself somewhere blissfully rewarding. This results in the two mutually exclusive beliefs — death is fearful and death is rewarding — battling for control of the believer's behaviour. Self-sacrificing believers usually give greater weight to some rewarding 'paradise' than to their instinctive fear of death; for everyone else it is the other way around. It can only be said that if it is surprising that brains are not more resistant to survival hindering beliefs, it is positively astonishing they are not more resistant to survival destroying ones. It is remarkable indeed that certain belief-sets are able to succeed in sacrificing some (but never all) of their possessors.

And as if sacrificing believers isn't bad enough, there is worse to come. Any belief-set capable of leading its possessors to think it is more valuable than they are can be expected to regard non-possessors as of even smaller value. If sacrificing believers is okay, sacrificing non-believers is better. Which opens the door to murder. What more obvious way for a belief-set to ensure its replicative success than to exterminate rival belief-sets, and what more obvious way to exterminate rival belief-sets than to exterminate the rival belief-sets' possessors? This behaviour again is utterly contrary to instinct, with the result that murder does not often accompany belief conflicts. Nonetheless, history is littered with examples of wars over belief-sets. Religions are particularly 'good' at this. Christian warriors in Europe spent a thousand years and more slaughtering fellow believers over minor doctrinal disputes, and Moslems warriors still do murder those who disagree with them even to this day. Political disputes, especially between racists and anti-racists, and between communists and anti-communists, also sometimes result in violence and death.

In describing and commenting on the ideological observations covered in this section, the distinction has been blurred between believers and their beliefs. Is it beliefs which try to replicate or is it believers who try to replicate them? Is it beliefs which exterminate rival beliefs or is it believers who exterminate rival believers? Is it belief-sets which create conflict resolving beliefs or is it believers who do that? If my beliefs control what beliefs I seek, what beliefs I reject, and how I behave, what role is left for me anyway? Are I and my beliefs distinct from one another at all? Given that I, as a conscious being, am born of scenario enactment, and that scenarios are beliefs, perhaps there is no distinction.

It is food for thought that subjective observers may each be no more nor less than the sum of their beliefs: an image of reality.

It is also an odd thought that human bodies are temporary survival machines built and controlled by genes in order that the genes can replicate indefinitely, and that human brains are temporary survival machines colonized and controlled by beliefs in order that the beliefs can do the same.

8.6 IDEOLOGY AND BIOLOGY

A NEW LEVEL OF COMPLEXITY

It has been stated above that encoded ideas do not become ideas until communicated to brains. This is because in encoded form ideas reside in a medium. Media do not possess the ability to generate movement and cannot therefore initiate communication interactions. As a result, ideas in a medium, so long as they are confined to it, cannot produce behaviour and so cannot display ideological properties.

This leads to the conclusion that brains are essential to the existence of ideas and that only cerebral ideas behave as ideas. This is something of a paradox in that observations always involve encoded ideas. It's a paradox which reinforces the point made in section 8.3 that it is actually communication *interactions* which are observed rather than encoded ideas in isolation. An encoded idea, in the absence of any interactions, would behave in a way quite unworthy of attention. Brains, then, play the active role in producing behaviour while media are passive. For any idea to interact with the world as an idea it must be communicated to a brain first.

Put another way, an encoded idea is just an arrangement of matter until it meets a brain. Then it is able to behave ideologically; that is, the arrangement of matter comes to be representative of the idea. The reason an idea can behave as such when in a brain is that it is stored there in a way that enables it to produce motor output. The motor output, when re-encoding the idea — or realizing it or acting on it — is unique to that idea, making translation and communication possible and generally giving ideas their properties.

In view of the central role of brains it would be worth reconsidering what qualifies as a brain or, more correctly, what qualifies as an idea translation system. (The comments on idea translators made at the start of section 8.3 may be worth reviewing here.) An idea translator needs three objective components:

- a sensory system enabling it to detect encoded ideas
- a signal processing system enabling it to store in a unique arrangement of matter each encoded idea detected by the sensory system
- a motor system enabling it to behave in a way uniquely determined by each stored idea, and enabling it to re-express each idea in encoded form.

Of the three components of an idea translator, the most crucial is the information store. The sensory and motor components can be comparatively simple, but the information store must necessarily be highly complicated — at least as complicated as the ideas it is to contain.

Ideas are arrangements of the material of a physical structure. So that structure, be it brain, machine, or medium, must have one essential property: some part of it must be capable of being arranged in a highly complicated way, such that the arrangement of material at one point in space or time does not determine the arrangement at any other. The arrangement of the material must be determined solely by whatever the information is that the material is being used to symbolize. Most crystals, for example, would be useless as an idea store because the location of each molecule they contain is largely determined by the position of neighbouring molecules.

Clearly, a material containing information cannot be simply described. Any simple description would reflect how the material interacts with the world but could not reflect how the ideas within the material interact with it. In other words, to completely describe an information containing material it is necessary to supply as much information as the material itself contains.

This is the second time a situation has been encountered in which large amounts of information are required to describe something. The first time was in section 5.7 and concerned living things and, in particular, DNA base sequences. The amount of information needed to describe what nucleic acids look like is much less than that required to specify the precise sequence of DNA bases that symbolize proteins. Similarly the information needed to specify an idea translator or a medium falls far short of the amount required to specify the detailed material arrangements that symbolize the ideas they contain.

As with living things it can be proposed that, because of this unavoidable informational requirement, ideas exist at a higher level of complexity, which I shall call the *sub-ideological level*. (The reason for the 'sub' prefix will become apparent shortly.) Again as with living things, the proposal is that the properties of systems viewed from this higher level look different from the way they look viewed from the lower physical level. The ideas within an idea translator, evaluated in terms of the lower level of complexity, are merely arrangements of matter behaving in predictable and inevitable ways. But viewed as single entities, without regard to their parts (i.e. from the sub-ideological level of complexity), idea translators and encoded ideas possess the property of *significance*; significance is conferred on encoded ideas by idea translators. And ideas also acquire all those properties listed earlier which permit them to be compared with genes (at the biologically level of complexity).

IDEAS, CONSCIOUSNESS AND MEANING

Regarding ideas as manifestations of a higher level of complexity, while giving a new slant on what ideas are, does not give any insight into what role, if any, consciousness plays in ideology. So far no requirement has been established for an idea translator to be conscious. If it can detect encoded ideas, store them uniquely, and behave in a way determined uniquely by the ideas, then it is an idea translator. There is no need for it to possess an internally sensed image of itself as well. It is indeed quite possible to think of encoded ideas interacting with a machine. If the machine has a data input device, a means of storing the data, internal programs instructing it on how to process the data, and a data output device, then it can clearly translate ideas from encoded form into internal objective form and back into encoded form. With its involvement, ideas can be replicated. If the machine's internal programs are sophisticated enough, it could even reject mutually exclusive ideas, thus enabling ideas to compete.

So is consciousness ideologically surplus to requirements? For matter to behave ideologically — that is, to display sub-ideological complexity — suitably complex forms of matter are required, but that would appear to be all. Ideology doesn't need consciousness. Or does it?

To address this issue, recall how consciousness arises. It is the brain's ability to enact scenarios that leads to its sensing of its own existence and its forming an I-image. The enactment of scenarios requires purpose — the achievement of the desired outcome in the scenario — and choice — the ability to select the best scenario. Choice and purpose are biological properties. So consciousness requires biological complexity. And as scenarios are one form of ideas, it requires sub-ideological complexity too. Consciousness can only arise in entities which are both biologically *and* sub-ideologically complex.

Next consider carefully what 'idea' actually means; as the concept 'idea' is generally understood, it means 'information in a *context*'. Unless an idea has a context it is not usually regarded as truly an idea. This is perhaps not an immediately obvious point to make. But think of a machine such as was described above, inputting some encoded idea into its data store, storing it there, and re-outputting it, thus replicating the encoded idea. If the machine does this automatically without knowing that's what it is doing — that is, it has no I-image — is what it replicates truly an idea? The machine may be a signal processing system, but without consciousness can it be said to understand the idea it has replicated? Is an idea an idea if it isn't understood? Or is it merely a mechanically stored and mechanically reproduced set of symbols? In order to understand something, it must be placed in context. Communication interactions between media and sub-ideologically complex machines are without context.

Now, for an idea translator to accord context to an idea, it must store the idea in a way that enables it to interact with other ideas entirely within the translator. Internal interactions with other ideas provide any given idea with its context. This is exactly the same process as takes place when scenarios are enacted and which leads to consciousness. Just as consciousness is a product of combining biological and sub-ideological complexity, so is putting ideas in context. Where consciousness is present, there also will be found ideas in context.

When an idea is given a context it can be said to acquire *meaning*. Ideas in biologically complex (conscious) structures have meaning. Ideas in structures lacking biological complexity, like computers and DVD recorders, have significance but they do not have meaning.

These ideological distinctions can be summarized as follows.

- An idea with meaning and significance exists in a structure which is capable of internally driven movement, and which possesses an I-image. Such a structure, which may be called a biological idea translator, must be both biologically and sub-ideologically complex.
- An idea with significance but not meaning exists in a structure capable of internally driven movement and which is sub-ideologically complex, but which has insufficient or no biological complexity. Such a structure may be called a mechanical idea translator in that it can translate ideas but cannot provide them with an internal context. Ideas in a mechanical idea translator, if communicated to a biological idea translator, then of course acquire meaning as well as significance.
- An idea with neither meaning nor significance exists in a sub-ideologically complex structure lacking the ability to engage in internally driven movement. Such a structure is a medium. Ideas in a medium are 'encoded' and acquire significance if communicated to a mechanical idea translator, and acquire meaning if communicated to a biological idea translator.

Thus what constitutes an idea depends on the point at which it is considered that an idea becomes an idea: when it is in a medium, when it has significance, or when it has meaning. One pays one's money and takes one's choice. And if one chooses the last of the three options, ideology needs consciousness after all.

THREE PROPERTIES OF IDEOLOGICAL ENTITIES

The combining of the two kinds of complexity in human brains can best be regarded as the superimposing of sub-ideological complexity on top of biological complexity. It creates a double level of complexity, and it is this which gives ideas their full potential. The double level will henceforward be called the (full) *ideological level*. The following scheme shows the relationships amongst the various levels.

Levels of Complexity: 0 Physical
                                 1 Biological OR Sub-ideological
                                 2 Ideological: Biological AND Sub-ideological

It was seen in section 5.7 that viewing entities from the perspective of the biological level of complexity leads to observations of properties of choice and purpose. It is natural to wonder what properties may be apparent when viewing entities from the perspective of the ideological (sub-ideological + biological) level of complexity. When ideological systems are considered as integral wholes without reference to their parts, do they have any properties that make no sense from the merely biological and physical points of view?

One ideological property, as just explained, is meaning. At a physical level things and events are because they are and happen because they happen, without significance and without meaning. At a biological level things and events acquire significance but not meaning. A DNA base sequence signifies a protein but it does not mean a protein. An organism's behaviour signifies its purpose but it has no meaning. At an ideological level things and events acquire meaning. Ideas and the behaviour produced by ideas have significance in the shape of realized ideas, but they also have meaning because they exist in a context, a context provided by internal interactions with all the other ideas in the biological idea translator.

A second property is, of course, consciousness. Consciousness arises from scenario enactment and scenarios are a form of ideas which involve choice and purpose. So any structure which is conscious must be ideologically complex; i.e. consciousness requires ideological complexity.

A third property that can be ascribed to ideological complexity is free will. Free will is an experience which accompanies scenario enactment when one out of a number of rival scenarios acquires control of motor output. The experience is one of 'I have chosen'. The choice is supposedly not predictable in any way in advance of its being made.

Viewed from the physical level of complexity there is no observational support for the existence of a property of free will. The brain is a collection of quarks and leptons, bound together in atoms and molecules, exchanging forces in a law-abiding way. The only unpredictability in this picture is that of quantum probability, and that cannot account for free will since quantum probability is random while free will is not.

Viewed from the biological level, the many interconnecting branching and merging cause and effect chains within the cortex give rise to the property of choice. The brain has the ability to choose what to do. For choice to become free will it simply needs to be experienced; that is, it needs to be associated with an I-image. As the I-image comes from scenario enactment, and scenario enactment requires an ideological level of complexity, it follows that free will is an ideological property.

The three ideological properties that have here been speculated on are all subjective. Observers can observe them in themselves but not in anything else. Individually they each attach meaning to ideas, they each have an I-image, they each have free will. In other people — other ideological structures — individuals can only observe assumed objective equivalents of these things; that is, they can only observe behaviour, and in particular the beliefs people communicate about their own experiences. It may thus be that subjectivity itself is the prime truly fundamental property to belong to the ideological level of complexity.

The beliefs that comprise this chapter are speculative. They conform to observation but do not constitute a theory and should be regarded skeptically. If they do correspond to reality (I naturally believe they do) then any difficulty people have in dealing with such abstractions may in part be due to people being, as observers, a collection of beliefs. We are in effect attempting to look at ourselves.

9

MORALITY

9.0 OBJECTIVE: To discuss the nature of morality and the bearing morality has on science.

We saw in chapter 8 that any view of reality, scientific or otherwise, consists of a set of beliefs. Those beliefs, residing in the brains of conscious observers, are images of reality, and they govern observers' behaviour in respect of that reality: beliefs determine behaviour.

One class of beliefs which is very potent as a determinant of behaviour is the class known as moral beliefs. These are beliefs about what is good and bad, and about what behaviour should and should not be undertaken. Because of their potency, moral beliefs are of great importance, both subjectively and in terms of the objective behaviour they lead to. They are thus a major part of reality and warrant attention accordingly.

We will begin by establishing how a moral code arises, and the circumstances to which it can be applied. Consideration will then be given to one notable characteristic of morally determined behaviour, namely, the value assigned to it in terms of goodness and badness. Next, since morality is a collection of beliefs, we shall ascertain to what extent moral beliefs show the kind of belief behaviour described in section 8.5. Finally, we shall examine morals which are associated with the pursuit of scientific method. These are morals which are attendant on scientific research, and morals which play a part in the search for, and communication of, scientific knowledge.

It should be noted that while this chapter is about morality, no attempt will be made to advocate any particular moral code. This is because, as we shall see, there is no systematic or repeatable measurement that can be made which would permit a scientific preference to be established for one moral belief-set over another. Since this book is concerned with scientific reality, i.e. measurable reality, it would be going outside its scope to enter an area where science's principle tool is inoperative. Only one exception will be made to this morally neutral stance, and that will arise when discussing scientific activities which are directly moral. (I refer here to experiments on objects which are conscious.) The nature of the discussion is such that my personal moral standpoint will be clear.

9.1 PREREQUISITES

NINE STEPS

If people are to acquire moral beliefs, they must first interact with their environment in particular ways. These interactions can best be described by listing them as a sequence of nine steps. By the time all the steps have been taken it will be possible to define morality and the small number of concepts crucial to the existence of morality. Each step will be described from the point of view of an idealized acquirer of beliefs, a moral agent, who will be called simply 'the agent'. I shall refer to this agent as 'he', though of course 'she' is equally valid; 'he' will be used rather than 'he or she' solely in the interests of brevity. The 'or she' should be understood throughout.

Step 1. The agent experiences reward and punishment. Imagine for a moment that the agent is in a simple environment which only interacts with him as a result of his behaviour. Everything that happens to him happens as a result of what he does. As was described in section 7.2, one of the functions of his subcortex is to evaluate what he does in terms of its outcome; that is, in terms of what happens to him as a result of his behaviour. The evaluation is one of continue/repeat or stop/avoid. In his simple environment the agent, by virtue of the nature of his brain, will experience these continue/repeat and stop/avoid evaluations whenever he undertakes any kind of motor activity. When an action on his part produces a continue/repeat experience, he will find the action rewarding. Conversely, a stop/avoid experience is punishing.

Step 2. The agent watches the behaviour of other people independently of his own. He observes their reactions to things that happen to them. He finds that he can interpret instinctively the meaning of, for instance, facial expressions displayed by other people in the course of their reacting. This enables him to infer what they find rewarding and punishing. The similarity between his reactions and theirs in comparable circumstances is so marked (talking here about simple situations where the rewards and punishments are crude rather than subtle) that he comes to believe that other people are conscious and experience reward and punishment in the same way that he does.

Step 3. The agent perceives that he can reward and punish others. This comes about because the agent knows what interactions with the environment reward and punish him, and can infer from step 2 that the same interactions have the same effect on others. It follows that by taking appropriate actions he can cause things to happen to other people that he believes will reward or punish them.

Step 4. The agent learns how to cause other people to reward or punish him. This follows from step 3. When the agent imparts a reward or punishment to others, there is a good chance they will behave to the agent in a reciprocal manner. The agent discovers that one of the surest ways to provoke a reward or punishment from other people is to do something to them first that he believes they will find rewarding or punishing respectively.

Step 5. The agent acquires the concepts of good and bad. When a person behaves in such a way as to reward the agent, one of the images that is caused to signal in the agent's cortex is associated with the rewarding feature of the experience. The same image is also caused to signal when the agent behaves in a way intended to reward another person. This image (or concept) is given the verbal label 'good'. Good behaviour, also called 'right' behaviour, is that which is rewarding to the agent, or expected to be rewarding to a recipient on the receiving end of the agent's actions. There are also corresponding concepts for punishing behaviour. These are 'bad' and 'wrong' respectively.

With this step the agent comes to believe that he does good when he gives an apparent reward to another person; that is, he causes the person to behave in a way which indicates that that person has received a reward. He similarly believes he does good when he behaves in a way that brings to himself a reward from someone else. The same applies to punishments: the agent believes he does wrong when he apparently punishes another person, or when he behaves in a way that brings a punishment onto himself from someone else. He also believes reciprocally that other people do good when they reward him and wrong when they punish him. (Obviously at this stage the terms good, bad, right and wrong are being used in a rudimentary sense.)

Step 6. The agent chooses whether to do right or wrong. It was described in section 7.2 how scenario enactment enables the brain to choose amongst the options open to it in any given situation. And in section 8.6 it was seen that free will subjectively accompanies choosing amongst scenarios. So when the agent confronts alternative courses of action, of which one is good and the other is bad (in the sense established in step 5), he experiences the ability to choose between them. Subjectively he decides whether to do right or wrong. Since he expects the former to bring reward and the latter to bring punishment (from step 4), he will want to do the right thing; that is, the outcome of enacting a scenario involving bad behaviour is likely to prove unsatisfactory because of its anticipated punishing consequences.

However, because the rewarding component image in the good scenario is only one of its components, the desire to do right rather than wrong is negotiable. The anticipated reward — which need not actually materialize of course — may be subordinated to other features entering into the rival scenarios. In other words, the choice between good and bad is very much a choice. It has no certain outcome.

Step 7. The agent concludes that other people also choose whether to do right or wrong. This step follows from step 2. If other people are conscious and have the same experiences as the agent, then if he experiences the ability to choose between right and wrong, so will they. The attribution of free will, of freedom to choose a course of action, to all people as well as to himself is one of the most critical steps in the development of the agent's morality.

Step 8. The agent learns to reverse roles with other people. This equally critical step goes hand in hand with step 7. The agent already assumes other people experience the world as he does and have free will. It is but a small extension of this line of thought to conclude that he knows what it is like to be in someone else's place. The ability to imagine — by means of scenario enactment — what it would be like to experience someone else's thoughts makes it possible to greatly expand the range of applicability of right and wrong.

From previous steps, the agent considers that good behaviour on anyone's part is behaviour which brings rewards to him, the agent, and that good behaviour on his part is behaviour which he expects to reward other people. Likewise for bad behaviour. At first sight, step 8 brings only a subtle change to this position. It is a change in perspective, and has two complementary effects. Firstly the agent continues to consider that good behaviour on anyone's part is behaviour which rewards him, but it now becomes good not only because he was rewarded but because the rewarder intended it to be rewarding. And secondly when the agent rewards other people, his behaviour is good not just because he expects it to be rewarding, but because the recipient of it actually is rewarded. The change in perspective comes about by means of the agent imagining himself — through *role reversal* — as doer of the behaviour as well as recipient in the first case, and as recipient of the behaviour as well as doer in the second.

The great expansion in the range of right and wrong behaviour that comes from this change in perspective is found by taking step 9.

Step 9. The agent, through role reversal, becomes involved in the behaviour of other people even when he is not himself a participant. Because he can reverse roles with both the doer of some action and its recipient, he does not actually need to play a part himself to decide whether what takes place is good or bad. He needs only to observe doer and recipient as a third party. With this last step he can have a belief about the behaviour of any other person in terms of whether it is right or wrong, regardless of who is affected by it. All human behaviour, insofar as it affects other humans, becomes right or wrong, depending on how it affects them.

These then are the nine steps people must take if they are to acquire a moral code. It should be noted that the steps are not to be regarded necessarily as occurring consecutively. In practice some may well occur concurrently or perhaps even in a slightly different sequence from that given here.

It can be remarked in passing, given role reversal is absolutely essential to morality, that there is evidence of objective neurological activity accompanying the subjective experience of role reversal. Computer based scans of the brain show distinctive activity in the frontal and parietal lobes, and also in the emotion-producing limbic system (see section 6.5), when human subjects passively watch people behaving.

SOME DEFINITIONS

With the concepts arising during the development of a moral code now elaborated, some terms can be defined.

Firstly: 'should' and its equivalent expression 'ought to'. Both these terms, when applied by an agent to his own behaviour or to the behaviour of anyone with whom he can reverse roles, derive from step 6.

Each scenario he enacts is aimed at achieving a goal. The achievement of the goal within the scenario is one of the factors which decide whether it is put into action or not. But there is an additional factor in his scenarios: the possibility of reward and punishment. It is a factor which has nothing at all usually to do with the goal under consideration. For example, if the agent wishes to obtain something belonging to someone else, the problem is one of how to separate the object from its current owner. Scenarios can be enacted accordingly. But the agent has learnt that the simplest solutions — stealing, basically — may bring punishment. The punishment has no bearing on whether any particular theft scenario actually works, but acts as an extraneous factor mediating against it. This factor can be quite divorced from any specific punishment.

When the possibility of punishment enters into a scenario, it functions as an intruding stop/avoid concept. But its effectiveness is not guaranteed. There are other amoral factors which may outweigh it. It is thus conditional. This conditional stop/avoid concept is given the verbal labels 'should not' and 'ought not to'. Where the concept is one of reward instead of punishment, the labels are 'should' and 'ought to'. 'Should' therefore in moral contexts is a concept meaning pursuit of reward and avoidance of punishment, for oneself or others, subject to other relevant considerations.

Behaviour which should be undertaken is called *moral behaviour*. Moral behaviour is behaviour which involves both a conscious, free-willed doer and a conscious recipient, both parties being individuals the agent can reverse roles with. (A qualification may be added here that some people do not require the recipient of the behaviour necessarily to be conscious at the time of the action. The recipient must, however, be conscious at some later stage, or else might be conscious at some later stage, depending on the effect of the moral behaviour. Think, for example, of behaviour involving human embryos.)

Another term to define is 'moral code'. The moral belief-set of any one individual can be seen objectively to be unique to some degree to that individual. It is this list of shoulds and shouldn'ts of a particular person that is called the *moral code* of that person. A moral code is the list of approved and disapproved behaviours that can be compiled from a person's moral belief-set.

A final definition is of a *moral*. That will be taken to be any one item from the moral code list. Hence a moral code is made of a collection of morals.

HEDONISM

The existence of morality opens the door to a new range of concepts whose meanings won't be defined here, but of which you will be aware. These concepts include guilt, altruism, heroism, justice, condemnation, righteousness, cruelty, dignity, etc. Every concept associated exclusively with the behaviour of people in relation to each other comes in this category.

Inasmuch as these moral concepts contribute to human behaviour through their formation as neuronal signalling pathways, there is no good reason to deny that human behaviour can be altruistic, cruel, righteous and so on.

It is sometimes asserted, though, that behaviour is ultimately selfish; that is, it is always aimed at maximizing rewards and minimizing punishments, a sort of hedonism, though of emotions rather than merely of physical sensations. Some of the nine steps listed above would tend to support this view since they identified good and bad with reward and punishment. And ideologically too, the hedonistic idea is probably tenable, given that the scenario enactments which precede voluntary behaviour are often associated with memories of a stop/avoid or continue/repeat nature. Stop/avoid memories — those associated with punishment — can be expected to put any scenario incorporating them at a disadvantage. The converse is true for continue/repeat memories.

However, this hedonistic notion is at heart an empty one. It reduces to a circular dialogue.

"I act as I do to maximize reward."

"How do you know that a given course of action was chosen in order to maximize reward?"

"Well, it's what I did, isn't it?"

This type of reasoning was commented on when discussing hypothesis 2 in section 2.3. It is a self-contained argument leading nowhere and will not be pursued further.

IRRATIONALITY AND ASSUMPTION IN ROLE REVERSAL

Role reversal is essential for the full development of the agent's morality. It is only through role reversal that the agent can be a bystander and yet still regard an action as good or bad. For example, when he sees somebody doing something unpleasant to someone else, he would say he believes the action to be wrong. And why does he believe that? Because he himself would not choose to behave in that way, and because if the action was done to him he would dislike it. In other words, if he was doer or recipient — if he was in their places — the action would be wrong. Without role reversal he could not imagine himself as doer or recipient and would not be morally concerned with other people's behaviour except when it affects him directly.

Yet the idea of role reversal is inherently irrational. The I-image that forms in people's brains, that leads people to be aware of their existence — to be conscious — is inseparable from the physical structure of the brain. I may be able to imagine that I have left my body; I may even come to believe that such an event is objectively real; but given the truth of chapter 7, nothing of the kind is possible. It can be no more than delusion. And if I cannot leave my body, I clearly cannot be anyone else. To say, as people do when making a bystander's moral judgements, 'if I was Mrs X' or 'if I was a dog' etc. is to talk nonsense. I am not Mrs X or a dog and never can be. Beliefs involving role reversal, as most moral beliefs do, depend on an identification which is completely unrealizable and unrealistic. (Moral beliefs involving behaviour in which the believer is a participant are, in contrast, in all respects eminently realizable and realistic.)

The conclusion to be drawn is that morality, except when experienced by a moral agent as doer or recipient, rests on imagining that something impossible, role reversal, can actually happen. This makes most moral beliefs irrational. However, morals are in this respect like any other kind of belief. They spread in the idea-environment on the strength of their replicative ability alone. Rationality is not of direct concern.

Another noteworthy point about role reversal is the way it relies on unverifiable assumptions (quite apart from being irrational). It is *assumed* other people are conscious and possess free will, and that they are in consequence candidates to reverse roles with. The assumptions are unverifiable because of the subjective nature of consciousness and free will. There is no objective measurement that can be made which could establish whether role reversal in any given instance is justified or not. As a result, role reversal lies completely in the eye of the beholder. Different individuals may confine their use of role reversal to other people, or they may extend it to all sorts of objects. Any object with whom an individual reverses roles is implicitly assumed to be conscious and, as the recipient of behaviour, to be subject to moral beliefs. For example, some people reverse roles with particular animals. Young children may even reverse roles with their dolls and teddy bears. In both cases, behaviour which affects the chosen objects becomes moral behaviour for the person concerned.

The degree to which these extensions of morality to non-humans are plausible will be discussed in section 9.4. Here attention will be confined to illustrating what difference it makes when it is assumed some given event comes within the bounds of morality.

Consider someone kicking a flower. If the agent reverses roles with the kicker but not the flower, the action is amoral. The agent is indifferent to it. If, on the other hand, the agent regards the flower as conscious and consequently finds himself reversing roles with it, he will regard kicking it as wrong because being kicked is painful. The behaviour is then no longer a matter of indifference but becomes something to be actively prevented. The bringing of an event within the moral domain can produce behaviour on the part of bystanders which the same event viewed amorally does not.

Another illustration, this time drawing on role reversal with a doer, concerns a motorist who knocks down a pedestrian. Reversing roles with both motorist and pedestrian the agent regards the motorist's behaviour as wrong. But if it is then discovered the motorist was incapacitated with a heart attack at the time, the assumption that the motorist, as doer, possessed free will would be invalidated (unconscious behaviour being involuntary). The motorist's action would cease to be moral. The agent's inclination to adopt a critical moral stance would disappear. Hence the way the motorist's behaviour is viewed depends both on role reversal and the attribution of free will. In the absence of free will, although role reversal can still take place, the motorist becomes a participant but not a doer. Behaviour in that case loses its moral significance.

9.2 VALUES

The nine prerequisite steps to the acquisition of moral beliefs in fact leave morality still short of its full elaboration. A tenth step is possible. This step, which is not essential to the existence of a moral code, is concerned with gradation. Moral acts tend not to be simply a matter of either right or wrong but of more right and less right, more wrong and less wrong. The tenth step, then, is the assignment of a value to each moral act. The value reflects the degree of good or bad in any particular piece of behaviour.

The immediate consequence of this moral evaluating is that if people face a situation in which the choice is between two good courses of action, they may choose the better course. The same applies to bad courses of action; if people have to do one of two bad things, they can choose the least bad, the one with the least negative moral value.

The assignment of value to each moral act can be pictured graphically. A scale can be envisaged rather like a distance scale found on a map. At one end of the scale is 'very bad'. Moving towards the other end entails passing points marked 'bad', 'slightly bad', 'neutral', 'slightly good', 'good' and finally 'very good'. Every moral act can be given a position on this scale, a position which enables its moral value relative to every other moral act to be established.

The discovery that there are values in morality raises the question of whether there is anything here for science to get to grips with. Science is concerned with systematic measurement, and all measurements are necessarily of values of one kind or another. Indeed, the observational side of science (as opposed to the theoretical side) is predominantly concerned with the measurement of values. What then has science to say about moral values?

To tackle this issue it would help to look first at the measurement of some values which are unconnected to morality. As an example, take distance. To make an objective measurement of a distance, it is necessary that an instrument of some kind exists whose response in any given situation is determined entirely by the distance as it enters into that situation. If, say, it was desired to evaluate the distance between two opposite walls in a house, a tape measure might be used. When extended between the two walls, the value indicated by the tape measure is determined only by the distance between the walls. Other factors, such as the temperature and the time of day, have negligible effects on the measurement. Similarly, if it was desired to evaluate the distance from the Earth to the moon, a pulse of light might be fired towards the moon and a clock used to determined how long it takes for the light to be reflected back to the Earth. Given a constant speed of light (see section 10.1), the time measured by the clock depends almost entirely on the Earth/moon distance alone.

So to make an objective measurement a device is needed which responds to the type of value being measured to the exclusion of all others. Evaluation of such quantities as time, pressure, temperature, mass, force, velocity, direction, rotation, and many more, is made possible by the existence of appropriate measuring devices.

Measuring devices have to work in a particular way. Of vital importance is that their response to what they are measuring is directly observable by human senses. If, for instance, a measuring device was invented which responded uniquely to temperature by producing a flow of electrons in a wire, that device would be useless unless the flow of electrons can be converted into a visible reading on, say, a meter.

It is clear from this that what a measuring device does is convert one quantity into another. It converts distance into a mark on a tape measure; it converts temperature into a reading on a thermometer. The quantity to be measured — which may or may not be observable — is converted into a different — and mandatorily observable — quantity.

It is essential therefore that if a scientific measurement is to be made of a quantity, that quantity must be convertible by some device into a different and directly observable form. By means of that conversion things that are themselves out of reach of human senses can be brought within range.

This is all very well but it begs the question of reliability. How can anyone be sure that the value shown on a measuring device is reliable? Say, for example, there are two weights, one noticeably heavier than the other. The weights are placed in turn on a weighing machine. The weighing machine will clearly identify the heavier one. But suppose it doesn't. Suppose it says they weigh the same. A weighing machine may get away with this on rare occasions — subjective assessments can be mistaken — but if it made a habit of disputing subjective experience, it would be regarded as faulty. In other words, a measuring device and subjective expectations need to be in close agreement with one another. It will then be concluded that the device is indeed reliable and is measuring what it is supposed to be measuring. The reliability of many measuring devices rests in the end on their ability to be in accord, either directly or via other measuring devices, with corresponding subjective evaluations. (See also the comments on the measurement of time in section 10.1, where another criterion for the reliability of a measuring device will be described.)

So far, the measurements being used as examples are all amoral and uncontroversial. But there are also amoral values which are far from uncontroversial. This time take beauty as a good example. No known device exists which can respond to the beauty in an object in such a way that its response is determined solely by the beauty as evaluated subjectively. There does not seem to be any quantity in a beautiful object for a measuring device to convert.

On the face of it this seems a little odd. The trouble is that my brain can evaluate beauty in just the same way as it can evaluate distance. And since I can subjectively evaluate beauty, it follows that the objective equivalent of that subjective evaluation within my brain is an objective evaluation of beauty. My brain can definitely measure beauty. So why can't a measuring device be found that does the same?

Part of the problem is that the assignment of beauty by individuals differs from one person to another; more so than does the assignment of, say, length. Because people's brains are similar, and because the concept of beauty is acquired through people having similar experiences, there is an underlying general consensus on what is beautiful and what is ugly. But this does not extend to detailed agreement. Hence even if a measuring device could be built, its readings would fail the close-agreement test. Some people would doubtless find their subjective evaluations of beauty match those indicated by the device. But only some people. Everyone else would find an unacceptable degree of mismatch. They would accordingly regard the device as not measuring beauty at all, because its evaluations would not be in close agreement with their subjective evaluations of the same quantity.

And the problem applies to far more than just beauty. One may list also humour, peculiarity, excitement, interest, challenge, and so on; and art, the most disagreed upon of all values. No measuring device exists for any of these things.

Scientifically speaking, the obvious answer to the question of why no such devices can be found is simply that there is nothing there to measure. There is no objectively existing characteristic of an object that is its beauty or its art, or whatever. The conclusion follows that those evaluations in the brain which measure these things are not measuring a characteristic of the object concerned as they might measure its distance; the evaluations are instead evaluations of signals as they pass through the brain. The unmeasurable values are not detected by the senses but are added *within the brain* to the sensory information. The brain adds purely internal evaluations to those derived from the external senses.

Evaluations can, then, be divided into two categories. Firstly there are those relating to objectively measurable quantities. These quantities are described as being *intrinsic* to the objects possessing them. Measuring devices can convert such quantities into other forms that can be directly observed. Secondly there are evaluations relating to objectively unmeasurable quantities. These quantities are described as being *extrinsic* to the objects possessing them. Objectively they are not characteristics of objects at all, but are added by observers to their subjective images of the objects.

Science is concerned with measurement and therefore with the intrinsic characteristics of objects and events. Chapters 1, 3, 4, 5 and 6 of this book are almost entirely concerned with that kind of characteristic. Science has far less to say about extrinsic characteristics. Questions arising over these latter characteristics are not resolvable scientifically because there is nothing objective to measure.

And that reveals the answer to the question that initiated this excursion into the topic of values. The question was: what has science to say about moral values? The answer: very little, for moral values are extrinsic. Observers add them to behaviour; they are not a part of it.

So science is silent on the matter of moral values. It can observe moral values through their communication interactions as beliefs; it can see where they come from and how they evolve, but it cannot evaluate them. Disputes amongst people about the relative moral value of this or that behaviour cannot be resolved scientifically. Such disputes must instead be settled by more time-honoured means. People — including scientists — sometimes have to fight for their moral beliefs.

9.3 MORALS AS BELIEFS

<u>THE NEED FOR DURESS</u>

Moral beliefs behave in the idea-environment much like any other kind of belief. They have a tendency to spread throughout the brain population; mutually exclusive moral beliefs compete with one another to that end; and they are communicated through the endeavour of their possessors. They do, however, differ behaviourally in some respects from amoral beliefs. The key difference comes from the association of moral beliefs with reward and punishment.

The nine steps an individual takes on the way to acquiring a moral code do not, of themselves, provide much in the way of moral beliefs. A few beliefs — mostly general in nature — are acquired through instinct. A good example is the moral that is derived from the experience of being aggressive with a submissive opponent. When an opponent behaves submissively it produces a tranquillizing effect on the dominant individual which makes it difficult to continue being aggressive. This effect of perceived submission is instinctive in nature. The instinct leads to the moral belief that striking a submissive opponent is wrong.

Further moral beliefs are acquired based on what individuals discover they like and dislike, or as a result of things they do which unexpectedly produce reward or punishment.

But the majority of moral beliefs do not arise from these simple sources. They are instead acquired through the active behaviour of those seeking to communicate them. This is a characteristic of beliefs generally, but moral beliefs are set apart by the duress which accompanies the communication. In the case of amoral beliefs, communication rests on beliefs merely being encoded, with duress as an ineffective optional extra (see section 8.5). In the case of moral beliefs, communication, at least in the early stages of moral code acquisition, relies on duress as an essential part of being communicated.

The need for duress, as opposed to the mere encoding of the beliefs in question, arises because moral beliefs are not just statements of what is supposed to be true; they are statements of what should be done. They are thus always realizable by way of appropriate behaviour — in principle if not in practice — and depend on the moral believer acquiring the concept of 'should'. This concept, as was seen in section 9.1, is derived from the ability to choose whether to do right or wrong. And knowledge of right and wrong depends on experiencing reward and punishment.

Thus it is clear, as the nine steps listed earlier revealed, that reward and punishment are essential prerequisites for morality. Indeed this point was made right at the start of step 1. The agent needed to experience continue/repeat and stop/avoid responses to his actions. The key fact here is that the response has only one characteristic: that it is rewarding or punishing. The *source* of the response is irrelevant. As long as the agent finds the consequence of an action pleasant or unpleasant, he will be able to learn that it should or should not be done.

This is something which can be taken advantage of when it comes to the active communicating of moral beliefs. Communicators of moral beliefs can intentionally supply a reward or punishment personally. They can, by backing up what they say with duress, increase the likelihood that the agent receiving the moral belief will acquire it. The agent is thus put in the position where he can acquire beliefs about what he should and should not do partly through the usual means of their being encoded for him to sense, and partly because the realized belief — the moral behaviour — really is rewarding or punishing, since the communicator of the belief directly supplies the reward or punishment. (Reward, of course, is given when the moral belief is realized by the recipient of the belief, and punishment when the opposite of the moral belief is realized.)

It must be emphasized here that reward and punishment are to be understood in a very wide sense. The infliction of physical pain is by no means the only form of punishment, any more than being given something desirable is the only form of reward. Rewards and punishments can be psychological as well as physical. For example, the very act of withdrawal by one person from another is something that can be singularly unpleasant. On the other side of the coin, praise can be very rewarding even when unaccompanied by any material benefit.

It is also worth pointing out that the use of duress in the communication of moral beliefs applies to people of all ages. Children, who have few existing moral beliefs for a newly communicated belief to come into conflict with, are most responsive to the use of duress, and in their case the duress often involves physical reward or punishment. But adults too are subjected to it. Even if they are less responsive to new moral beliefs because they already have a moral code, and even if they are less likely to be subjected to physical forms of duress, the duress is still there, albeit not as effective and not as obvious.

MORAL BELIEFS IN CONFLICT

Imagine the agent of section 9.1 possesses a particular *amoral* belief. If he encounters people espousing a rival belief, he might well be inclined to defend his own belief and attack theirs. If he encounters people actually acting on a rival belief, provided the action is itself amoral, he might consider them misguided or mad but he would be unlikely to exert himself much to stop them.

Now imagine the agent with some *moral* belief. His response to someone espousing a rival belief would be little different to his response had the belief been amoral, except that he might be more easily angered. But if he encounters people actually acting on a rival moral belief, he would consider them evil rather than mad and would feel a strong urge to physically cause them to stop. (Evil is a concept associated with people who choose to do what is, from the agent's viewpoint, wrong.) This is because, by role reversal with the recipients of the moral action, he feels that he is personally a victim of the wrongdoing. His instinctive response to being attacked, which is to defend himself aggressively, is activated, and he is thereby impelled to intervene.

It can be concluded that moral beliefs are particularly able to produce behaviour whereby they are defended by their possessors, and whereby their possessors fight against rival moral beliefs in the idea-environment. The abilities of moral beliefs in this respect greatly exceed those of amoral beliefs. And fascinatingly enough, the very act of defending their moral beliefs is something possessors find rewarding, though usually with the proviso that the defence is successful.

There is a corollary to this. If the possessors of a moral belief prevent the holders of an incompatible rival belief from acting, then the rival believers are caused to act from their own point of view immorally. They are accordingly inclined to resist. Unless the rival believers on one side or the other are prepared to acquiesce in the perceived immorality of their opponents, physical conflict is difficult if not impossible to avoid.

This proneness to conflict is made worse by the marked tendency of humans to form themselves into groups. Human groups form in the first place because of an instinct which produces gregarious behaviour. It is an instinct which works to the advantage of beliefs of all kinds, since it makes their proliferation that much easier. The formation of groups affects moral conflicts because one of the groups most people consider they belong to is that group of people who share a broadly similar set of moral beliefs. Religious sects and political parties are examples of such groups. (A person can of course belong to more than one group at a time.)

A group defined by its moral code easily gets into disputes with rival morally defined groups. Apart from it usually being considered good to support members of one's own group, each group member, by role reversal, feels personally attacked by the immoral behaviour of the rival believers in the opposing group. If both groups collectively take up an aggressive stance in response to the immorality each perceives in the other, inter-group conflict results. Potentially such conflicts can be murderous. Moral disputes are one of the causes of human warfare.

Scope for conflict arising from incompatible moral beliefs, whether between individuals or between groups, is increased by the fact that a moral code is not just a collection of morals; it is a collection of moral *values*. It thereby becomes possible for two people or groups to agree that two different actions are both wrong, but to disagree on which is the least bad. For example, if a woman whose life would be endangered by pregnancy becomes pregnant, it may be agreed (and then again it may not) that both to kill her foetus and to endanger her life by not killing it are wrong. But which is the least wrong is open to debate. The potential for conflict is easy to appreciate.

THE SOURCE OF MORALS

How morals are acquired as a result of intentional communication has been described. But how do morals come into existence in the first place?

As was seen at the start of this section, some morals are instinctive. Aside from the example given there, mention can be made of moral beliefs such as that murder, incest and rape are wrong. These beliefs can be assigned to instinct because any species organized along the lines of human tribes whose members were not restrained from these activities would find that they imperilled successful biological replication.

But what of moral beliefs acquired through learning? These beliefs are mostly passed from one generation to the next. Since they are not instinctive, it follows they must have been created at some time in the past by individuals. What presumably happened in most instances is that a particular social problem arose which needed a solution. The problem was solved through creative thinking and presented to the society as a belief about what should be done (i.e. as a moral belief) to solve the problem. The belief was then adopted by enough people to ensure its survival in the idea-environment.

This notion of where most moral beliefs come from is often at odds with the source the believers themselves claim for their beliefs. They claim their beliefs come from a god. (I shall leave this word undefined for now. The scientific view of gods will be explored in section 10.4.) This claimed source for moral beliefs has the effect of altering the supposed status of the beliefs. Moral beliefs created by humans to suit the circumstances of the times can obviously become obsolete as times change. It follows directly that right and wrong also depend on circumstances and can be expected to vary as societies progress and the idea-environment evolves. And it is observably the case that beliefs about right and wrong do change in this way, so much so as to be hardly worth mentioning. But by claiming morals are god-given, a believer asserts that they are absolute and not open to change or obsolescence.

Why would a believer claim a god-given origin for a moral belief? How does a god get involved? When a new moral appears, the person espousing it has to persuade at least some of the populace to adopt it voluntarily. This may not be difficult when a society is under some kind of external threat, but a moral created to deal with internal threats is much harder to communicate successfully. A struggle is likely between those posing the threat and those resisting it. Those who are undecided in the struggle tend to be influenced by the supposed source of the new moral. A claim that the moral comes from a god helps to get waverers to accept the communicated moral as a belief.

The reason why a belief in the god-given origin of a moral facilitates that moral's communication is that it interacts with beliefs people already have about gods (again see section 10.4). The problem faced by the creator of a moral which supposedly comes from a god is not primarily one of communicating the moral belief, but one of communicating the additional belief that the moral is indeed god-given. Success in this latter communication depends very much on what people believe about the person doing the communicating. Not many would-be creators of morals or moral belief-sets have succeeded in convincing their audience that they have been blessed with a revelation from their god. Only a select few have managed to do it with lasting effect in the whole of human history.

This is not to say that a god couldn't pass a moral code to someone — it is self-evident that such a being could — but that a moral code does not *need* a god-given source to account for it. It is quite able to arise in the idea-environment just like all the other beliefs found there. Humans can make their own moral beliefs.

The conclusion from a scientific, observationally based point of view is that instinct is the primary source of morals. To this, secondarily, are added a large number of beliefs created by individuals in response to social problems that have arisen in past times. The whole of non-instinctive morality is an accumulation of these problem solving beliefs.

And this is an on-going process. As new societal problems arise, new morals are invented to deal with them. And as old societal problems pass away, old morals are made redundant. As such, morals are very much a personal matter for each individual. There is no absolute good or bad. There are no moral certainties. There are only beliefs. Those beliefs, wherever they come from, have a great influence on human behaviour. And therein lies the value of morality. Morality is valuable or harmful, not because of its source, but because of the behaviour it leads to.

9.4 SOME MORAL ASPECTS OF SCIENCE

THREE MORAL CONSIDERATIONS IN SCIENTIFIC WORK

In general the practice of science is a pretty amoral business. A question is asked, observations are made, a theory is constructed. The scientist deals with objects to which role reversal does not apply. Neither observing the stars nor colliding particles together nor most other scientific activities can be thought of as good or bad.

Four ways will be mentioned in which scientific activity acquires a moral dimension. Three of them, which will be dealt with here briefly, are ways science indirectly becomes a moral matter. The fourth, whereby science becomes directly moral, will be discussed at length shortly.

The three indirect ways are all reflections of the fact that although most science relates to objects which are not considered conscious and so fall outside the moral domain, the practice of science takes place within human society. Inasmuch as it affects other members of society, it is moral behaviour. Specifically, it affects society by consuming resources; by creating ideas whose realization can physically change the world; and by creating ideas which may alter people's perceptions.

Consumption of resources. Whenever scientists make observations or perform experiments, they spend their time in a way that is not of immediate benefit to anyone. They could have utilized the time rendering assistance of one kind or another to people in need of it. In addition, observations and experiments use equipment which has to be made by other members of society. They, too, could have spent their time making things to use in good causes instead. Expressing it financially, scientists spend on the pursuit of knowledge money which could be put to moral uses. They might for instance build an observatory with money that could have bought an irrigation system for a drought-stricken community.

By taking up society's resources, scientists may deprive their fellow citizens of benefits they would otherwise enjoy, and at worst may add to the injuries of society's weakest members. To resolve this issue society must allocate to science, or more correctly to each individual science, a position on its scale of moral values. Each science can then be given a level of priority in terms of resources. The act of scientific study, for all its amorality, thus has to have allocated to it a moral value.

Technology. Science is at heart an elaboration of the need of people to learn about their environment so as to enhance their chances of survival. It is an exploration of objective reality. Consequently the ideas that are created during the pursuit of scientific investigation, precisely because they are realistic, can often be incorporated into realized ideas. These realized ideas are then able to affect the very reality from whose investigation they spring.

Scientific knowledge, when it is used to make objects whose purpose is to affect reality, is known as technology. While science is concerned with discovering truth, technology is concerned with putting what has been discovered into practical use.

Because technology affects the world, it also affects the world's inhabitants. The use of any given technological device — scientific invention — can be rewarding to people or can have punishing consequences for them. Its use may therefore be either good or bad. The choice of whether to put an invention to good or bad use rests with whoever comes to be in control of it. Whichever way the choice goes, the scientist, as an indispensable link in the chain leading to the making of the invention, cannot avoid a little of the praise or blame attendant on how the invention is employed. Although science's discoveries are amoral, the uses to which they are put are not.

Changing people's world-views. When scientific ideas are communicated they can, like any other kind of idea, affect people's perceptions of themselves and of the world. They are thereby able to cause people subjective pleasure or displeasure, and may even bring about conflict between those who believe some new and iconoclastic scientific idea to be true and those who refuse to believe it and insist it is a lie.

The moral issue in this case boils down to a simple question: is knowledge necessarily good, or can it, without regard to any technological applications, be bad as well? There are often facts about themselves that people would be distressed to know. A scientist discovering such a fact cannot deny that, while the discovery itself is amoral, to communicate it is a moral act.

It is no less true that knowledge can be good or bad for society as a whole. Scientific discoveries, especially when they are incompatible with existing widespread beliefs, can lead to communal unpleasantness without producing any offsetting benefits; the communication of scientific beliefs can, through their effect on human behaviour, have good or bad consequences.

SCIENTIFIC STUDY OF RECIPIENTS

This leaves the fourth moral aspect of science.

When observers reverse roles with some non-human object, they implicitly assume it to be conscious. They may also assume it possesses free will, though this is a minor consideration because people do not in general attribute an awareness of morality to non-humans. The results of these perceptions by a role reversing observer are two-fold. Firstly, any apparently voluntary action carried out *by* a non-human moral object will be exempt from moral considerations because the object lacks a sense of right and wrong. Secondly, in contrast and more pertinently, any voluntary action carried out *on* a non-human moral object by a human will be a moral act. Thus if scientists concern themselves with an object that an observing bystander reverses roles with, anything they do to it will be seen by the bystander as moral behaviour. The practice of science, when it concerns itself with things that an observer can reverse roles with, becomes a direct moral issue.

Two groups of objects can be identified to which role reversal applies. One group consists of other people. Because most if not all observers reverse roles with other people, it is accepted by scientists that actions involving people are moral actions. The rules governing experimentation on humans are very strict, so that when breaches occur they are covert. Only in special circumstances is such experimentation openly acceptable, and then usually the experimentee's consent is mandatory.

The other group consists of a range of animals. Here the validity of role reversal is less generally agreed. And even when it is agreed, the relative moral value of an action involving an animal is likely to be disputed. Since there is the potential in this for direct scientific immorality, the question of role reversal with animals needs detailed examination.

It must be made clear from the outset that what is of concern is the application of experimental techniques as opposed to passive observation; the active doing of wrong is accepted widely as a greater immorality than is the passive doing of nothing to prevent it. Also the discussion here is about animals as objects of specifically scientific investigation rather than as objects targeted by people in pursuit of non-scientific activities. Any conclusions drawn will nevertheless be relevant for all human behaviour involving animals.

CRITERIA OF CONSCIOUSNESS IN ANIMALS

To draw conclusions about the immorality or otherwise of animal experimentation it is necessary to decide, firstly, whether role reversal with animals is plausible. Secondly, if the answer is taken to be affirmative, consideration must be given to the criteria by which moral values are allocated to experiments involving animals. Each of these problems will be looked at in turn.

The view of consciousness developed in chapter 7 associated it with a structure. The structure had to be able to store information derived from sources external to itself; it had to be able to emit signals (i.e. behave) in ways reflecting the stored information; and it had to allow the stored information to interact with itself in a manner leading to the formation of an I-image. For animals possessing a brain — not all do — their brains are structures which demonstrably possess the first two of these abilities. The question mark hangs over the third.

On a quantitative level, chapter 7 provides no clue as to whether animals are conscious; that is, whether they possess an I-image. It is the question of how much ideological complexity is 'sufficient' — a question not addressed in section 8.6 — which holds the key to this. Unfortunately it is a key that is not yet within reach. There is no knowledge at all about how many intermediate neurones an animal needs for the neurones collectively to have that third ability and so give rise to consciousness. The number of neurones any given animal possesses thus currently provides no guidance about whether or not it is conscious (though the relative number of neurones may provide a criterion of the *likelihood* of consciousness, see below). Are there any other criteria by which animal consciousness might plausibly be judged?

An alternative approach used by many observers is to attribute consciousness to animals on the basis of the ease with which they can reverse roles with them. This is unsatisfactory because it puts the attribution of consciousness in the eye of the observer, much as the beauty-value discussed earlier was observer-added; it treats consciousness as an extrinsic property. As such, the criterion does not depend on the properties of animals' brains but on signals in the observer's. It is accordingly an implausible criterion to use. In fact it tends to rest on the response of the observer to animals' physical characteristics. Animals with large eyes, especially at the front of the head, those with hands shaped like human hands, those that walk or sit upright, all are easy to reverse roles with. This criterion of consciousness, while understandable, is not rational and cannot be expected to carry any intellectual weight in disputes over the moral or amoral nature of animal experimentation.

Animal behaviour is a characteristic that offers much more hope. The behaviour of an animal is the external manifestation of what is taking place in its brain, which is in turn a manifestation of the images stored there. It can be proposed that the more nearly human-like an animal's behaviour is, the more likely it is that the animal is conscious. If it acts like a human, it probably thinks and feels like one as well. Behaviour thus provides an objective criterion against which consciousness can be judged.

To what extent do animals behave like people? At first glance, hardly at all. They don't make or use tools, don't worship a god or know they are going to die, don't possess any awareness of a moral code, don't have a sense of humour, don't fight wars, don't produce works of art, and so on. Above all else, they don't use language.

A list of this kind would seem to cast grave doubt on the possibility that animals are conscious. There are, however, a number of criticisms that can be raised about the kinds of behaviour listed.

Firstly, the behaviours with which the list is concerned are associated with the most sophisticated information processing properties of the human brain. Concepts of the highest abstraction (self-death, morality, art, etc.) are involved. But it was proposed in section 7.2 that the formation of the I-image occurs, not in conjunction with abstract concepts, but during the course of scenario enactments relating to concrete situations. None of the abstract behaviours listed above is necessary for the enactment of scenarios dealing with practical matters. It therefore seems likely that I-image formation and the concomitant attainment of consciousness arise at a lower level of neurological organization than that needed to support abstract concepts.

Secondly, following on from the first point, when consideration is given to practical as opposed to abstract human behaviour, there is much that humans and some animals have in common. This is mainly the case with animals possessing comparatively big brains, and particularly those which are social. Maternal or parental care of the young, the establishment of hierarchies amongst individuals within the social group, the use of gesture to convey information about subcortical neuronal signals which, in people at least, are matched with emotions, all are behaviours shared with humans.

Thirdly there is the issue of language. It is sometimes claimed that language — speech or sign language or some equivalent form of complex symbolic communication — is either essential for consciousness or is an inevitable accompaniment to it. If a thing can converse, *then and only then* does it know it exists. So far as is known no other animals are capable of communicating in the way humans are. If the equating of speech with consciousness is correct, the conclusion is that only humans are conscious and that all other animals are amoral objects.

Needless to say, the equation and its ensuing conclusion are highly questionable. Doubt arises because not all humans can speak. It is quite possible for the speech areas of a person's brain to be destroyed, or even for the entire speech hemisphere to be lost. Patients in cases of this kind will still be able to display a degree of responsiveness sufficient to convince most observers that they remain conscious; they continue to be regarded by other people as moral objects despite their being speechless. It seems probable that a brain which has lost its store of words, including the word 'I', has not lost the concept of 'I' — its I-image — but simply no longer has access to a verbal label for that concept. It only follows that absence of language (in humans or animals) implies absence of consciousness if the I-image is confined (in humans) to the speech areas of the brain, which is very unlikely.

It can be concluded that abstract and linguistic behaviours are less plausible criteria of animal consciousness than are behaviours of a more practical kind. So is there any aspect of practical behaviour that might be particularly indicative of consciousness?

One possibility is the displaying of behaviour which is complex enough to suggest the presence of free will. It will be recalled from section 8.6 that free will is an ideological property; that is, it requires consciousness. Because of its subjective nature, free will cannot be observed. But the ability of an animal to enact scenarios of a complexity suggestive of free will can be. If an animal can be observed making a deliberate choice in a complex situation, this could be considered a good indicator of consciousness.

But how complex must the situation be? Would it be sufficiently complex if the choice has as one of its options a biologically disadvantageous course of action, and that the animal, at least some of the time, chooses the disadvantageous course? If the animal's behaviour is

424

controlled by biology, it would choose the advantageous course automatically every time. But if ideology controls its behaviour, scenario enactment, and only scenario enactment, could bring about the choice of a disadvantageous course.

Acting in biologically disadvantageous ways is very much a characteristic of human behaviour (remember the ideological subversion of biology discussed in section 8.5) and is clearly associated with free will and thereby consciousness. Thus it is suggested that an animal whose behaviour is so sophisticated that it can choose to do something biologically disadvantageous will possess an I-image and so be conscious.

As an example, consider the training of dogs not to eat their food without first being given some command word. The training need not involve the infliction of physical pain. Some dogs, but not all dogs, can be trained not to eat even to the point of starvation in the absence of the appropriate command. In the circumstances it seems very unlikely that a well-trained dog has learnt not to feel hungry, and indeed the smell of the food can be expected to produce all the usual behaviour short of actual eating. The dog wants to eat but chooses not to. It has in essence acquired a moral belief from its trainer, a belief that it is wrong to eat without permission. Certainly a human behaving like the dog would be regarded as probably being motivated by a moral belief. This correspondence between humans and dogs can be further emphasized by pointing out the resistance of some dogs to training. This has a human counterpart in the variability amongst people to do as they are told; that is, to do what it has been made clear to them they should do. And therein lies the key point. The dogs' behaviour resembles human behaviour. It is reasonable therefore to consider that dogs are conscious.

Unfortunately, very few observations have been made to identify which animals can learn to reject biologically advantageous courses in favour of disadvantageous ones. The size and organization of a brain that can do this is correspondingly unknown. Is a cortex required? And if so, how much of one? All mammals possess a cortex, so are all mammals capable of being conscious? My own view, drawn from anecdotal evidence, is that apes and monkeys, elephants, some if not all cetaceans, and the aforementioned dogs and their carnivore relatives, plus equines and pigs, are able to carry out scenario enactments and can be presumed conscious. They would appear to have 'sufficient' ideological complexity.

In future these animals will be referred to collectively as the higher mammals.

The conclusion that the higher mammals are conscious cannot be the end of the matter. Biologically disadvantageous behaviour implies consciousness, but failure to exhibit such behaviour does not imply no consciousness. Lower mammals, whose behaviour is firmly rooted in biological advantage, may simply have a less sophisticated conscious awareness. Experiments on rats have shown their brains have a 'pleasure centre'. Does that make sense unless rats are conscious? And what about some or even all of the other vertebrates? Crows especially can readily be observed behaving in ways which compare strongly to conscious human children. Nor can it be considered safe to exclude some invertebrate animals. In this last category there are, for instance, the super-brains of the molluscan world, the octopuses, whose ability to handle information (and enact scenarios?) is far in advance of their primitive physiology.

CONSCIOUSNESS BY DEGREE

By now it must be woefully obvious how few observations there are that can be used as a guide to attributing consciousness to animals. In view of this disturbing ignorance it is not possible to achieve any consensus as to whether experimentation on this or that species of animal comes within the moral sphere. Nor is lack of observation the root of the problem. The real problem is lack of agreement on what it is that should be observed. It has been advocated above that specific types of behaviour may provide a useful criterion, but other rival criteria such as possession of language also have their supporters.

And it is not even accepted that behaviour — of whatever kind — should be the determining factor. A completely different alternative arises from the following proposal: the amount of consciousness present in any animal is a function of the number of its intermediate neurones multiplied by the extent of their interconnectivity. (See Bibliography: Rose for advocacy of this proposition.) The criterion of consciousness in this case is the anatomy of the animal's brain rather than the way it behaves.

This proposal clearly implies that all animals with interconnected intermediate neurones are conscious, and raises new fundamental questions about the nature of consciousness. It also embraces a full solution to the problem of allotting a relative moral value to animal experiments. How does it do this?

To say that consciousness is a function of the number of neurones and their interconnectivity is in effect to say that there is such a thing as a *degree* of consciousness; speaking very broadly, the larger the brain, the more conscious it is. Thus humans are more conscious than chimpanzees, which are more conscious than dogs, which are more conscious than worms. From this it is defensible to regard all animal experimentation as wrong, and to go on from there to assign a value to the wrong proportional to the amount of consciousness thought to be present in each case. It then becomes possible to argue that while experiments on humans, chimpanzees, dogs and worms are all wrong, it is less wrong to experiment on chimpanzees than on humans and more wrong to experiment on dogs than on worms. Given the necessity of choosing, it is better to save a human's life than to save a chimpanzee's, and better to save a dog's life than a worm's. The opposite applies to the sacrificing of a life. The moral dilemma in scientific experimentation on animals is thus resolved. The potential indirect good that can be done to people by acquiring a given piece of knowledge can be weighed against the bad done to an animal if that knowledge is obtained at its expense.

This way of settling the morality of animal experimentation is so easy and convenient that it is widely, if not explicitly, used. Unfortunately it must be severely questioned. In the first place, while brain anatomy undoubtedly gives an indication of what the brain is capable of, behaviour alone can reveal its actual abilities. And in the second place, the idea of degree of consciousness is compatible with the account of consciousness provided in chapter 7 only with difficulty.

The difficulty arises because the formation of an I-image is an all-or-nothing event. Either a brain contains an I-image or it doesn't. It is thus either conscious or it isn't. Degree enters into this picture only in a limited way. It does so, not as degree of consciousness in itself, but as degree of what the brain is conscious *of*. The brain (of both humans and animals) may be conscious of what its senses are detecting to a variable degree, and it may be conscious to a variable degree of the signal processing taking place inside itself. In terms of the I-image, this is equivalent to the extent of I-image activation at any given moment. (Think, for example, of alcohol intoxication here.) But it remains the case that while what a brain is conscious of may vary, it is still either conscious or it isn't.

From a subjective point of view the idea of consciousness by degree is not easy to deal with either. What I am conscious of in the way of sensory signals and what I am conscious of in the way of scenario enactment may vary according to whether I am wide awake or drowsy or drunk or meditating, but I talk of being conscious of a lot or a little, not of being more conscious or less conscious. It is hard to see how I can half know something or know it more or less than at other times. It is equally hard to see how a conscious animal can know what it knows less than I know what I know.

When dealing with purely subjective ideas such as those of the preceding paragraph it is important to be aware that different people may experience consciousness differently from each other. Degree of consciousness may seem perfectly sensible to some people. Accordingly it would be unwise to be dogmatic. I shall however stick with the account of consciousness given in section 7.2; namely, that the I-image is an all-or-nothing affair.

This is not, though, to reject consciousness by degree completely. It can still be a useful concept. Consider, for example, frogs' brains. These very small, cortex-less brains have been shown to respond only to a limited range of visual inputs; namely, food, predators and other frogs. Visual stimuli falling outside this very select list produce no response in the vision-processing part of frogs' brains. What frogs are visually aware of — should they be conscious — is therefore highly circumscribed. This shows that what an animal is fully conscious of depends fairly directly on the size of its brain; i.e. the number of neurones multiplied by their interconnectivity. Provided it is understood in the limited sense of 'consciousness of', consciousness by degree can be a credible adjunct to an all-or-nothing I-image.

ALL-OR-NOTHING CONSCIOUSNESS

If consciousness is fundamentally an all-or-nothing phenomenon, then the conclusions about moral values that follow from consciousness by degree must be rejected. Where does that leave the morality of animal experiments?

As a preliminary to answering this question it would be reasonable, having criticized consciousness by degree, to address two criticisms of all-or-nothing consciousness.

One criticism is this. I was not conscious when I was conceived (obviously enough) but I am conscious now. Without consciousness by degree, it follows that there must have been a moment of transition when I became conscious for the first time. Such an event might be thought to have dramatic behavioural consequences. Yet there is nothing in what has been said so far about consciousness to give any clue as to what those behavioural consequences might be. And observationally there is no obvious discontinuity in the behavioural development of a child that might clearly be associated with the onset of conscious awareness.

In defence of all-or-nothing consciousness against this criticism it can be remarked that many of the ideas created by people during scenario enactment are produced in a flash of inspiration. It is not implausible that perception of the self could also arise in this way. But beyond that it has to be admitted that the behavioural changes which might accompany this most profound of inspirations can only be speculated on in a way that borders on fancy. Perhaps a child's first temper tantrum marks the moment when, in the course of its being frustrated, the interacting signals in the synthetic cortex form the crucial image of themselves, and the child perceives in a flash that 'I' am being frustrated. It is only a wild guess. Against this particular suggestion is the fact that any one child's first temper tantrum occurs over a highly variable timeframe; anything from a few months of age to about three years. And also, parents may well believe their baby is conscious long before they witness its first tantrum. Could they be mistaken?

The other criticism of all-or-nothing consciousness — and one which is of direct relevance to the morality of animal experimentation — is concerned with threshold brain size. See figure 9.1. Take it, as was done when considering consciousness by degree, that the number of neurones times their interconnectivity can again be used as a quantifiable definition of brain size. Clearly some animals will have brains too small to contain an I-image, others will have brains of sufficient size — of sufficient ideological complexity — and yet others will have brains that are borderline cases; that is, their brains will be on the threshold of consciousness.

In terms of figure 9.1, some animals will have a brain size below n_1.

They will not be conscious. Some animals will have a brain size above n_2. In their brains the chance of the I-image flash of inspiration occurring is so high as to be more or less certain. These animals will be conscious. Animals whose brain size lies between n_1 and n_2 are the threshold animals. Being on the threshold they may or may not individually acquire an I-image. Some will, some won't. The probability of their doing so will increase with brain size from no chance at n_1 to near certainty at n_2.

The criticism which arises out of these considerations is that for any one species whose brains are on the threshold, only some individuals will have the flash of inspiration leading to consciousness. Again, it would be expected that the absence of an I-image in the rest of the species will have behavioural consequences. Those endowed with an I-image will behave differently from those not endowed with one. (It is assumed both here and in the previous criticism that consciousness does, of itself, affect behaviour, although this does not follow directly from the account of it

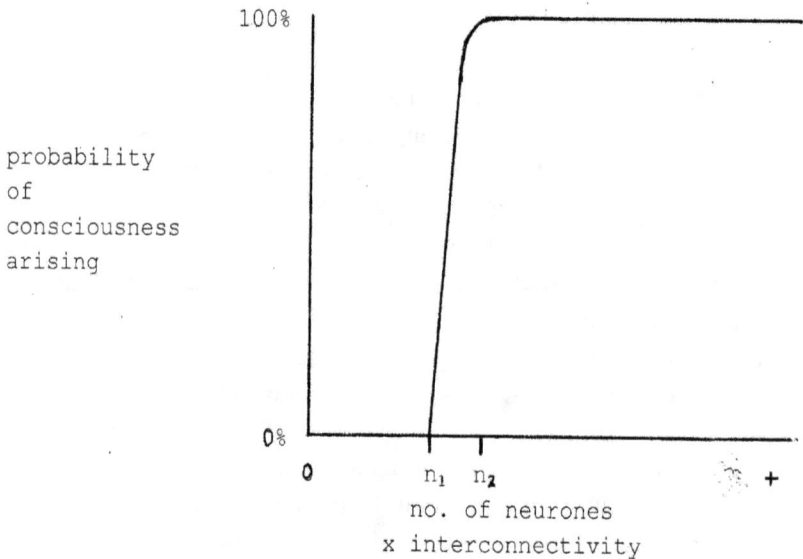

Figure 9.1 A speculative graph of how the probability of consciousness arising in a brain varies with the number of neurones and their interconnectivity if consciousness is an all-or-nothing quantity.

given in chapter 7.) The difference in behaviour should be observable. And yet no such observation can be cited.

The trouble, as with the previous criticism, is lack of knowledge. The sizes of n_1 and n_2 are entirely unknown, as is the difference between them. It can only be said that n_2 is less than $n_{\text{human brain}}$. Consequently it cannot be known whether diversity of behaviour should be sought in spiders, or frogs, or monkeys. It cannot even be said whether there are currently on the Earth *any* species with brains between n_1 and n_2 in size. After all, if there are behavioural advantages to acquiring consciousness, then when a few individuals in a threshold species cross the threshold, those individuals should be better at surviving and replicating than their unconscious companions. Within a few generations all the members of the species would be expected to acquire consciousness, the unconscious variant becoming extinct. In evolutionary terms it would not be a stable situation for a species to straddle the consciousness threshold.

The problem, to restate it once more, is that it is not known what behavioural changes consciousness produces in individuals or species. (The human-like behaviour mentioned earlier as a way of identifying a conscious species is actually a behavioural *dis*advantage, and so can't be the behavioural change being sought.) As a result, all-or-nothing consciousness hinges on one fundamental and wholly unanswered question. When going up the scale from primitive unconscious animals to advanced conscious ones, or when following a human baby from its unconscious conception to its conscious childhood, a transition from unconsciousness to consciousness *must* occur. Subjectively that transition is beyond evaluation: the difference between nothing at all and something priceless. But objectively? How is the transition manifested? *What objective behavioural difference does consciousness make?*

It has to be admitted that answers to criticisms of all-or-nothing consciousness are nicely insulated from theoretical and observational support, at least for the time being. It can only be said that any idea about something is better than no idea at all, so long as the shortcomings in the idea are openly acknowledged.

CONCLUSIONS CONCERNING ANIMAL EXPERIMENTATION

Bearing this catalogue of ignorance in mind, what conclusions about the moral value of animal experiments can be drawn from all-or-nothing consciousness? There are two principally.

The first is that since the size of the speculated n_1 and n_2 are unknown, it is only possible to assign a vague likelihood to the members of any given species being conscious. That likelihood decreases as the actual size of n — number of neurones times interconnectivity — decreases.

Fortunately the argument advanced earlier about the presence of consciousness in the higher mammals was concerned with behaviour resembling human behaviour, not brain size. The view that higher mammals are conscious is therefore unaffected by ignorance about the behavioural changes that accompany crossing the threshold of consciousness. Only when proceeding to animals with smaller and smaller brains does brain size become an issue. Except for higher mammals, the question of whether a given animal is a moral object thus becomes a matter of decreasing likelihood as brains get smaller.

The second conclusion is that if consciousness is an all-or-nothing affair, it cannot be argued that bad behaviour affecting conscious animals is less wrong than bad behaviour affecting people. If it is wrong to experiment on a person then it is equally wrong to experiment on a conscious animal. Both classes of objects are equally conscious. Role reversal is fully valid in both cases. The argument that experimenting on an animal may save much human suffering at the expense of the animal's suffering is certainly still morally sound. Adding up all the good and bad values gives a total of net good. But if conscious animals and humans are equally conscious, the same argument justifies experiments on unwilling humans. And by the same token, the arguments against this latter behaviour also apply with equal strength to conscious animals.

These two conclusions make it possible to be quite definite about the morality of animal experiments. Consider the field of brain research. It is undeniably true that a large amount of the knowledge set down in the neurobiology chapter of this book has been obtained from experiments on higher mammals, particularly apes, monkeys, dogs and cats. Many neurological experiments rely for their rationale quite explicitly on the assumption that these animals' brains are sufficiently human for the results to be applied to humans. That assumption casts a very dark shadow over the scientists concerned, since it denies them even the excuse of ignorance. They can be in little doubt that their victims (for such they are)

are conscious. Despite this, they have in the course of their experiments destroyed, at one time or another, every part of these near-human animals' brains. Animals have been blinded, crippled, paralysed, have had electrodes cemented into their heads, have been subjected to repeated electric shocks, have been dosed with chemicals toxic to neural tissue. The catalogue of tortures is large and horrible.

It would be an insult to neurobiologists who experiment on higher mammals to accuse them of being callously indifferent to the wrong they do. Yet it is hard to see how it can be otherwise, since if they were not indifferent to their wrongdoing they would not continue with it without very good reasons. With few exceptions, very good reasons are not forthcoming. Most of their experiments are primarily conceived out of the usual scientific curiosity or out of a desire on their part to contribute to their field of knowledge. Theirs is a case of direct scientific immorality with little if anything to excuse it. Yes, the brain is a marvellous structure and knowledge about it is fascinating; but no, it is evil to obtain that knowledge by experimenting on higher mammals.

Nor in fairness are neurobiologists alone to be singled out. Animal biology in general is a fertile field for animal experimenters. Although they tend to deal with animals which are less likely to be conscious than are the higher mammals, there is often enough doubt about the status of consciousness in the animal under investigation to warrant circumspection. In far too many cases that circumspection is not evident.

9.5 THE MORALS OF SCIENTISTS

When scientists carry out typical amoral scientific research, their moral beliefs enter into their behaviour to some extent. This is because they do not undertake the research in isolation. Their work entails them interacting with other scientists before, during and after they carry it out.

The community of scientists engaged in these mutual interactions can be discerned collectively to subscribe to two principal moral beliefs. These are that the communication of knowledge, except possibly about people, is good, and that dishonesty about what is observed is bad.

As far as communication of knowledge being good is concerned, there is not a lot to add to what was said in the previous section. Suffice it to say that this particular moral belief is an essential member of any belief-set, including the belief-sets of scientific knowledge, which is to stand much chance of surviving the deaths of its possessors. This is because survival demands communication. And if beliefs are to be communicated it is necessary that believers think their beliefs are worth communicating. That means in turn that either believers must receive some benefit from the communicating, or else they must consider the communicating to be morally good; that is, something that should be done regardless of material benefit.

Communication of scientific beliefs is not generally something that brings much in the way of benefits to the communicators; at least not that they couldn't get equally well by communicating non-scientific beliefs. So for scientific belief-sets to replicate in the idea-environment, (most of) their adherents have to regard their communication as morally good.

On the matter of dishonesty, science is to some extent self-policing. All scientific work, observational and theoretical, is liable to scrutiny by other scientists, thereby ensuring that dishonesty is detected. Accidental errors are also brought to light by such scrutiny. Neither dishonesty nor error is likely to endure for any great length of time.

(It is interesting to note, though, that in practice scientific scrutiny is actually a rather hit-and-miss business, tending to be done haphazardly and not on a systematic basis. This is perhaps because there is not much excitement and kudos to be had from repeating other scientists' work.)

The importance of honesty in the practice of scientific method can be appreciated by considering what would happen if most scientists were dishonest about their observations. Scientific data would become value-less, and science would be reduced to chaos. Fortunately dishonesty is not widespread.

When it does occur, dishonesty often takes the form of a scientist falsifying the results of observations or experiments in order to give apparent confirmation to a pre-held theory. Because this dishonesty is rooted within the scientific community, it can easily become accepted scientific knowledge. It may not be until perhaps many years later that someone engaged in scrutiny uncovers the deception. As the consequences to science of fraud becoming commonplace are so severe, the penalties for getting caught are also severe. Scientists who report observations they have not made risk being regarded with a degree of

434

opprobrium proportional to their seniority and to how successful their lies were during the time they went unexposed.

As with scrutiny though, practice and principle are not wholly in accord with one another on this issue. A scientist who makes up observations to support a theory which transpires in due course to be borne out by genuine observations may well be lauded for possessing great insight rather than condemned for jumping the gun. Morally this is rather dubious.

Aside from the two morals of communication of knowledge being good and dishonesty being bad, there are few other morals that play a big part in governing the practice of scientific method. It might be wished that science could provide a moral code of a more embracing nature for its adherents, but attempts to construct such a scientific morality would fail to be convincing. Moral values cannot be measured. In the absence of measurement, scientific method simply doesn't work. Any scientific moral code would lack the very foundation on which science depends. It would be missing the key characteristic — measurability — which makes science what it is.

10

TIME AND LIVES

10.0 OBJECTIVE: To put the objective reality science describes, and the observer science encompasses, into perspective as a part of human experience.

Perhaps the most all-pervading feature of reality, the one that we are aware of throughout our lives, is the passage of time. Time defines a beginning and an end for each of us, and passes inexorably as we travel from the one to the other. There is no way to escape its limits, nor to exert any control over where we are within those limits. For these reasons time is a very powerful component of reality. We shall begin therefore by looking at time objectively, finding out how it is measured, and what the passage of time amounts to when expressed in scientific terms. In doing this, arrows of time will be encountered. These arrows can point in one of only two ways, and so make it possible to distinguish two directions. How arrows of time give rise to the subjective experience of time — our sense of past, present and future — will be described. We shall then see what science has to say about the future of various major parts of reality, from human life to the entire universe. Earlier chapters having described how things got from their beginning to now, this look at the future will complete the scientific account of the key components of objective reality.

With the whole picture at our disposal we shall next confront a second pervasive feature of human perceptions, the question of purpose: what is existence for? We shall see that the answer provided by science is terse and, for the many people who need — or at least would strongly prefer — an answer which is grand and inspiring, distinctly disappointing. For these people, this lack of emotional satisfaction may lead to an openness to additional 'truths'. Non-scientific sets of beliefs are thereby enabled to thrive which propose a purpose for human life and the universe which lies outside the reach of observation. These unscientific belief-sets are called religions, and we shall investigate what science has to say about them.

The last two sections of this chapter are predominantly philosophical in content. As with earlier parts of the book of a philosophical orientation, these sections should not be thought to be advocating a standard view of scientists to the issues under discussion.

10.1 THE PHYSICS OF TIME

THE MEASUREMENT OF TIME

The first thing to establish about time is how it is measured. Such an endeavour may seem rather lacking in profundity, but it brings out some important preliminary points.

Time is, of course, measured by clocks. A clock is some device within which an event occurs repeatedly at regular intervals. The swinging to and fro of a pendulum is an example of a suitable regular and repeating event. The rotation of the Earth with respect to the sun is another. Basically, anything which vibrates or rotates, from atoms to pulsating stars, can be pressed into service as a measurer of time. So can objects in which the repeating event is more contrived; for example, candles with marks at regular intervals down their sides, and hourglasses. In each case, what matters is that the measurement made by the clock solely reflects the amount of time as it enters into the measurement situation. Other factors, such as what the weather is like and the speed the clock is travelling at, should have negligible effects. This much was established in general terms in section 9.2.

There is a particular problem associated with time measurements. Normally, subjective evaluation of the measured quantity is compared with the evaluation given by the measuring device to see if the two closely agree with each other. If they do, it can confidently be believed that the measuring device is measuring the chosen quantity correctly. Unfortunately, subjective evaluations of amounts of time are often at odds with the readings given by clocks. People who are bored tend to think more time has passed than is indicated by a clock. And people in a hurry frequently can't believe how quickly the limited time available to them has been used up. How can someone, in these circumstances, be confident that clocks are indeed measuring the time accurately?

Since clocks involve repeating events, it is inevitable that time will be a major component in the evaluations clocks provide. But other factors might also be present. These would take the form of distortions to the time component. So the problem in effect comes down to a matter of regularity. How can it be established that the repeating event associated with a given clock is repeating free from extraneous distortion (i.e. regularly)? And when an event is taking place which subjectively seems to be regular, such as the audible ticking of a clock, how can it be established that the event is occurring at the same rate now as it was a while ago?

When doubt over subjective observations arises, the scientific response is to make measurements. The trouble with clocks is that the only way to establish by measurement that a clock contains a regularly repeating event is to time the event, which naturally requires another clock. The regularity of the repeating event taking place in this second clock is itself open to doubt. To establish objectively that the second clock is, in turn, regular needs yet another clock. And so on, ad infinitum. The conclusion seems to be that there is no absolute standard of regularity on which to base measurements of time.

To resolve this dilemma it is necessary to *assume* that one of the clocks — it doesn't matter which one — contains a regularly repeating event. Now, every type of clock uses some particular property of matter to produce its supposed regularity. Different types of clock use different properties. The clock which is assumed to be regular can be compared with clocks using other properties of matter, and the question asked: is the chosen clock regular with respect to the others? If it isn't, then the initial assumption was wrong. But if all the other clocks are regular with respect to the chosen clock, and thereby with respect to each other also, the assumption can be considered justified. The chosen clock, and all the others too, can be regarded as measuring time in a regular fashion.

In practice it is (helpfully) found that each type of event incorporated into clocks occurs regularly with respect to all the others. This correspondence amongst different clocks gives firm grounds for believing that their measurements of time are free from extraneous influences. As such, their agreement can be taken as definitive of regularity, since if it was proposed that all clocks are identically irregular, it would have to be asked: "irregular with respect to what?" If all clocks are identically irregular, the question clearly has no meaning.

By using the correspondence between different types of clock to define regularity, it is possible to establish an objective unit of time arbitrarily. This can be a unit which is equivalent to, say, k rotations of the Earth, l vibrations of a particular atom, m swings of a particular pendulum, and so on, where k, l and m are fixed numbers whose ratio, $k:l:m$, never varies. This is the foundation on which the objective measurement of time rests.

THE STRANGE EFFECT OF SPEED

Consider a clock based on the following mechanism. A flash of light is produced at a point in space. The light travels to a mirror, which reflects most of it back to its origin. The return of the light to the origin causes 1 to be added to a counter and also triggers off the next flash. The cycle continues indefinitely. Obviously, the value shown by the counter will increase regularly. Knowing that the speed of light is $2\cdot998 \times 10^8$ m s^{-1}, if the distance from the source of the light to the mirror is $149\cdot9$ m, say, the value on the counter will go up by one million every second.

This kind of clock can be expected to be regular with respect to every other kind, and observation shows that it is indeed so. For instance, in the same period of one second a given pendulum may oscillate once and a given candle may burn down by 10^{-2} mm. The numbers 10^6, 1 and 10^{-2} are fixed, and their ratio, $10^6 : 1 : 10^{-2}$, never varies. The three kinds of clock thus establish each other's validity in a reciprocal fashion.

Now perform an imaginary experiment. Take the Earth, on which observers and the three clocks are located, and cause it to travel through space very fast at a steady speed. The pendulum, which works through the effect of the Earth's gravity, will be unaffected by the motion, since it is travelling with the Earth. The candle, which works by means of chemical reactions, will also be unaffected. The flashing light clock though is more doubtful. Picture the clock being oriented at right angles to the Earth's motion. Its light will still have to travel the same distance from source to mirror and back again, but it will also have to travel a certain distance forwards due to its need to keep up with the speeding Earth. The forward distance will be added to the normal source-to-mirror distance, thereby increasing the total distance the light has to travel. As a result, the flashing light clock will count more slowly than the other two kinds of clock when travelling at speed. It will get out of step and cease to qualify as a good clock.

There is nothing especially disturbing about this revelation. It's just a shame that the flashing light clock isn't such a good idea as it at first seemed. But, of course, to be scientific about it, this conclusion must be checked against observation. Fortunately this can be done in practice as well as in imagination, for the Earth is already travelling fast through space. Its velocity as it orbits the sun is about 30 km s^{-1}.

Suppose the flashing light clock is set up at right angles to the Earth's motion and the ratio of its counter to the pendulum and the candle is established at that time. Then move the clock so that the flashing light to mirror direction is aligned parallel to the Earth's motion. The distance the light has to travel to the mirror and back will be different from the distance it has to travel when the clock is set at right angles. That difference will be manifested in an alteration in the rate at which the flashing light clock counts. The ratio of its counter to the pendulum and the candle will differ between the two orientations. (In practice, far more accurate clocks than a pendulum and a candle would be needed to detect the predicted difference.)

The experimental check that the flashing light clock gets out of step with clocks working by other means yields an unexpected surprise. No irregularity can be detected. No matter what the clock's orientation, and no matter which direction the Earth is heading in as it orbits the sun, the clock remains in step with the others. The flashing light clock is a good clock after all.

The predictions about how and when a flashing light clock should prove to be irregular are clearly wrong. That in turn means the theory on which the predictions are based must also be wrong. The trouble is that the theory is scarcely a theory at all; it is more a set of elementary and obvious assumptions. What are these assumptions? Three spring readily to mind.

The first is that the Earth is moving through space. Could it be instead that the Earth is actually stationary and the rest of the universe moves round it? That would explain the regularity of flashing light clocks, but such a solution is incompatible with the rigorously observation-based cosmological model (described in section 4.3) that has been constructed over the last hundred years. The assumption that the Earth moves is really very safe.

The second assumption is that the speed of light is independent of the speed of the emitter of the light. Perhaps instead, it depends on the emitter's speed. Experiment shows that if the emitter moves while the observer keeps still, the speed of light is unaffected by the emitter's velocity. The second assumption is accordingly supported by observation and is therefore valid also.

The third assumption is that measurements of distance and time are fixed, while measurements of the speed of light vary depending on how fast an observer is travelling. Could it be the other way round? Could it be the speed of light which is fixed and distance and time which vary? The consequences of this possibility are described by the Special Theory of Relativity.

THE SPECIAL THEORY OF RELATIVITY

Observation shows that the speed of light is indeed fixed — or 'invariant', to use the technical term. In that case, if a flashing light clock is to stay in step with other kinds of clocks when moving fast, something odd has to happen; specifically, when travelling at speed, all clocks — not just flashing light clocks, but all clocks — must become equally irregular with respect to comparable stationary clocks. The moving flashing light clock *does* get out of step, ticking more slowly than *stationary* clocks. But it doesn't get out of step with clocks travelling along with it. Only clocks travelling at different speeds or in different directions get out of step.

This is a neat explanation which accounts for what has been observed thus far, but it creates a problem worse than the one it solves. Why should a pendulum swing more slowly, and a candle burn more slowly, just because they have a speed? Since the speed affects all clocks equally, the implication is that speed affects the regularity of time itself. Time passes at different rates according to how fast observers and their clocks are travelling. Even the neurological clock inside an observer's brain, which gives rise to the subjective experience of time, will be affected. Neurological clocks, flashing light clocks, pendulum clocks and candle clocks slow down at high speeds because time itself slows down.

This is a revolutionary idea, and to come to terms with it an attempt must be made to quantify it mathematically. How much speed produces how much slowing down of time?

To begin to answer this question, imagine two observers, X and Y — a wife and husband perhaps! They get together and build two identical flashing light clocks. They each have identical rulers with which they carefully measure the distance between the flasher and the mirror, setting it to be 149·9 m. They designate one million flashes to equal one second. They also build two other clocks, two mechanical watches, say, which tick exactly once per second. Observer Y takes one of the flashing light clocks, one of the rulers and one of the watches and leaves observer X with the second ruler and the other two clocks.

So far so good. Both X's and Y's time measuring devices perform identically and stay in step. And, of course, their rulers measure the same distances. (The rulers will turn out to play an important part in what follows.) Now, imagine separating X and Y and causing them to rush towards each other at v m s^{-1}. Assume that when they pass one another both are travelling at a constant velocity so that there are no forces to confuse things. Get them both to make some measurements.

X sets her two clocks going and finds that they are in step. She also checks the distance between flasher and mirror in her flashing light clock using her ruler. The distance is unchanged. She concludes that the speed of light is $2 \cdot 998 \times 10^8$ metres (as indicated by her ruler) per second (as indicated by her watch). She finds that this is so no matter in what direction her flashing light clock is oriented. It seems reasonable for X to think she is stationary and that Y, who is rushing past her, is moving.

Y makes the same measurements using his own equipment. What does X expect to see? She expects to see Y's measurement of the distance between flasher and mirror give a result identical to her own, which it duly does. Taking account of the slowing of time due to Y's speed, she also expects to see his watch tick slower than her own, which it does. But the flashing light clock leaves X in a quandary. Certainly it counts slower, but it does so at a constant rate irrespective of its orientation. A straight mathematical analysis would lead X to expect the rate at which Y's flashing light clock counts to depend to some extent on its orientation with respect to the direction Y is travelling in. X can observe Y conclude that from his own point of view the speed of light is $2 \cdot 998 \times 10^8$ metres (as indicated by Y's ruler) per second (as indicated by Y's watch) irrespective of the orientation of Y's flashing light clock. Y concludes that it is he who is stationary and X who is moving.

As these observations stand, they are irrational. The speed of light cannot be the same in all directions for two observers moving with respect to one another. Not, that is, unless the different rate at which time passes is compounded by having lengths undergo a matching change. Suppose, from X's point of view, Y's ruler shrinks in addition to his watch ticking more slowly. In that case X can rework her mathematical analysis to enable Y to measure a constant speed of light regardless of the orientation of his flashing light clock. It is then a direct consequence of this shrinkage of lengths as well as of time that the slowing of Y's flashing light clock is independent of its orientation.

The requirement that no matter how fast observers are travelling they always measure the same speed of light in all directions puts stringent restrictions on the way time and length alter with velocity. These restrictions enable the mathematical relationship between velocity and time to be formulated. If X has a clock — of any kind — that registers an event repeating regularly, then if she observes an identical clock travelling with Y and possessing a velocity v, she will see the value shown by Y's clock, t_y, will be $(1 - v^2/c^2)^{1/2}$ times that shown by X's; i.e.

$$t_y = t_x \left(1 - \frac{v^2}{c^2}\right)^{1/2}$$

where c, as usual, is the speed of light.

X will also observe Y's ruler to be unaffected by his velocity providing it is held at right angles to the direction of motion. But if it is aligned parallel to the direction of motion, it shrinks. The distance between two unit marks on Y's ruler, z_y, will decrease so that in terms of units of X's ruler

$$z_y = z_x \left(1 - \frac{v^2}{c^2}\right)^{1/2}.$$

This shrinkage is, like the slowing down of time, a shrinkage of space itself rather than merely a shrinkage of rulers. From X's point of view, the faster Y travels, the more he is affected, the effect being confined to distances aligned with the direction of motion. If Y travels head first, he gets shorter; if he travels stomach first, he gets flatter. The same applies to all material objects that are travelling with him.

This apparent connection between time and space will be considered further in a moment. For now, look at the equation relating X's and Y's measurements of time. This quantifies how time passes as viewed by X. The equation shows that the faster that objects, including people, are observed to travel, the slower is time observed to pass for them. The nearer their speed gets to that of light, the slower do clocks travelling with them tick. Were they to reach the speed of light, time would stop passing for them altogether. All events would be suspended. (In fact, objects possessing rest mass-energy cannot reach the speed of light.) Because of the slowing of time, a traveller measuring time at speed will be observed to get values which are too small. Relative to observers who self-designate themselves to be stationary, the values will get smaller and smaller as the traveller's speed rises. This is the effect of speed on the measurement of time.

SPACE-TIME

The complementary shrinking of distances and times with increasing speed suggests there is a close connection between these two seemingly unrelated quantities. Now, the relating of distances — but not times — to one another is a branch of mathematics called geometry. One geometrical relation in particular is worth reviewing here because of the revealing light it casts on how science now views times as well as distances. It is the property of space described by Pythagoras's Theorem.

Picture a triangle, one of whose angles is a right angle. If the length of the largest side, the one opposite the right angle, is h, and the lengths of the other two sides, which have the right angle between them, are a and b, then

$$h^2 = a^2 + b^2.$$

This is Pythagoras's Theorem. For any one given value of h^2, the values of a^2 and b^2 can vary. But whatever a is measured to be, and whatever b is measure to be, when they are combined in the form of $(a^2 + b^2)$ the answer will always be exactly h^2.

Space and time are connected in a similar way to Pythagoras's a and b. It has been seen that times contract at speed. The same is true of distances aligned with the direction of motion. This can be expressed mathematically in an equation resembling Pythagoras's:

$$l^2 = c^2 t^2 - x^2$$

where t is the time between two events, x is the distance between them, and I is what is known technically as the *interval* between them. The key point about this equation is that the value of I^2 for any given two events never varies, no matter whether an observer is stationary or moving. The values of t^2 and x^2 do vary depending on the relative speeds of different observers, but all observers can be sure, when they each combine their measurements of t and x in the form of $(c^2t^2 - x^2)$, that the answer will always be exactly the same I^2.

It is amazing how rich in meaning a simple equation can sometimes be. Here are eight facts the $I^2 = (c^2t^2 - x^2)$ equation reveals.

Fact 1. Obviously, a time and a distance can't be added or subtracted. But multiplying time by speed converts it to a distance. The natural speed to use is the speed of light. This shows that the speed of light is not merely a constant but is a fundamental part of the structure of the universe.

Fact 2. By revealing precisely how time and distance are connected, the equation establishes that time — when multiplied by c — is another kind of distance. This is a unification of sorts between space and time. In particular it enables time as well as distance to be described by geometrical equations. Physicists generally refer to space-time rather than to space and time separately.

Fact 3. Pythagoras's Theorem only works for triangles drawn on flat surfaces. Because the interval equation has the same form as Pythagoras's Theorem, it says that the geometry of the space-time to which it applies is also flat. (Recall figure 4.1.)

Fact 4. The big difference between Pythagoras's Theorem and the interval equation is the minus sign in the latter. Whereas a and b are both distances, on an equal footing in all respects, the same cannot be said for ct and x; ct may be a kind of distance, but the minus sign makes it different from a distance too.

Fact 5. Because Pythagoras's a^2 and b^2 are added together, h^2 is always positive and h always has a realistic value, something that can be marked off on a ruler. In contrast, when x^2 is bigger than c^2t^2, I^2 is negative and I, being the square root of I^2, is not a realistic number. This reveals a powerful truth hidden in the equation.

To spell this truth out, the equation can be altered slightly to the equivalent

$$I^2 = t^2\left(c^2 - \frac{x^2}{t^2}\right)$$

where the quantity x/t is a velocity, and is how fast something would have to travel in order to cross the interval I between one event and the other. In other words it reveals how much time is available to cover the distance. Clearly, only when x/t is greater than c does I^2 become negative (and unrealistic). Since I^2 *must* be positive to be realistic, this means in turn that x/t *must not* exceed c. The speed of light is the maximum speed any material object can attain.

Fact 6. When I^2 is positive, two events can be causally connected; that is, one can cause the other. What is more, because of the way the measured values of x and t vary, all observers, no matter how fast they are travelling — and they cannot, of course, travel faster than the speed of light — will agree on the sequence of cause and effect. If event 1 causes event 2, it is impossible for any observer to see event 2 happen before event 1.

Fact 7. What a negative I^2 really means is that the two events separated by the interval I cannot be causally connected. Event 1 cannot cause event 2. Nor can event 2 cause event 1. The amount of time (t) available to cross the distance (x) between them is simply insufficient. And again, all observers agree this is so.

Fact 8. Because all observers agree on the sequence of events when I^2 is positive, they all agree in which direction time is flowing. The past for one observer is the past for them all. Fascinatingly enough, the same can't be said when I^2 is negative. The order of events which are not causally connected can be in one time direction for one observer and in the opposite time direction for a second observer travelling fast with respect to the first. Thus the direction in which time flows is ambiguous when $x^2 > c^2 t^2$.

All these pronouncements about reality from a simple geometric equation!

THE EQUIVALENCE OF +t AND –t

There is a very significant ninth characteristic of the interval equation that remains to be mentioned. For causally connected events — that is, those for which $I^2 \geq 0$ — there is something curious about the values t can have. Because c^2, t^2 and x^2 are all squared numbers, even if c, t and x are negative, c^2, t^2 and x^2 will always be positive. Now c is just a value. It lacks a direction, so to talk of $-c$ has no significance. In contrast, t and x do both have a direction. If I take a large step forwards, I move forward by $x = +1$ m. If I take a large step backwards, in the opposite direction, I move forward by $x = -1$ m. In either case $x^2 = +1$ m^2. What this means is that in calculating the interval between two events it doesn't matter whether one is spatially in front of the other or behind it.

The same applies to time. It does not matter whether an event is ten seconds later or ten seconds earlier than some other event. The time part of the interval calculation will be the same either way. It is therefore possible to put a negative value of t into the expression $(c^2t^2 - x^2)$ and still get an interval permitting a causal relationship. This suggests that one event may cause another separated from it by negative as well as positive values of time. An event may cause things in the past as well as in the future.

At first sight this is a plain unrealistic idea. It does not accord with observation. Effects always come after their causes, never before them. The equation that led to this unrealistic consequence must, despite its richness, surely be wrong. But it would be unwise to dismiss the interval equation so hastily. It might be casting an unexpected light on cause and effect.

Imagine an electron, call it e_2, resting in a vacuum. Next imagine another electron, e_1, which is caused to start travelling towards e_2. (The cause of e_1's motion will be ignored for the moment.) When e_1 approaches e_2 it repels e_2 and is itself repelled. The result is that e_1 changes direction and e_2 acquires a velocity. The near collision of e_1 and e_2 is an event, E_2, which is causally related to the acquiring by e_1 of its initial motion in event E_1. So E_1 causes E_2. As is obvious, E_2 happens after E_1 and is separated from it by a positive value of t.

Suppose a film is made of this encounter between e_1 and e_2 and replayed backwards. In the reversed film, E_2 is seen happening first and is the bringing to rest of e_2 by a near collision with e_1. Event E_1, the return of e_1 to its starting position, happens second. In observational terms, reversing the film is akin to reversing time. And the most important thing about this reversed time sequence of events is that what appears to happen is not impossible. The reversed sequence of events could have taken place quite realistically. To prove this, all that is needed is to reverse the directions of motion of e_1 and e_2 after their near collision. They will retrace their paths and event E_2 will 'unhappen' so to speak and be followed by a similar unhappening of E_1.

Now, the only difference between the sequence of events E_1 then E_2 and the sequence E_2 then E_1 is either the direction of time (as in the reversed film) or the direction of motion (when the electrons retrace their paths). Furthermore, the reversed sequence brought about by apparently reversing time is exactly identical to the sequence brought about by reversing direction. The conclusion follows that there is no more reason to frown on the idea of reversing the direction of time, that is, changing $+t$ into $-t$, than there is to frown on the idea of reversing the spatial direction, that is, changing $+x$ into $-x$. Both have equivalent effects. Hence it is every bit as valid to regard E_2 causing E_1 with time reversed as it is to regard E_2 causing E_1 when the direction of the electrons is reversed. Both reversal options are equally realistic.

Let this point be spelled out. Because of the way time is perceived (see the next section), E_1 is observed causing E_2 with time increasing between E_1 and E_2. But it is equally valid to regard E_2 as causing E_1 with time decreasing between E_2 and E_1. This is the implication of the possible $-t$ in the interval equation.

The process of reversing the direction of the particles involved in causally related events, or of playing a film of those events backwards, or more likely of merely substituting $-t$ for $+t$ in an equation, is called the time reversal operation. If an event sequence occurs equally possibly with time reversed as it does with time running normally, it is said to be symmetric under time reversal. Such events are called *time symmetric*.

The equation $I^2 = c^2t^2 - x^2$ is time symmetric. It is quite unaffected if $(+t)^2$ is replaced by $(-t)^2$. The laws encountered in chapters 1 and 3 are also time symmetric. For example, mass-energy is conserved whether time is passing one way $(+t)$ or the other $(-t)$. Because the laws of chapters 1 and 3 lie at the very heart of physics, it is not surprising to discover that almost all of physics is time symmetric. It follows that the universe in the vicinity of the Earth — where the physical laws have been discovered — must be time symmetric too. Films of events taking place on and around the Earth look just as realistic when they are run backwards as when they are run forwards. For instance, a film of the moon orbiting the Earth appears perfectly reasonable regardless of whether the film is run normally or in reverse.

So according to the laws of physics, cause and effect are fully reversible. The only constraint on this is that (from the interval equation) all observers will always agree on the sequence of events, whether time is passing one way or the other.

10.2 ARROWS OF TIME

LIKELIHOOD

Unfortunately, confidence in the time symmetry of physics is totally unwarranted. Almost nothing that is observed on a large scale, except for simple things like planetary orbits and colliding billiard balls, looks realistic in a reversed film. Physics may say that the universe is locally time symmetric, but that is patently contrary to observation. It is wrong. It follows that the laws of physics encountered so far cannot be a complete description of objective reality. Something must be missing from them. To find out what, the nature of the asymmetry needs to be examined more closely.

Reconsider the collision between the two electrons. It was easy to envisage how it could be reversed in practice by means of reversing the electrons' directions of motion. Yet such ease of reversibility is something of a special case. If it is now asked what set the electron e_1 moving in the first place, it may be found that it was ejected from a heated wire filament as a result of many collisions with other electrons and atoms. To time-reverse this event is easy enough with a film played backwards, but to do it by reversing directions of motion is much more difficult. If e_1 is to be neatly reabsorbed by the filament in the reversed event, all the electrons and atoms in the relevant part of the filament must be in the correct places and travelling with the correct velocities to absorb the kinetic energy of e_1 as it strikes the filament. If the electrons and atoms in the filament are not correctly located, then the event E_1 reversed by means of reversing directions of motion will not look the same as event E_1 reversed by means of playing a film backwards.

All the events observed in the world at large are events like E_1. To reverse them in practice, it is necessary to reverse the directions of motion of enormous numbers of particles, all of them to an arbitrarily precise degree of accuracy, so that they exactly retrace their steps. This is what is observed to happen when films of such events are played backwards. But to achieve the same thing by reversing directions of motion is impractical to say the least. Because backward events cannot be observed in practice by reversing directions of motion, backward events observed in principle by reversing the direction of films look unrealistic. Hence, backward and forward processes are distinct from one another, and the events concerned are therefore time asymmetric.

The cause of time asymmetry is thus quite easy to spot: it arises out of the practical difficulty of reversing the motions of large numbers of particles accurately. But more thought is needed before this can be expressed as a law of physics. Further consideration needs to be given to E_1, which can be taken as a typical event.

To limit the size of the event, assume that the system containing E_1 can be completely isolated. This is an assumption traditionally made about the kind of systems under consideration, and it enables such systems to be thought of as complete in themselves and of limited extent. (Unfortunately the assumption is questionable when considering time reversal, as will be seen shortly.)

Given complete isolation, there is no reason why all E_1's particles cannot retrace their paths. After all, each particle individually can move through space as well in one direction as in any other, including in an exactly opposite direction. If each particle individually can retrace its path, then obviously all the particles collectively are equally able to do so. Consequently, the time-reversed event E_1 is not impossible. When a film of E_1 is watched being played backwards, what is seen is not something which cannot happen. Since the individual particle motions remain normal, it is an event which is definitely possible. It is not a viable proposition, therefore, to build time asymmetry into physics on the basis that reversed events involving many particles are impossible. The practical difficulties of reversing events could in principle be overcome.

If the lack of ease of reversibility cannot be translated into impossibility, there is still the next best thing to fall back on: improbability. Every particle in event E_1 is following a certain path. In the time-reversed E_1, every particle is following an exactly opposite path. Now, the location and velocity of each particle at any instant has a huge range of values open to it. When E_1 occurs (in the normal time direction) it is not specified in advance which electron is to be emitted from the filament nor, except within very broad limits, what its speed and direction are to be. Hence, a wide range of particle paths can lead to the emission of an electron.

In contrast, the reversed event is much more tightly constrained. For example, it is required that one specific electron must travel on a specific path which will lead it to strike the filament. A slight deviation and it will miss its target. Similar considerations apply to other aspects of the event. Such factors make the time-reversed event more unlikely — more improbable — than the unreversed event.

Probability it is then that brings time asymmetry into physics. Event E_1 with time passing forward is a comparatively probable event. The reversed event is comparatively improbable. If E_1 looks normal when time passes in one direction, then it will look abnormal under time reversal; that is, when a backward film is watched, or all the component particles' directions of motion are (miraculously) reversed. Probable, under time reversal, becomes improbable.

In order to express this as a law, it is essential to be clear what it is that makes things probable or improbable. To illustrate this, imagine taking an animal and converting it into a set of basic molecules (NH_3, CO_2, etc.) dissolved or suspended in the water that the animal contained. Both the chemical mixture and the animal from which it was derived will consist of exactly the same number of atoms of each type, since the former was made solely by turning the latter into its constituent parts.

Suppose some biochemists wish to signal to someone on another planet how to make the mixture. They need to specify how much NH_3, how much H_2O, and so on, to mix together and, for completeness, the shape and size of the container, its temperature and sundry other similar things. With these few bits of information, the biochemists will be able to ensure their extra-terrestrial counterpart will make an identical mixture, taking the mixture as a whole.

It is significant that the mixture is described 'as a whole'. If the mixture is viewed as individual molecules, then the particular molecule that might be found at some precise location in the biochemists' container may well not correspond to what the alien on the other planet finds at the same location in the container there. The two mixtures are thus not identical in detail. Indeed, the precise arrangement of the molecules in one container is very unlikely to bear any resemblance to the arrangement in the other. The two mixtures are only identical when looked at in their entirety.

Inside the biochemists' container are enormous numbers of molecules. The number of possible arrangements of these molecules is nearly infinite. The vast majority will look the same (as a whole). But clearly a few of the arrangements will correspond to the distribution of the molecules in the original animal. These will not look the same; they will look like the animal and not like the mixture in a container.

Now, when the biochemists told the extra-terrestrial how to make the mixture, they assumed that the actual arrangement of molecules that their counterpart would fortuitously come up with was one of those that looked like the mixture. The extra-terrestrial could by sheer chance have hit on one of those resembling the animal; the result in this case being that the mixture would then have had very little similarity to the biochemists'. The biochemists' justification for assuming the extra-terrestrial will not accidentally make the animal while preparing the mixture is that almost all the arrangements of the molecules look like the mixture. Extremely few look like the animal. The alien on the other planet will create one arrangement at random. A random arrangement is very likely to be a typical mixture and very unlikely to be an animal.

Deciding what is probable and what is improbable thus rests on the properties of systems as a whole and on the position of each of their parts. A system (like the mixture) which has many equivalent arrangements of its parts, a system whose individual parts can be widely swapped around without affecting its overall properties, is probable. A system (like the animal) which has few equivalent arrangements of its parts, a system whose individual parts cannot be widely swapped around without affecting its overall properties, is improbable.

This definition of probable and improbable systems can be applied to events. When an event occurs, the participating particles will be rearranged to some extent. The new arrangement may be one of the few that constitute an improbable system, or it may be one of the many that constitute a probable system. Because of the number of available arrangements in each case, the odds favour the latter. Hence there is a tendency in all systems which are to any degree improbable — as almost all systems are — to alter in ways that make them more probable as time passes.

The observed time asymmetry can now be put in purely physical terms. As events occur, improbable arrangements of matter will turn into probable arrangements. The reversed-direction process — replacing $+x$ with $-x$ — in which probable arrangements turn into improbable ones does not occur. That is why the equivalent reversed-time process — the backward film which replaces $+t$ with $-t$ — looks abnormal and hence is time asymmetric.

The position is clear enough for a law to be formulated to introduce time asymmetry into physics. The law is this: whenever an isolated event occurs, the configuration of the constituent particles after the event will probably be more probable than before; that is, the greater the unlikelihood of the before configuration, the greater will be the probability that the after configuration will be a more likely one.

This law is known as the *Second Law of Thermodynamics* and it can be stated in a variety of different ways according to the context it is being applied to. By being added to the other laws of physics, it introduces the time asymmetry that is observed.

It will be noted that because the Second Law of Thermodynamics has its origins in improbability rather than impossibility, it is a law which doesn't actually forbid anything. It doesn't say that improbable arrangements of matter do not arise as time passes, but only that they are very unlikely to. Improbability is not impossibility.

(It is remarkable that both large-scale physics, as being discussed here, and micro-scale quantum physics are probability-based. This may not be a coincidence.)

That improbable arrangements of matter are not forbidden from arising by the Second Law of Thermodynamics encompasses an important characteristic of the universe; namely, that complicated (i.e. improbable) structures can arise in spite of the general trend towards probability. Living things are pertinent examples of this. They are of necessity very complex, as was discussed in sections 5.7, and owe their existence to

455

chance events in which matter becomes more complicated. Biological species only exist because complex DNA base sequences were — and still are — able to form from simpler ones. The gradual increase in the complexity of living things over the last three and a half billion years is the story of the gradual chance increase in the complexity of DNA, as permitted by the Second Law of Thermodynamics, and as preserved thereafter through the mechanism of replication. We humans are ourselves a demonstration of the fact that increases in the complexity of matter are improbable but not impossible.

It must be pointed out, though, that the growth of *individual* organisms from conception to maturity is *not* an instance of improbability arising from probability. That's because organisms are not isolated systems. The growth of a living thing involves it in interactions with its environment. Those interactions, while permitting the organism to grow in complexity as it grows in size, produce increases in the probability of the organism's surroundings. This more than compensates for the increases in the improbability of the matter in the living thing itself. Individual living things, when considered in conjunction with their environment, are part of the general trend towards probability.

THE THERMODYNAMIC AND COSMOLOGICAL ARROWS

It is clear that time can flow in one of two directions which have arbitrarily been designated forward and backward. Disregarding probability, physical processes are generally time symmetric, in that they can take place with time flowing in either direction. Physical laws — which are just formal descriptions of physical processes — are also time symmetric; their validity is unaffected by the direction of time. This is why the forward and backward designation is arbitrary. Almost all physical processes can detect no difference. One direction in time is as good as the other.

Because, despite this physical time symmetry, the universe is plainly not time symmetric, an explanation has been sought in the Second Law of Thermodynamics: improbable arrangements of matter turn into probable arrangements as time passes in the forward direction. This tendency of matter defines the forward direction of time, distinguishing it from the backward direction. For this reason it is called an *arrow of time*, pointing

456

unambiguously to the future and making the future quite different from the past. As other arrows of time are about to be encountered, the arrow derived from the Second Law of Thermodynamics will be referred to, not unreasonably, as the thermodynamic arrow.

But there is a puzzle here. Probability cannot increase without limit forever. Eventually a most-probable arrangement of the matter of any given system must arise — an arrangement in which any change in its probability must be to less probable states. If some piece of the universe is taken and completely isolated, then according to the Second Law of Thermodynamics it will, after a while, almost certainly reach its state of maximum probability. Then what? What direction will the thermo-dynamic arrow of time indicate then?

To discuss a specific example, suppose the isolated piece of the universe contains nothing more than a warm ball of gas in a box. Suppose further that gravity is too weak to affect it. Suppose finally that it begins its isolation already in a most-probable state. As the gas molecules move about, they go from one maximally probable arrangement to another. The appearance of the gas does not change significantly from one moment to the next. If a film of this gas is played in reverse, the backward film is indistinguishable from the same film played forwards. Events in the gas are completely time symmetric. There is no thermodynamic arrow of time. Backward is no different from forward.

As a development of this argument, it can be imagined that every so often in the course of moving about, the gas molecules will come to be arranged in a way which is less than maximally probable. This will be a purely chance business, occurring at random. After this less than maximally probable state has arisen, further molecular movements will almost certainly restore maximum probability fairly soon. The sequence of events can be written:

probable → event 1 → less probable → event 2 → probable.

The arrows in this sequence define a direction in time, but on investigation the direction can be seen to be trivial. Again, if the sequence of events is played backwards, which is equivalent to reversing the arrows above, no difference is found between one direction and the other. With the arrows as shown, event 1 increases improbability and event 2 reduces it. With the arrows reversed, event 2 increases improbability and event 1 reduces it. Either way, what begins with maximum probability ends with maximum probability. The process does not destroy time symmetry. In fact, whatever happens to the isolated ball of gas, time symmetry is preserved. In isolated, maximally probable systems, the Second Law of Thermodynamics does not define a direction in time. It is as time symmetric as the rest of physics.

The puzzle is now manifest. How can the Second Law of Thermodynamics be time symmetric in some cases and time asymmetric in others? It is a puzzle which calls into question the use of this law to explain the universal time asymmetry.

The chief distinction between the isolated ball of gas and the time asymmetric physical processes occurring everywhere on the Earth is that the Earth is not isolated. Perhaps violating the isolation of a system introduces the time asymmetry.

Consider taking the ball of gas and, just for an instant, permitting something outside it to disturb it, so that it briefly ceases to be isolated. At the moment of the disturbance, only some parts of the ball of gas will be affected. That's because the disturbance can affect all of the gas simultaneously only if it travels infinitely fast, which has been seen above to be impossible. To have the disturbance concentrated in limited parts of the gas is to make the formerly isolated system less than maximally probable. Only when the disturbance has spread evenly throughout the system will the gas once again be in its most probable state.

Violating the isolation of a system can thus be seen to provide a source of improbability for the thermodynamic arrow to work on. Given an outside source of improbability, thermodynamic processes will lead to a state of maximum probability with time passing forwards. It can be concluded that it is indeed loss of isolation that keeps the thermodynamic arrow in operation.

However, this is not yet a complete answer. Yes, violating the isolation of a system provides a source of improbability, but no, it hasn't been explained yet why an improbable system becomes more probable in only one of the two time directions. After all, the 'less probable' state encountered above that arose by chance in a completely isolated system could turn into a probable state via event 1 with time passing in one direction or via event 2 with time passing the other way. Why can't that happen similarly when improbability is caused by breaching the isolation of a system?

The source of the disturbance to the previously isolated ball of gas is itself a system, as is the agency which brings about their connection. The three systems together — that is, ball of gas, disturbance and connecting agent — can be regarded as three components of a single isolated super-system. The isolated super-system will itself gradually approach a state of maximum probability, with subsequent loss of the thermodynamic arrow of time. To retain the arrow within the super-system, yet further outside disturbances must be invoked. These new disturbances are themselves systems which can be aggregated with the super-system to make an even bigger isolated system needing yet more outside disturbance. Eventually, taking this process of ever larger isolated systems to its limit, the whole universe is involved. The whole universe is the ultimate isolated system.

Now, if the contents of the universe as a whole were in a maximally probable configuration, this ultimate universal isolated system would be time symmetric like the isolated ball of gas. There would be no thermodynamic arrow of time. But this is not the case. The contents of the universe collectively are not arranged maximally probably. So it seems that it is cosmology which holds the key to a complete account of the thermodynamic arrow of time. The fact that less probable systems become more probable in only one time direction has a cosmological cause.

Because there is no system to which the universe can be connected, the improbable arrangement of its matter must be taken as something arising in the course of the Big Bang. A key factor here is the direction of motion of all the universe's particles. At any chosen moment since the Big Bang it is found that there is a pronounced trend for particles to move in a particular way. In one direction of time — designated forward — they move away from each other; in the other direction — designated backward — they move towards each other. If it was possible somehow to create a large, static universe in such a way that all its particles were moving randomly, it would be most unlikely there would be any trend amongst them to move apart or together. In this imaginary, maximally probable universe there would be neither expansion nor contraction. And no arrow of time.

This contrast between the real, time asymmetric universe and an imaginary, maximally probable one is very informative, not least because of what happens to the latter. It cannot remain static. Gravity would begin to produce a contraction. Random motion would be supplemented by a trend amongst particles to approach one another.

Assume for the sake of argument that this trend in the imaginary, initially static universe develops in both directions of time: moving away from the moment of creation in either time direction, the universe contracts. To spell this out, if one direction in time is arbitrarily designated as forward (i.e. with time increasing) and the other is designated backward (i.e. with time decreasing), then the picture can be described like this: with time passing forwards, the universe expands until the moment of its creation when it is at its maximally probable state, after which it begins to contract; with time passing backwards, the universe expands until the moment of its creation when it is at its maximally probable state, after which it begins to contract. Clearly, the only difference between the two sequences is that time is increasing in one and decreasing in the other. A film of the first, played backwards, becomes the second. This imaginary universe thus remains time symmetric.

But now suppose there are observers who begin their observations after the contraction has begun (in either of the two time directions). Although their universe is time symmetric, it certainly won't appear so. As these observers look towards the moment when the universe is maximally probable, they will see it is larger at that time. As they look away from the moment when the universe is maximally probable, they will see it is smaller. The observed expansion and contraction, due to gravity, would make their universe look time asymmetric.

The initially large static universe which has been described here is purely imaginary. It does not, except superficially, resemble the actual universe, and the reasoning applied to it cannot be translated easily into reality. However, it does bring out the source of the time asymmetry in the real universe. When the Big Bang occurred, it could conceivably have done so in both time directions. In that case, a film of the event would show everything crashing together to a point and then exploding apart. The reversed film would show exactly the same thing. The account of section 4.6 given with positive values of $\mathbb{T}$ (the age of the universe) would be equally valid with negative values, and the universe would be time symmetric. But it is in the nature of the Big Bang that if it has two sides (one with $\mathbb{T}$ increasing from zero, one with $\mathbb{T}$ decreasing from zero) observers on one side cannot see the other. They can only see their half of the film. Hence the universe looks time asymmetric. Play the available half of the film one way and an explosion from a point is seen. Play it the other way and a collapse to a point is seen. (It must be stressed here that there is no observational evidence for there being another half, a $-\mathbb{T}$ half, to the universe.)

The time asymmetry of the universe as a whole defines an arrow of time, which can be called the cosmological arrow. Time passes forwards as events move away from the Big Bang, and backwards as they move towards it. Use of the symbol $\mathbb{T}$ below will signify that the cosmological arrow of time is being used to define the direction in which time passes forwards.

How does the cosmological arrow of time give rise to the thermodynamic arrow? The simplest assumption about the state of matter when the Big Bang occurred is that the matter was randomly distributed. It was chaotic rather than ordered. This is a way of saying it was arranged in a maximally probable way. However, as the universe expanded, vast amounts of space were created, thereby providing a matching vast increase in the number of possible arrangements of matter (since the matter had more space to be arranged in). The matter, in a sense, has been unable to keep up with the expansion. It has not yet taken up the number of possible arrangements available to it. As such, the expansion has left it in a less than maximally probable state. So now, still trying to catch up with the expansion, matter becomes increasingly probably arranged in the same time direction as the expansion takes place in. Hence looking towards the Big Bang is also looking towards times of greater improbability, and looking to the cosmologically defined future is also looking towards times when matter is arranged in more probable ways. The thermodynamic arrow of time — the tendency of matter to distribute itself more probably as time passes forwards — is thus determined by the very expansion of the universe, which is itself an arrow of time.

(Incidentally, given that the surplus of matter over antimatter in the universe was created during the Big Bang, then the mechanism by which that surplus was produced — whatever it was — must also be an arrow of time, producing more matter than antimatter as $\mathbb{T}$ increases, and thus producing more antimatter than matter if the time direction is reversed. A $-\mathbb{T}$ half to the universe would consequently be made of antimatter!)

THE ELECTROMAGNETIC ARROW

Because of their close relationship, the cosmological and thermodynamic arrows always point in the same direction: away from the Big Bang is always away from improbability. There is a third arrow which is equally consistent with these first two in always pointing in the same direction as they do. It is called the electromagnetic arrow of time.

When an electrically charged particle accelerates, it emits electromagnetic waves. If the physics of the emission was completely time symmetric, it would be expected that waves would be sent out in both time directions, towards (cosmologically) earlier times as well as to the later times that observation actually reveals. Waves that a particle emits to earlier times would be observed, as $\mathbb{T}$ increases, to converge on the particle. An observer should see waves from the past converging on the particle, arriving at the moment it starts to accelerate, and spreading out from it into the future.

No waves are ever observed to converge on a particle when it accelerates. Only outgoing waves are seen. This means that in a reversed film the waves would only be converging ones. There would be no outgoing waves. Hence electromagnetic phenomena are time asymmetric.

The electromagnetic arrow of time is defined by the direction in which waves spread out. Time increases as waves dissipate (in the absence of lenses, concave mirrors, etc.); time decreases as waves converge. Because it always points in the same direction as the cosmological arrow, the electromagnetic arrow must, like the thermodynamic arrow, ultimately be of cosmological origin: heading towards the Big Bang, electromagnetic waves converge; heading away from the Big Bang, electromagnetic waves dissipate.

Given the cosmological connection, it is interesting to note that a mathematical analysis of the electromagnetic arrow of time indicates that the asymmetry requires the universe not to be hyperbolic; i.e. it would appear to rule out the model 3 option of section 4.3.

THE SUBJECTIVE ARROW

A position has now been arrived at in which the two directions of flow of objective time can be distinguished. One direction, with time increasing as defined by the three arrows, is to the future; the other direction is to the past. This objective viewpoint must be capable of providing a framework for the conscious experience of time; it must be able to produce, or at least accommodate, the subjective distinction between past and future.

Of the three physical arrows, one can be discounted as far as subjective time goes, and that is the cosmological arrow. The passage of cosmological time is not noticeable on the scale of a human life; the universe does not change perceptibly in a mere hundred years. That leaves the thermodynamic and electromagnetic arrows with which to account for human perceptions of time.

What is the human perception of time? Time is perceived as a sequence of events which is divided into three: the past, the present and the future. Some of the events that lie in the past are ones individuals are fairly certain of the reality of; they say they know what happened. All other past events can be found out about in principle, if not always in practice, simply by making a sufficient effort. And of course past events cannot be reversed; they are irrevocable. The present, usually referred to as 'now', is the border between past and future. Because the past does not gradually merge into the future but does so abruptly, now is a moment — an interval of time of arbitrary smallness. Finally, the future consists of events whose reality can only be guessed at and which can be influenced to some extent.

The fixity of the past and the vagueness of the future provide one of the two components of the subjective arrow of time. Events that are unalterable are in the past, at lower values of time than now; events that are potentially alterable are in the future, at higher values of time than now.

The subjective arrow, by defining a past and a future, is not obviously different from the various objective arrows. But there is a second component to the subjective arrow: perception of time includes an additional characteristic absent from physics, and that is that 'now' moves. Events are constantly being transferred from future to past. Future events are not only alterable; they are also transferrable to the past side of 'now'. Past events are not only fixed, they are also not transferrable to the future side of 'now'. The subjective perception of time to be accounted for thus has two closely related aspects: the future can become the past as well as being alterable; the past cannot become the future and is not alterable.

The main reason for thinking an account of subjective time is possible is that the subjective arrow always points in the same direction as the three objective arrows. The subjective past and future are always respectively towards and away from the Big Bang (time travel fantasies notwithstanding). This implies not only that the subjective and objective arrows are related, but that the relationship is comparatively straightforward.

On purely thermodynamic considerations it might seem that such confidence is misplaced. It has been seen that with increasing T the probability of arrangements of matter also increases. Now, every piece of information added to the brain makes the arrangements of its molecules increasingly improbable. Increases in improbability accompany decreasing T, so the amount of information in a brain — the amount a person knows — should increase towards the Big Bang, which is towards the time of a person's youth. On this reckoning people should know most when they are born and increasingly little as they age. Clearly, thermodynamics on its own cannot provide a simple account of the subjective arrow of time.

What of electromagnetism? Recall chapter 1 for a moment. It will be remembered that all forces passing between objects (more than 10^{-15} m apart) are resolvable into gravity and electromagnetism. Gravity, being extremely weak, can be discounted. Interactions between the objects in the brain, neurones, must therefore ultimately be resolved into disturbances in the lines of electromagnetic force — i.e. electromagnetic waves — passing between the electrons and ions that the neurones are made of. A similar statement can be made about interactions between sensory neurones and the environment.

This fundamental involvement of electromagnetic waves in neurological events has two implications for the brain's arrow of time. Firstly, it means that all interactions between the environment and sensory neurones will be affected, directly or indirectly, by the electromagnetic arrow. Specifically, all events detected by the senses will relate to electromagnetically earlier times. Secondly, within the brain all the interneurone and intraneurone events will also be obliged to follow the direction of the electromagnetic arrow.

Because of the constraints imposed by the electromagnetic time asymmetry, the information formed in the brain can only relate to electromagnetically earlier times. The information in the brain is therefore not time symmetric. As a result, the subjective past will look very different from the subjective future.

The thermodynamic arrow has no role in this asymmetry, except in a mildly negative way. The thermodynamic considerations mentioned earlier actually conflict with the electromagnetic ones. As time passes forwards, the brain does indeed lose information, or conversely gain information as it gets younger. However, thanks to electromagnetism the information concerned always relates to earlier times, and is much more than offset by the acquisition of new information arriving from the electromagnetically defined past. The new information, which increases the improbability in the detailed structure of the brain with time passing forwards, is paid for thermodynamically by a compensatory increase in the probability of the brain's surroundings.

The close connecting of the subjective arrow of time with the electromagnetic arrow means that the past will be known and the future will be unknown. As such, it accounts for perceptions of past and future. It does not explain, though, the movement of 'now'. The explanation for this subjective aspect of time is, however, not hard to make out.

The three objective arrows of time define past and future relative to any given event. If some event takes place at time $\mathbb{T}_r$, then relative to that event other events taking place at times $\mathbb{T} < \mathbb{T}_r$ are in the past. Those taking place at times $\mathbb{T} > \mathbb{T}_r$ are in the future. The reference event at $\mathbb{T}_r$ divides past from future and is in effect an arbitrary present.

Conscious experience is a matter of events taking place in the brain, events which affect the I-image. Each event registered consciously by an observer is regarded as a reference event dividing past from future. In other words, there is a subjective reference time $\mathbb{T}_r$ selected by an observer: it is the time $\mathbb{T}$ at which the event corresponding to a given thought arises in the observer's mind. Thus each thought has its own reference time. This means that an observer's subjective reference time is not static. Every thought occurs at its own particular moment, and that moment is called 'now'.

Added to this is the electromagnetically produced time asymmetry in the brain. At any given moment a thought can relate to earlier 'now' thoughts about which observers have information, but cannot relate to later 'now' thoughts about which they have no information. Hence, 'now' thoughts appear to follow a sequence from past to future. And with each revision of the subjective reference time to higher time values, part of what was the future is transferred to the past. The subjective movement of 'now' towards future times is due to the non-static nature of observers' subjective reference times and the restriction of information in their brains to the electromagnetic past. It is out of this that comes the experience of a moving present and of the passing of time.

10.3 FUTURE TIMES

THE STORY OF HUMAN LIFE

The story of each human life is a purely biochemical process. A fertilized egg cell divides inside the mother. The two cells resulting from the division divide in turn, and an enormous series of cell divisions then ensues. As the mass of cells grows, differentiation takes place, and gradually a human body forms. After nine months, that body is capable of an independent existence and emerges from its mother to become a fully separate human being. This is how each life begins.

From birth onwards, human development proceeds on two levels, the biological and the ideological. Biologically, under the control of biochemical events determined by DNA, the human grows to maturity. It takes about twenty years for this to happen. Ideologically, too, the human matures. The brain comes to be able to extract from its sensory inputs the many subtleties and nuances found in them, to achieve almost perfect control over its muscular outputs, to acquire awareness of its own existence, and to build a world-view enabling it to match its behaviour to its environment in highly sophisticated ways. The timescale for this latter maturation is about the same as for the biological.

The steady attaining of maturity can be looked on as the fulfilling of the potential of the human. This expresses itself in the capacity to survive and deal with all the challenges the world may throw up. On reaching maturity, a human is as resistant to disease, as fit, as agile, as adaptable, as quick-witted, as cunning, as intelligent as he or she will ever be. In this state, humans are well-equipped to realize their primary biological potential of raising offspring, and their primary ideological potential of acquiring, communicating and acting on their ideas.

But then, having matured, a strange thing happens. Biological faults begin to develop. The skin and muscles become less flexible, and the body becomes more prone to diseases of various kinds. A form of degeneration sets in. At the same time, the ability of the brain to embrace new ideas begins to decline, as does its general ideological adaptability. The speed with which the brain can solve problems and react to events decreases.

Biological degeneration is not at first glance what would be expected of organisms. It reduces their ability to survive, to replicate and to protect their immature offspring. Yet it seems to be the rule for all multi-cellular creatures, and humans are no exception. It is puzzling, for there is no apparent reason why DNA should not specify structures capable of indefinite self-repair.

Ideological degeneration goes hand in hand with biological degeneration. Given that the latter is going to occur, it is not hard to explain why ideological degeneration accompanies it. A human brain with its quantity of neurones fixed at birth is ideologically viable for well over a hundred years. This is longer than its containing body remains biologically viable. Hence there is no evolutionary advantage to DNA building a brain capable, by way of neurone regeneration, of being ideologically immortal; the human brain lasts long enough as it is.

As all adult humans are aware, biological degeneration is not something that can go on forever. Eventually it reaches such a stage in some part of the body that that part fails to perform the minimum function required of it to sustain overall biological viability. The elaborate marvel that is the human body then falls to pieces. The cycles within each of its cells, the protein-sugar cycle and the protein-DNA cycle (see sections 5.3 and 5.4 respectively) break down. The body as a whole, and each of its individual cells, dies. (Degeneration is not, of course, the only way that faulty biology can lead to death.)

When biological death occurs, the cells of the brain are not exempt. The patterns of neuronal interconnections that have formed during the individual's life are lost when the neurones which contain them become incapable of signalling. It follows that ideological death accompanies biological death. And if the view of consciousness advocated in chapter 7

467

is correct, or at least on the right lines, the destruction of the images in the cortex, including the I-image, will correspond to a complete loss of consciousness and the end of subjective reality.

This then is the whole story of human lives. Both as organisms and as ideological structures, we are born, grow to maturity, degenerate and die. It is perhaps unfortunate that our intellectual capability enables us to see all this, but there is no escaping it. Human lives have an end every bit as definite as their beginning.

(As a crumb of consolation it can be remarked here that none of us will ever die subjectively. As consciousness is something that requires time — for neurones to transmit their signals at the very least — the transition from life to death is not consciously evaluable. No one can ever know that they are dead. Subjective reality, consciousness and being alive are inseparable.)

What applies to individual people does not apply necessarily to the human species. While all people will die one way or another, replication makes it possible for the numerical losses to be replaced indefinitely. There is no evidence that species ever degenerate the way individuals do. On this reckoning humanity could go on forever, though changes in the human DNA base sequences might in the long term alter human anatomy to a speculative extent.

There are, however, three external hurdles that humanity would have to overcome in order to endure permanently. While two of them might conceivably be surmounted, the third cannot. Consequently humanity — or its descendant life-forms — has an end as well as a beginning, just as we do as individuals. The only doubt about humanity's demise is the form it will take: whether it is to be home-made planetary devastation or (being extremely optimistic) one of the three hurdles.

THE STORY OF THE UNIVERSE

The three hurdles that humanity faces are found in the future of particular parts of the universe: the future of the solar system, the future of the galaxy, and the future of the universe as a whole. As the past of these objects, their beginning, has already been described in section 4.6, it remains to complete their temporal description by seeing what can be said about their futures.

The fate of the solar system is largely determined by the star at its centre. The sun will remain in its present state for $5\cdot5 \times 10^9$ years, give or take a few hundred million, getting slowly hotter and brighter. This increase in temperature will inevitably drastically alter the climate of the Earth. However, it is unlikely to do so at a rate, or to an extent, that would imperil the existence of complex organisms in the short and medium terms, astronomically speaking. In the long term, however, prospects are not hopeful. The Earth's surface will become hot enough to boil water in about $3\cdot5 \times 10^9$ years' time.

Once the $5\cdot5 \times 10^9$ year mark is reached, events will begin to move a bit faster. The sun's central parts will run out of hydrogen and start to contract. The contraction, taking place over roughly a billion years, will have the effect of gradually increasing the sun's energy output by several hundred times, as hydrogen nuclei continue to combine to make helium in a shell surrounding the spent solar core. The sun's surface temperature will drop to perhaps 3000 K, but that surface will expand to be 120×10^6 km across. To well-refrigerated observers on the Earth, the sun will then appear one hundred and fifty times bigger than the moon, a red globe so large that it will have an angular diameter of nearly a right angle (as against its current value of $0\cdot5°$). At this stage the sun will be what is known, with some justice, as a red giant star. Needless to say, no carbon-based living things will be able to survive the terrestrial heating that will go with this phase of solar evolution. And the Earth's atmosphere will be blasted away by the intense solar energy output just for good measure.

Thereafter, although the details are not completely clear, the sun will start fusing helium nuclei in its core, turning them into carbon and oxygen (see section 4.6). This stage in the sun's life will last for about 10^8 years and ends when the core runs out of helium. The sun's surface, having contracted considerably during the helium fusing stage, will blow out to an even greater extent. Its outer layers of gas will stream into space, leaving behind a remnant star with about half the mass of today's sun, but so compressed by its own gravity that it will be only slightly bigger than the Earth. This star, a type known as a white dwarf, will initially emit tens of times more energy than the sun does at present. Gradually, though, it will fade until after a further 10^8 years or so it will be barely bright enough to cast a shadow on the Earth's desolate surface.

And yes, the Earth will still be there. An interstellar visitor will find it orbiting further from the sun than now because of the sun's reduced mass — a solid, rocky, airless little world, shrouded in darkness, sterile and frozen, illuminated by a sun no brighter than present-day Venus.

And after that? All that remains is for the sun to grow colder and redder and dimmer as the billions of years pass, until it can no longer be seen.

This is the first hurdle.

To escape the death of the sun, humanity or its successors will need interstellar mobility. Such a thing is not beyond the bounds of possibility. They would then face the second hurdle.

The Milky Way was born out of a condensation of the gas emerging from the Big Bang. Inside that condensation, comparatively small concentrations of gas formed stars. There remains today much galactic gas that has not yet met this fate. But gradually this situation will change. Stars are continually forming from the residual gas, reducing the quantity remaining. As a result, the rate of star formation in the galaxy will slowly decrease, while the existing stars burn themselves out one by one just as has been described for the sun. The number of active stars in the galaxy will consequently begin to decline. Over many thousands of millions of years the Milky Way will lose its population of big, bright stars, only partially replacing them with small, long-lived, dim ones. By 10^{11} years from now the process will be well advanced and the dead Earth's sunless sky will be much less starry than it is at present. In this manner, the future of the Milky Way is to fade slowly like the embers of a once blazing fire as, one after another, its stars reach the end of their lives.

As the number of active stars in the Milky Way declines, the number of habitable planets will also decline. The same state of affairs will be found in all other galaxies; living things and complex ideological structures will gradually run out of places to live.

This is the second hurdle.

Quite how bad this situation will eventually get depends on the timing of the third and final hurdle. As was seen in section 4.3, there are three simple models of the universe and model 1 was chosen as the one to use. Now, the model 1 universe can only exist for a limited amount of time, its longevity depending on how near to being flat it is. Flatness, put in terms of figure 4.1, is a matter of the size of the model 1 balloon relative to the objects on its surface. The bigger the balloon is in terms of its surface objects, the flatter the surface appears to be to observers on those objects.

The connection between lifespan and flatness is reasonably uncompli-cated. The model 1 balloon expands for a while, reaches a maximum size and then contracts. This sequence of events is brought about by the force of gravity between the objects on the balloon's surface. If the balloon's surface is comparatively highly curved (small balloon), it means the force of gravity is strongly felt and the events take place rapidly — the universe has a short lifespan. If on the other hand the balloon's surface is nearly flat (large balloon), it means the force of gravity is weakly felt and the events take place slowly — the universe has a long lifespan.

It will be recalled from section 4.3 that our real three-dimensional universe is nearly flat (i.e. near the critical density) and that the degree of flatness remains imprecisely known. As a result, it is impossible to say when gravity will halt the universal expansion. It cannot therefore be predicted how far galactic decline will get before it is overtaken by the universe's contraction.

To pick an arbitrary value, say the universe reaches its maximum extent in 10^{12} years and begins to contract after that time. In 2×10^{12} years it will have returned to the size it has at present. It will look rather different, however, for galaxies will be approaching one another instead of receding, and they will be but shadows of their former outward-bound selves, masses of dim and dead stars.

As the contraction proceeds, the galaxies will collide and merge. Also the wave mass-energy density of the universe will once again gradually become predominant (see section 4.4 for the reason), causing an equivalent rise in temperature. Within a second of the end, temperatures will exceed 10^{10} K, matter will be converted back into neutrons, protons and electrons, and pairs of electrons and antielectrons will be made in vast quantities. Then, in another second, it will all be over. The soaring temperatures will unify the fundamental forces, the balance between matter and antimatter will be restored, and the universe will return to the quantum gravity state that it began with at $\mathbb{T} = 10^{-43}$ s. This will mark the end of space and time. The final collapse will be, to all intents and purposes, a Big Bang in reverse.

This is the third hurdle.

It might be thought this last hurdle could be avoided if it turns out that the universe is actually a model 2 or 3 universe, or the current favourite, a practically flat model 1 universe with self-repelling Dark Energy. All these model universes expand forever. That changes the nature of the third hurdle, but unfortunately doesn't get rid of it. As the uncountable ages pass, 10^{12} years, 10^{22} years, 10^{32} years, the galactic decline mentioned earlier runs its course. The endlessly receding galaxies will slowly become dimmer and dimmer. The skies of the long dead Earth, once filled with myriads of stars, will become completely dark, with only an occasional burst of light produced by the collision of stellar remnants to relieve the blackness. Eventually proton disintegrations (assuming protons really are slightly unstable) will reduce the quantity of protons in the universe towards zero. The antielectrons resulting from proton disintegration will annihilate with electrons, and all that will remain are neutrinos and photons (and Dark Matter?). Space and time in these models do not so much end as fade into emptiness. The hurdle may be in a different form, but it is every bit as insurmountable.

Nor is the third hurdle avoided by adopting one of the more imaginative cosmological models. Mentioning just one example, it is possible that within that final fraction of a second of the model 1 universe's collapse, some process takes place which resets the universe and causes it to engage in a new Big Bang. The universe would then have an infinite duration consisting of a series of Big Bangs, expansions and contractions. It hardly needs saying, though, that the conditions in which each resetting happens are equivalent to an end followed by a new beginning. Each end is every bit as final as in the simple model 1 cosmology.

10.4 RELIGION

THE NEED FOR PURPOSE

It is possible to set down with confidence the past and future of each of our lives, the past and future of humanity, of the solar system, of the Milky Way and, in very broad terms, of the universe. There may be ignorance about some of the details but the broad outlines are clear enough. These are the key components of objective reality and they have now been described in a temporally complete way.

It cannot escape notice that the one thing all these differing aspects of reality have in common is impermanence. Everything has a beginning and an end. Things are born, they exist for a time and they die. It is possible that the universe as a whole is an exception, engaging in an unending series of reincarnations, but that doesn't alter the fact that the present incarnation will come to an end, one way or another. For observers here and now, the present incarnation is the only one that counts.

This is an area where biology and ideology clash with a vengeance. At the level of individual human lives, biology commands staying alive, and for the great majority of people that means endeavour and struggle. Life is, in effect, a series of goals to be achieved. Some goals are mandatory, such as finding sufficient food each day; and some goals are optional, such as achieving ascendancy over rivals. But the objective, in every case, is to reach the goal, to succeed.

Yet the ultimate biological goal, that of staying alive indefinitely, is unattainable. Ideologically this is anathema. Does anyone ever choose to pursue a goal if they know in advance that there is no chance whatever — none at all — that they might succeed? There simply isn't any point striving after a goal which cannot possibly be reached. To attempt the absolutely impossible is ideologically insane. Biology demands a fight for survival; ideology says don't bother.

Of course, the deaths of the universe, the galaxy and the solar system can in the main be discounted. They are at least hundreds of millions of times further away than the death of a long-lived ideologically mature adult. The universe, the galaxy and the solar system can thus be considered, in terms of human lifespans, to be near enough permanent. Their demise has consequences for life, hopefully including the far distant descendants of some of the Earth's current living things, but not for people on the Earth today.

The impermanence of human lives, on the other hand, is not so easily dismissed. And since the biological imperative to survive is so strong, ideology is forced to come to terms with it: to find reasons to bother, to find a point in something pointless. Viewed objectively, people are born, raise children, wait about for a while and die. Human life completely described in a brief sentence! How can you find a purpose in that?

The scientific answer to this question is threefold. At the lowest level of complexity, the physical level, there is no purpose in human life; there is no purpose in anything. At the biological level of complexity the purpose of human life is to produce children. At the ideological level it is to assimilate, create and pass on ideas. This amounts to saying in the final analysis that the universe *is*, life *is*, ideas *are*. If this response is unsatisfactory, then so be it. It is the only answer science can give.

For some people — perhaps the majority, perhaps not — this scientific fatalism isn't good enough. They ask: "What's the point of having children? What's the point of passing on ideas? There has to be more to human existence than that." Those who insist on believing that this 'more to human existence' is real, and then can't resist attempting to spell out what the 'more' is, inevitably find they have no choice but to solve the ideological conundrum by unscientific means: introspection maybe, or plain fantasy.

If someone is determined to create new problem-solving beliefs that science can find no observational justification for, they can only do this by abandoning evidence-based rationality. Once that line is crossed they are free to say and believe whatever they like. Outside the jurisdiction of the Court of Observation, anything goes.

The importance of this point cannot be overstated. *Once it is accepted that ideas can be true without observational evidence, it means nothing is ruled out.* It opens the door to a way of acquiring beliefs that underpins, and leads directly to, religion.

THE INGREDIENTS OF RELIGION

A religion is a belief-set which includes beliefs about:

- life after death
- objective reality
- a moral code
- a god or gods.

All these beliefs are connected in one way or another to the view that observed reality is somehow inadequate when it comes to providing a purpose for (or giving meaning to or justifying) human life.

Religion gives purpose to the universe and all it contains in the simplest possibly way: the ultimate purpose of everything is to enable people to exist. And following on from that, the ultimate purpose of human existence is to serve the god. Thus the purpose of the universe is to be a home for humankind, and the purpose of human life is to do what the god wants.

(In the interests of convention I shall refer to the god(s) henceforward as 'God'. I shall also refer to God as 'He' rather than the more appropriate 'It'. These are not conventions worth defying. Incidentally, the reason for the tradition of regarding God — or the top god if there's more than one — as 'He' is probably that when a religious belief-set initially appears its creator is also always masculine.)

What God wants people to do is spelled out by the moral code. This might seem to restrict God to providing a purpose only for moral behaviour; that is, behaviour affecting moral objects. However, God is Himself conscious and therefore a moral object. On the understanding that all behaviour affects Him, all behaviour is thereby brought within the compass of moral behaviour and is covered directly or indirectly by the moral code. Hence God can, if desired, provide the purpose for every piece of behaviour. The concept of God is a complete answer to the need of many humans to assign purpose to what they do and what they are, and to fix any problems that objective reality seems to present them with.

Religions thus contain a set of unscientific ideas formulated to address issues where the scientific view is considered unsatisfactory. At the heart of these unscientific ideas is an entity, God, who transcends the finite limits of human life; who intervenes in human affairs, at the very least to communicate the moral code; who defines, by way of that moral code, what behaviour should and should not be carried out; and who provides both an explanation for why the universe exists, and a reason why people should bother. If each of these facets of religion was fully insulated from observation, science could say nothing about them. But in fact none of them are. They all have an impact on objective reality. To see how this is so, each facet needs to be looked at in turn.

LIFE AFTER DEATH

It doesn't seem much of a reward for a life spent in the service of God to die at the end of it. It would be far more rewarding to go on living forever. This is perceived as a problem for a religious belief-set, since no one has ever managed to live forever, no matter how obedient he or she has been to the moral code.

The solution to the problem has been found by creating a new concept: the concept of life-after-death, also known as eternal life. The concept finds expression in a belief that people who serve God do not die when they die (!) but continue to live endlessly in some other rewarding place. Conversely, and more ominously, those who do not serve God also continue to live endlessly, but for them the endless life is one of eternal punishment. Following on from this, the claim can be made that people serve God in order to have the reward of a happy eternal life; and also, obviously, to avoid an eternal life of agonizing torture. A happy eternal life becomes a goal which gives purpose to the service of God.

Before getting into how science views this, it can be pointed out that there are several things wrong philosophically with the concept of eternal life, happy or otherwise.

In the first place, the concept is rather odd because it isn't logically necessary. The purpose of people's behaviour is to serve God. The reason why people should serve God is because that is their purpose, not because it is rewarding. They should serve God quite regardless of whether they are to be rewarded with eternal life or not.

In the second place, the concept of eternal life is virtually incomprehensible. Infinities in space and/or time are well-nigh unfathomable, and that certainly applies to imagining a life that goes on and on forever.

Thirdly, the concept leads to a behavioural circle. The purpose of serving God is to have eternal life; the purpose of eternal life is to serve God. (What else could it be?) It is quite unnecessary to replace the simple goal of serving God with a more complicated goal of serving God/having eternal life. Especially since the latter confers no additional philosophical benefits.

Coming now to the science, eternal life requires either reincarnation or that there is some way in which human consciousness can continue in the absence of neurological events. Leaving the former aside for a moment, the latter gives rise to the notion of spirit or soul. (As far as I am aware, these two words are synonyms.) Spirit is consciousness disconnected from the objective world and yet somehow attached to it. Applying science to this speculation requires that it does not contradict any secure theory, and that it preferably avoids requiring the invention of new universal laws. If the concept of separate consciousness — spirit — conforms to these two constraints then it is scientifically viable. Otherwise it is not.

Making a definite proposal, let it be supposed that my brain, or perhaps only a part of my brain, is in communication with an immaterial spirit. Furthermore this spirit, which is separable from my physical self and capable of an independent existence, is the thing which is conscious.

Now, I know subjectively that if I consciously decide to raise my arm it duly rises. The raising of my arm is objectively observable as, in theory at least, is the activity in the neurones of my motor cortex which control the muscular event. Because the activity is consciously initiated, the equating of consciousness with an immaterial spirit suggests that the cause of the output of signals from my motor cortex lies outside my physical brain; that is, the cause is my immaterial spirit rather than incoming signals from other neurones. But spirit has, by definition of 'immaterial', no physical existence and so is not detectable by measuring instruments. Consequently there must be, at the interface between spirit and brain, events occurring which have no observable cause.

It is here that the concept of spirit comes into conflict with science. Suppose it was possible to locate the part of the brain where the spirit initiates motor activity. The hypothesis of an immaterial spirit predicts, recalling section 6.3, that the sodium channels of neurones in that area of the brain would be observed opening spontaneously. The act of opening involves movement of the atoms constituting the channel protein from a closed to an open configuration. That implies that the atoms acquire a quantity of kinetic energy during the movement. The law of conservation of mass-energy requires that that energy must come from somewhere. An immaterial, unmeasurable spirit cannot supply the energy unless it violates the conservation law by creating the energy out of nothing. Science therefore insists that a choice be made between the hypothesis of spirit and the law of conservation of mass-energy. One or other can be true but not both.

As a way out of the dilemma it may be wondered if the definition of spirit could be modified. Perhaps the relationship of spirit to brain is akin to two entities on opposite sides of a window. On one side is the material world going about its business according to physical laws; on the other side is the spirit, which experiences what is going on but does not influence it. Without influence, spirit is placed out of range of scientific investigation. It is then able to co-exist happily with any and every scientific law. However, spirit as it is now defined doesn't actually *do* anything, so there is nothing to be gained by postulating its existence. It contributes nothing. It has been reduced to no more than an empty word.

As one final last ditch effort to rescue spirit, suppose there are indeed places in the brain where energy is created, and that a sufficiently ingenious neurosurgeon could find them and witness the creation. In that case, it would be possible to detect and measure the creation taking place, much data could be collected, patterns could be sought therein, new laws could be proposed (if necessary replacing conservation of mass-energy) and a theory constructed. Spirit would then become a manifestation of a hitherto unsuspected property of the universe. Explained in this way, it would be put on a par with all other scientific knowledge. Spirit could no longer be considered truly immaterial. It would cease to be what it was defined as. There is no saving spirit this way either.

As if these arguments aren't damning enough, reference to section 6.7 provides even more ammunition to shoot spirit down with. It was shown there that memories and personality are functions of the brain, as are sensory perceptions and behaviour. It follows that the destruction of the brain, such as accompanies physical death, brings with it loss of sensory perception, ability to behave, memory and personality. Which leaves very little left for the spirit to be. It has also been argued in chapter 7 that consciousness itself is a purely neurological process which therefore depends on a functioning neuronal system. All things considered, the concept of spirit is scientifically absurd.

Without spirit, eternal life comes to rest on literal magic: a physical reconstruction of the body at some future time or a reincarnation taking place in some other non-universe realm of existence. Such things are not inconceivable (obviously!) but it is impossible that any observer could ever have come across objective evidence for them. By far the simplest position to take is that the concept of reincarnation is unrealistic. Lacking any theoretical or observational support, it must be regarded as imaginary.

If life after death is philosophically unnecessary and scientifically silly, why does it remain a widespread concept in the idea-environment? The answer is that it brings with it a solution, not only to the problem of 'why bother to be good?', but to a much tougher conceptual problem. Life is finite in duration. It is also asymmetric in time. We know we did not exist before our lives began. Given the Big Bang theory is correct, we were in fact dead for well over ten billion years before we were born. This causes us no conceptual difficulties at all. But taking the same approach to non-existence after our lives end is another matter. It is very difficult to conceive of the personal non-existence that ensues. Far easier to imagine continuing magically to be conscious. Hence eternal life spares us from having to come to grips with future personal oblivion. If eternal life is as difficult to comprehend as non-existence, that is not a problem here. The 'eternal' part of the concept can be considered separately from the 'life' part, enabling the difficulty with the concept of eternity to be dealt with later. There will, after all, be plenty of time....

INTERVENTION BY GOD

God is required to inform people about which of their actions serve Him. This information must be conveyed either through the mind or through external events, or both. As each of these options involves Him in interacting with objective reality, both can be examined by scientific method.

Consider, first, communication via the mind. God has a couple of ways open to Him to do this. The cleverest way would be to use DNA to wire the knowledge directly into the human brain. With this option, God's communication would take the form of instinct (see section 8.1). People would know instinctively how God wants them to behave. Some moral beliefs are indeed instinctive (see section 9.3) so this is not immediately in conflict with science. Even better, variability in human moral values could be explained in terms of errors in DNA replication. The original DNA specified by God would occasionally acquire errors during the replicative process so people would, as the generations pass, diverge to a degree in moral behaviour. Some would possess the full complement of God-specified DNA, and others would possess error-containing variations and so have fewer or different moral instincts.

The flaw in this line of approach is that at some time in the past God would have had to implant the requisite DNA into human eggs and sperm. This could not have been done by natural means, since if it had been done that way, the moral instinct would simply be a natural process requiring no God to bring it about. And if it was done by unnatural, interventionist means, geneticists should be able to find in the human genome a large section of DNA having no trace of a counterpart in any other living thing, including close human relatives such as chimpanzees. Since chimpanzees and humans share more than 98% of their DNA, and the less than 2% which is unique to humans has nothing in any way odd about it, the evidence does not support the idea that God intervened by this means.

The other way God might communicate via the mind is to implant thoughts. These thoughts might perhaps be heard as a disembodied voice — words that seem to come from nowhere — or perhaps they might take the form of a feeling or emotion. The thoughts, whatever their form, may precede behaviour or follow after it. It can be noted too that there is no need for this communication to be something everyone can experience. It is sufficient that chosen people — prophets — hear it and pass it on.

The subjective thoughts or emotions that constitute God's communication with a person must cause neuronal signals to pass through that person's brain. This signalling must necessarily not be caused by normal external events or by neuronal activity, but only by God. Just as in the discussion of spirit, this requires uncaused events to occur in certain neurones. In this instance they would be neurones at the interface between God and brain rather than between spirit and brain. And of course exactly the same objections apply. Indeed, everything said in connection with spirit interacting with the brain applies equally to God interacting with it. The idea that God speaks to people is thus not scientifically defensible.

The success of the idea in the idea-environment can, as with eternal life, be accounted for quite plausibly. It is a matter of common experience that thoughts can sometimes spring into people's consciousness apparently unprovoked. Such thoughts tend to be particularly pointed when they are doing something wrong — or contemplating doing it. Conversely, feelings of well-being often accompany doing something right. It is subjectively mysterious where these thoughts and feelings come from, and God provides an easy answer. (The neurological explanation will be found in section 7.4 in the discussion of spontaneity.)

It is interesting that so real is the presence of God in the minds of some people, they state categorically that they know He exists. Furthermore, believing that God speaks to them, there seems no good reason why they shouldn't speak back. Speaking to God is called praying. There is no doubt that for many people the sense of the closeness of God as mentor and confidant is as subjectively real as the world they live in. How can this be accounted for scientifically?

It will be recalled from section 9.2 that when the brain evaluates what it perceives, the evaluations fall into two categories; namely, values which are objectively measurable, and values which are added within the brain. These latter values are subjectively real but have no objective existence. They are values which lie in the mind of the beholder rather than lying in the object being beheld.

The similarity between the many subjective values added within the brain and the experience of God is so pronounced as to indicate a common identity. People who experience God find He is subjectively real just as people who experience drama or frightfulness find these qualities are real. There is no incompatibility with science in this situation providing no assertion is made that these internally added values have an objective existence. A subjective God added to subjective reality — i.e. created by the brain — is every bit as reasonable as, say, subjective loveliness when that is added to subjective reality. The reality of God to believers thus comes from within their brains, not from some intervening outside source. If their subjective God talks to them, or they talk to Him, they are only talking to themselves. The evidence for the truth of this verdict is that God never tells a believer anything it would be impossible for the believer to know or guess by any other means. Nor does He answer prayers, as will be seen in a moment. Believers may believe their subjective God is objectively real but, as has often been remarked, there is no guarantee that subjective beliefs have any correspondence to objective reality.

If God doesn't communicate with people via their minds, there is still the second option; that is, He could communicate by means of events external to the brain. (Prayer, incidentally, assumes explicitly that God controls external events. The whole point of prayer is to get God to bring external events about.)

What sort of events would an observer expect to see if God uses them to signal to people? Basically, there should be a correlation between certain types of behaviour and good fortune. If God helps people or societies who serve Him, then those who happen, by chance or intention, to behave as God desires will be better off as a result. Similarly, if God answers the prayers of those He approves of, then the recipients of those prayers will also be better off than people not prayed for. As to what

481

constitutes 'better off', one, some, or all of the following seem reasonable: above average levels of fame, wealth or power; greater longevity; absence of disease or suffering; increased likelihood of recovery from illness; foreknowledge of natural disasters and consequent escape from them; exceptionally robust happiness; victory in war. (This last is routinely cited by warlords and religious warriors: 'I'll win because God is on my side.') Out of all the putative indicators of God's help, at least one should correlate with the behaviour that God is served by. Indeed, a correlation *must* exist if God uses this method of communication.

No such correlation can be found. There is no evidence that the universe distinguishes between people or societies, any more than it does between quarks or galaxies. Events are totally impersonal. It can be stated with complete confidence that there is no observational evidence whatsoever that God uses events to communicate with people.

Despite this overwhelmingly negative verdict, beliefs that God intervenes, like beliefs about eternal life, are very successful ideologically. It is widely held, especially in less scientifically knowledgeable societies, that God physically helps those who serve Him and, more pointedly, physically punishes those who decline to. Diseases, earthquakes and plagues are his stock-in-trade. As with other beliefs, it would be nice to suggest a reason why this one flourishes.

The answer here is to do with powerlessness. A review of the huge catalogue of awful things which can happen to people can be a cause of fear and despair. A few misfortunes can be avoided through personal endeavour, but most are a matter of luck. People are powerless to protect themselves from the potential horrors of life. It is comforting to think that God, who can protect anyone He chooses, might actually do so. Obviously, God is more likely to extend his protection to those who serve Him than to those who don't. And if God uses his power in this way, controlling events to the advantage of his servants, then it is but a small extension to believe He also controls events to the detriment of those who oppose Him. Believing these things brings it within a person's power, via service to God, and by way of prayer, to avoid through God-approved endeavour all the awful things that might happen. A belief that God has control of events brings with it a negation of fear. This is the key to the belief's success.

(To digress briefly, the belief that God looks after those on his side has an interesting ideological characteristic. Because good and bad moral behaviour do not match up with good and bad fortune, the God-on-my-side belief, if it is to spread in the idea-environment, needs to attach itself to other beliefs which render it immune to objective testing. For example, if something bad happens to a bad person it gives observational support to the notion that God manipulates events as a means of registering his disapproval. And if something bad happens to a good person? Well, that isn't evidence against God's involvement in events. No, it only means God is testing that person's moral steadfastness. (See the Biblical book of Job for a classic example of this line of argument.) With beliefs combined in this way, no matter what happens to whom, it is impossible to disprove that God takes account of human behaviour when controlling events. All contrary observations can be explained away. Such beliefs are, of course, utterly unscientific.)

RELIGIOUS MORAL CODES

How would scientists set about devising a moral code? Their first step would be to define what is meant by good and bad. Since these things are not measurable their definition would be arbitrary but also necessarily reasonable. In addition, the definition would have to contain within it something that can be measured. Say, for example, the scientists choose to equate moral good with being liked. A measure of being liked could be how many people smile at, seek the company of, and offer help to the supposedly liked person.

The point here is not that taking these steps would necessarily lead to a fine moral code, but that the process is rational. A definition of good is arrived at, a way is formulated of measuring success or failure in achieving goodness as defined, and then, and only then, are rules drawn up. These rules — the moral code — must logically follow from the definition of good and from the chosen measure of success. Providing general agreement can be reached on the definition and the measure to be used, there is a high chance that the moral code will achieve its aims. Adherence to the moral code will be good for most people.

A further point to be made about this is that if circumstances change, the moral code is flexible. If a new, better measure of being liked is discovered, or should people's requirements for being liked change, then the rules can be amended if that proves to be logically required. At every step along the way, reason and common sense would be in control.

Turning now to religions' view of moral codes, a totally different situation exists. It is just possible, being charitable, to concede that a religious moral code, when first drawn up, arose in a rational way as described in section 9.3. But any rational origin is rapidly hidden by the assertion that the resulting rules are actually dictated by God, and as such are fixed and unalterable. This absence of flexibility is then compounded by an insistence that the rules *define* what is good. Instead of what is good defining the rules — the rational approach — religions put it the other way round. This unavoidably entails dismissing reasoned argument in favour of dogma, and leads to some obvious moral absurdities — obvious, that is, except to people who prefer obedience to actually thinking.

Examples of absurd religious moral rules are easy to come by. If the rules state that believers who renounce their beliefs must be killed by other believers, then these terrorizing acts of intolerant murder are good. If the rules state that terminally ill people who are suffering greatly and want to die must be denied any assistance to that end, then these appalling acts of gratuitous cruelty are good. If the rules state that babies must have their genitals mutilated when they are too young to defend themselves, register an opinion or give their informed consent, then these grotesque acts of child sex abuse are good. Even more ridiculous, if the rules prescribe how individuals living on their own must behave in private, then their violations of those rules, which affect no one else and are therefore not moral in any way, are nonetheless morally bad.

The root of these immoral religious moral rules lies in their being ascribed to God. But it has already been shown that there is no observational evidence that God intervenes in human affairs. If God neither speaks to people nor gives them signs by way of events, the connection between moral codes and God is broken. Religious moral codes, in the absence of God's guidance, are no better than non-religious ones, and can be observed to be far more concerned in practice with their own survival and dissemination as beliefs in the idea-environment than with bringing tangible benefits to believers. This is not surprising. As was described in section 8.5, belief-sets generally behave in this way — replication is what counts. Being of benefit to believers is only a consideration inasmuch as it enhances a belief-set's ability to replicate.

Since other aspects of the connection between God and moral codes have already been discussed in section 9.3, this component of religion will not be considered further here.

GOD

The concept of God, though every adult seems to know what it means, is actually very hard to define. Here are four vague answers to the question of what God is.

- The traditional answer. God is the creator of, and so is greater than, the universe. He communicates with people and actively intervenes in human affairs by way of supernatural happenings; that is, by way of things which could not happen through the operation of the laws of the universe as discovered by science. These God-caused events reveal the existence of God and his personality.
- The creator-only answer. God is the creator of, and so is greater than, the universe. He created a perfect universe which therefore never needs to be 'fixed'. Thus He never intervenes in human affairs or in any other aspect of the universe. You don't fix things unless they're not going according to plan, and a perfect universe must always proceed according to plan precisely because it's perfect. God's existence (and personality?) is revealed by the fact of the creation and nothing else.
- The equivocal answer. God may be the creator of the universe. Or He may not. Either way He is outside the reach of science and observation. It is therefore impossible to know anything about God, including whether He actually exists or not.
- The atheistic answer. God is an invention of the human mind. There is no such thing as an objectively existing God. Given the two propositions 'God invented people' and 'people invented God', the second is so much more likely to be true than the first that it can be taken as practically certain.

As far as science is concerned, the traditional answer above is the one that matters, as it is open to observational verification. The other three answers are scientifically indistinguishable. Thus the argument is between

485

advocates of a traditional, interventionist God on the one hand, and either a non-intervening or a non-existent God on the other. The differences between the creator-only, equivocal and atheistic answers make no observational difference and so are not objectively real differences.

Before going into the scientific verdict on the traditional answer it would be informative to comment philosophically on the creator-only and equivocal answers, as these can be disposed of quite easily.

The creator-only answer holds that the only way the universe could come into existence is if God created it. No God, no universe. But if God was required to pre-exist first and 'cause' the universe, the obvious and unavoidable next question is what pre-existed God and 'caused' God? The question seems to have the answer that God didn't need to be caused because He is self-creating. In which case, if God can be self-creating, why can't the universe be? Invoking the existence of a creator-only God is pointless, wholly unnecessary and contributes nothing to an understanding of reality.

The equivocal answer starts from the view that it is impossible to know anything about God. It follows logically that it is also impossible to define in any way what the word 'God' means, since a definition must rest on knowledge. In the absence of a definition, the word 'God' becomes meaningless. As does any statement about God, including the statement 'God may or may not exist'. Thus the equivocal answer, which is agnosticism, is irrational. Either agnostics believe the word 'God' is definable, in which case they belong in one of the other three camps, or they don't, in which case they literally haven't a clue what they're talking about and so can say nothing intelligible on the subject.

So much for philosophical comments. What about the scientific view? As remarked above, science is limited to passing judgement on an interventionist God and the concomitant belief that supernatural events reveal God's existence and personality.

It has already been seen in the course of discussing interventions by God that the universe shows no trace of God-caused supernatural events. The scientific verdict is thus that God's existence is not revealed by supernatural events, because there aren't any.

Which leaves God's personality — a question of who God is rather than what God is. The historical God personality has variously been portrayed as being that of: someone who is the ultimate dominant tribal male; someone who is brutally vengeful towards anyone who tries his patience; someone who is insanely jealous of false gods (to be jealous of rivals who don't exist is insane); someone who is merciful; someone who is just; someone who is sadistically vindictive towards non-believers, especially after they've died; and — the current Christian favourite — someone who is loving.

Since God does not intervene, the only way the universe can reflect God's personality is for the natural laws to reflect it directly. And again, the observational verdict is negative. The universe does not behave, either at large or locally on the Earth, in ways that reveal its progress is governed by so much as one of God's various disparate (and contradictory) supposed personality traits. If the universe reflects the personality of God, the kindest that can be said is that He's totally dispassionate, totally amoral. And someone not disposed to kindness, noting that none of the assertions about what or who God is can claim the slightest observational support, would ask the pertinent question: "Can any object be said to exist if it is indescribable and in all respects unobservable?" The scientific answer is no.

Given this verdict, an explanation is needed for why beliefs about the objective existence of God are so widespread in the idea-environment. Why is it that every adult knows, albeit often in an undefined way, what God is?

Partly it comes down to cultural history. Put simply, children are told about God by their parents and teachers. Knowledge of God is thus perpetuated from one generation to the next ad infinitum.

It is also possible beliefs about God have evolutionary roots. Believing an interventionist God exists gives people the power to alleviate their fears about what could happen to them, as has already been commented on. From an early stage in human evolution the creation of ideas about God, or about gods, would be compelling for this reason. What if a dissenter arose among the ranks? In an ideologically cohesive tribe, the dissenter would be handicapped in finding a mate, and would receive below average support from fellow tribespeople in raising offspring. Thus natural selection within a social species would make dissent rare. Furthermore, in the struggle for limited resources, a cohesive tribe is more likely to be successful than an ideologically divided tribe, thereby enabling each individual in the former to produce more children than individuals in the latter. So stamping out dissent not only makes sense from the point of view of beliefs-in-competition (see section 8.5), it also makes sense genetically. In this way, ideological evolution — the way ideas spread in the idea-environment — may have influenced human biological evolution. Ideology might well favour a biological, genetic predisposition to conform to the existing beliefs of the tribe. To suggest that humans have an instinct to conform accords well with observation, both objectively and subjectively. To go further and suggest they have a specific conformity-based instinct to believe gods exist, whilst more questionable, is not incompatible with observation.

IN SUMMARY

Religion was invented to provide a grand purpose for human life, an explanation for life's caprices and vicissitudes, a way out of its distressing transience. It does this by offering life after death (without any evidence), a conscious spirit separate from the brain (despite the evidence), a God-dictated moral code (without any evidence God was involved), and a God who intervenes in human affairs (despite overwhelming evidence He doesn't), who has a range of awe-inspiring characteristics (without any evidence) and is unobservable (the only fact about God that fits the evidence). This is independence from observation at its most extreme!

And to a religious believer all these beliefs are true; that is, they correspond to objective reality. They are true irrespective of, and where necessary in defiance of, observational evidence. A religious believer holds to his or her beliefs come what may and calls this steadfastness 'faith'. To have faith is one of the most emphasized rules of religious moral codes (what a surprise!) and is therefore especially good.

Sometimes these faith-based religious believers claim they *do* have evidence. The refutation of their claims is simply the question: 'Then why is there more than one religion?' Science is riddled with disputes, but those disputes are resolved — or will be resolved — by observation; that is, by the evidence. If religious believers genuinely possessed evidence, then their differences would be resolved by appeal to that evidence. But this *never* happens. Religious believers argue endlessly; religions are rife with dissent, schisms, factions, sects, heresies. When a religion claims it has evidence, why is it that that evidence mainly only convinces people who already subscribe to that particular religion? Why does the evidence not convince adherents to other religions? The answer is that the evidence is not objective; that is, it exists only subjectively, like God Himself, in the minds of believers. It is not real in the scientific sense.

Religion — every religion — is a collection of beliefs inhabiting the idea-environment. To that extent religions are objectively real. But beyond that, because they have no observational support, there is no reason to believe they correspond to reality. Indeed, the complete absence of supporting evidence makes a scientific mockery of religious belief-sets. From an objective standpoint, religions are no more than wishful thinking: mere assemblages of empty fantasies, irrational superstitions, neurotic rituals, and groundless myths. Adherence to such beliefs through the fraudulent, self-delusional con-trick of faith constitutes the most effective and implacable psychological prison any person could possibly lock themselves inside; a prison which is the very antithesis of human curiosity, intelligence and honest intellectual freedom.

What it comes down to ultimately is that either you believe observation determines truth, in which case you're a scientist (at heart if not professionally); or you don't, in which case you can believe anything you like, including any religion which takes your fancy.

10.5 SCIENCE IN PERSPECTIVE

To end this book's account of a scientific view of reality and its inevitable consequent dismissal of the religious view, it would be interesting to look at three mysteries, to discuss them from a rational, science-constrained perspective, and then to see how science and religion differ in their approach to tackling them.

Here are the three mysteries.

Mystery number one. The universe permits observers to exist. There are many physical quantities governing the way objects interact, and the size of these quantities is frequently found to be of vital importance to the existence of life. As an example, there is the strength of the hydrogen bond (see section 5.2). If this physical quantity had been much greater than it is, DNA replication would have been too accurate to permit the beneficial replicative errors to occur that led to the formation of complex

489

organisms (such as ourselves). On the other hand, if the quantity had been much smaller, replication would have been so inaccurate that life would have been unsustainable. Many other examples of vitally important quantities are known. Given that observers are necessarily complex structures, it is curious that the physical quantities relating objects to one another should all happen to fall within the narrow range permitting complexity to exist.

Mystery number two. The universe is governed by laws. It has been seen that all objects, from tiny quarks to huge galaxies, from simple balls of gas to complex brains, behave in a way that can be described by a set of laws. These laws are never observed to be violated. The import of this is that the universe is fundamentally ordered. Why? Why isn't it utterly chaotic, with the laws changing from one moment to the next and from one place to another in a random fashion? A chaotic universe would not be an observable universe. It is curious that the laws needed to make the universe stable enough to contain observers are the ones found to be in operation.

Mystery number three. The universe is describable by mathematics. The study of mathematics is an intellectual game played by mathematicians without regard to objective reality. Mathematics is completely abstract. Yet this same mathematics, a creation of human logical thinking, can be used to describe the workings of the universe. Why should this be so?

These three mysteries might be taken to suggest that the universe was designed for life and, ultimately, for observers; that it was built rather than just happened. This is the religious view: that the purpose of the universe is to be a home for humans. Or, to put it more scientifically, to be a place where complex biological structures — living things — and complex ideological structures — observers — can exist. And the amazing, subtle intricacy of the universe as revealed by physics generally, and especially by particle physics, tempts one to concede it might indeed have been designed. But....

For a start, it is an extreme non sequitur to go from speculating that the universe was designed, to asserting that the universe was designed *for us*. Knowing what something is for — knowing its purpose — requires that the designed object be viewed in a context external to itself. Since the universe is everything, it cannot be placed in such a context. Hence its purpose is unknowable in principle as well as in practice. Life, including we observers, could simply be an insignificant by-product. It is extraordinary anthropocentric arrogance on our part to seriously suggest this whole immense universe was set up just so we could be in it!

Moreover, the scale of the universe does not support the notion that its purpose is to enable life to exist. A simple calculation can illustrate this point well. The furthest distance in the universe that can be observed is roughly 3000 Mpc away. In a flat universe the volume equivalent to 3000 Mpc is given by $4\pi r^3/3$:

$$\frac{4\pi(3000 \times 10^6 \times 3 \cdot 25 \times 10^{16})^3 \text{ m}^3}{3} \sim 10^{78} \text{ m}^3.$$

The volume of a planetary biosphere is the product of its thickness, say, about 20,000 m, being generous, and the planet's surface area, $4\pi r^2$. Taking the Earth to be a typical size for planets with a biosphere, this gives a volume of roughly

$$(20 \times 10^3) \times 4\pi(7 \times 10^6)^2 \text{ m}^3 \sim 10^{19} \text{ m}^3.$$

Assume, being generous again, that the average number of habitable planets is one per star, that there are 10^{11} stars per galaxy, and 10^{11} observable galaxies. The total volume of biosphere in the observable universe is then $10^{19} \times 10^{11} \times 10^{11} \text{ m}^3 = 10^{41} \text{ m}^3$. To spell out what these figures mean, if you were to take a carbon-based organism — something which can only exist in a biosphere — and place it at a random location in the universe, the chance that death would not follow very rapidly upon arrival at the location is one in $10^{(78-41)}$; that is,

1 in 10,000,000,000,000,000,000,000,000,000,000,000,000.

That is a universe designed for *life*?

And even where life does exist — on the Earth (and elsewhere?) — organisms survive for the most part despite their environment, not because of it. Survival is a constant battle against the forces of nature. And for non-photosynthetic organisms, survival involves killing other living things. Life is lethal even to itself. Thus the universe is implacably hostile to life. It permits life to exist, but only barely and under sufferance. This strongly suggests the universe was not designed for life, far less for observers.

And what about conceding the 'design' claim in the first place? Bearing in mind how observer-biased we observers are bound to be, is that just another prejudice? In fact, to regard the three mysteries listed above — physical quantities, laws and mathematics — as suggesting that the universe was designed is to beg the question of what the universe would have been like had it not been designed; it is to assume, in particular, that the universe would have been different in such a way as to render it observer-less. There is no observational or theoretical basis for such an assumption. There is no certifiably undesigned universe available to be examined.

Consider for example the first of the mysteries listed above. It is not known whether any or all of the various physical quantities crucial for life are arbitrary and independent, or not. It is quite possible they have to have the values they do. Perhaps if you change one you have to change them all — and still end up with a universe fit for living things. Similar arguments apply to the other two mysteries. It is thus impossible for anyone to establish that an undesigned universe wouldn't be identical to our own. There is no way to choose between 'designed' and 'inevitable'.

The mysteries of physical quantities, laws and mathematics are but three facets of the overarching mystery of the totality of existence. Such deep mysteries are something science is completely comfortable with. They are, indeed, the reason science exists. To admit some facet of the universe is either unknown or cannot be explained is not, for science, an admission of failure; it is a challenge, a work in progress, a spur to observation and the application of logic and creative intelligence.

This sets science apart from religions, for religions have no interest in mystery. They already have an explanation for everything. There is nothing to think about, nothing to puzzle over, nothing to explore (other than God Himself). If this or that facet of the universe cannot be explained, it is trivially attributed to God: God did it. Mystery solved! Quite apart from the fact that 'God did it' is not actually an explanation at all (*how* God did it is an explanation; *that* God did it is not), the statement is either a restatement of the mystery or an additional complication. In the former case, why call the mystery God? Why not call the mystery a mystery? In the latter case, as has already been made abundantly clear, God is Himself a mystery: unobservable, indescribable, etc. Why add a second mystery to the first? What is gained by doing that? However it is viewed, invoking God to 'explain' mysteries leads to nothing but self-imposed ignorance.

Their conflicting attitudes to mystery graphically demonstrate the incompatibility of the religious and scientific views of reality. Religion rests on passive acceptance of claims that there are other non-observational ways to know about reality; that some truths are determined by historical hearsay and tradition; that unsubstantiated religious revelatory claims take precedence over observation whenever a conflict arises. Science rests on active pursuit of observational evidence. Observation is paramount. Revelation, which has its place in scientific theorizing, fails if it finds no hard evidence to support it.

And that highlights one of science's strongest features. It is human curiosity and intelligent enquiry formalized into a set of rules and procedures. It liberates people to believe what they see with their own eyes and to make of it what they can, and then to test what they make of it by looking ever further into the objective reality within which they live their subjective lives.

Science's role, its true place in an observer's reality — technology aside — is as a creator of images. Within each observer's mind, it enables images of objective reality to be formed which correspond far better than any unscientific ones to the objective reality that is actually 'out there'. Science functions something like a cartographer whose images collec-tively make a number of maps. Maps describe the lie of the land, and scientific maps are quite simply the finest there are. They reveal better than anything else the landscape around us, and guide us to a deeper appreciation of what we observe. For example, when our ancestors looked at the starry night sky, they doubtless saw the same beauty that we do. But their stars were little lights a few miles away. Ours are mighty suns or even entire galaxies stretching across unimaginable distances of space and time. When we look at the night sky we can see not only beauty but staggering grandeur as well. And when our ancestors contemplated consciousness, they doubtless valued it the same way that we mostly do. But their consciousness had to be accepted as a given. Ours resides in a brain of a hundred billion neurones, a brain in which chemical reactions of immense subtlety and precision, occurring continuously, somehow create an amazing structure which can know of its own existence. When we contemplate consciousness we are aware not only of its value, but of the mind-boggling complexity that underpins it. For us, to be conscious is not only a given; it is (except when suffering badly) an infinite privilege.

And yet there is another side to science's maps — a purely subjective side. When we are in pain, or hungry, or lonely, what science says about the nature of these things, some neurological happening, is quite irrelevant to our perceptions. Similarly, when we are enjoying ourselves we do not regard the experience as a mechanical series of signals passing through our brains, but as a pleasure. It serves as a reminder that science enriches our experiences; it does not make them.

Which puts science (and all observationally acquired knowledge generally) clearly in perspective. Scientific maps, like all maps, can only show us where everything is. They cannot tell us where to go, nor even how best to get there. These are decisions we must make for ourselves. Science is a far sighted guide but a blind master. It is a treasure to be used, not a god to be worshipped.

The scientific knowledge that we possess today describes a universe whose components, be they quarks, brains, or galaxies, are ordered and law-abiding. It describes a universe which is awesome and full of immense beauty; a universe where we observers can each live out our one and only life free from the primitive tyranny of a Big Brother God watching our every move, monitoring our every thought and terrorizing us with hell-fire and damnation. Science has set us free. It is up to us as individuals whether, and how, we use that freedom.

Such is a scientific view of reality.

APPENDIXES

A.1 LESS FAMILIAR UNITS USED IN THE TEXT

Length	pc	parsec	$3 \cdot 0857 \times 10^{16}$ m
	Mpc	megaparsec	10^6 pc, $3 \cdot 0857 \times 10^{22}$ m
Temperature	K	Kelvin	identical to the temperature in degrees Celsius +273; e.g. 30° C = 303 K
Energy	J	Joule	a 1 kg weight falling 10 cm at the Earth's surface acquires very nearly 1 J

A.2 VALUES OF THE THREE FUNDAMENTAL CONSTANTS

c	speed of light	$2 \cdot 9979 \times 10^8$ m s^{-1}
G	gravitational constant	$6 \cdot 6726 \times 10^{-11}$ m^3 kg^{-1} s^{-2}
h	Planck's constant	$6 \cdot 6262 \times 10^{-34}$ J s
$\hbar$	Planck's constant / 2π	$1 \cdot 0546 \times 10^{-34}$ J s

A.3 GREEK LETTERS USED IN THE TEXT

γ	gamma	electromagnetic waves = photons
Δ	delta	amount of (uncertainty in) the symbol following
λ	lambda	wavelength
μ	mu	muon
ν	nu	neutrino
π	pi	pion; also the constant $3 \cdot 14159....$
ρ	rho	rho-meson
τ	tau	tauon
Ω	omega	cosmological density parameter

A.4 PRONUNCIATION GUIDE

Technical words whose pronunciation may not be obvious to you are listed below. The problem is mainly with knowing where to place the stress and how to pronounce the vowels. Each word is accordingly accompanied by a phonetically re-spelt equivalent using the following rules.

- Stressed syllables are indicated by capital letters.
- Vowels, as in bag, beg, big, bog, and bug are unmarked.
- Vowels, as in fail, feel, file, foal, and fuel are underlined.
- Vowels, as in park, perk, and pork are indicated by à, è and ò.
- The vowel, as occurs in 'shout', is rendered ou.
- Indistinct vowels, as for example the *a* in infant are indicated by ə: infənt.

acetylcholine	ASitilKOLen	adenine	ADinen
adenosine	aDEnəsen	amino	əMEno or əMIno
baryon	BAreon	Cepheid	SEfeid
corpus callosum	KÒpəs kaLOsəm	covalent	koVAlənt
cytosine	SItəsen	eukaryote	uKAreot
glyceraldehyde	glisəRALdəhid	guanine	GWÀnen
guanosine	GWÀnəsen	Heisenberg	HIzənbèg
kilometre	KIloMEtə (NOT kiLOmitə)	medulla	məDUlə
meson	MEzon	myelin	MIəlin
Noether	NÈtè	parietal	pəRIitəl
pion	PIon	prokaryote	proKAreot
Schwann	shwon	synapse	SInaps
tauon	TOUon	thymine	THImen
uracil	URəsil	Wernicke	VÈnikə

BIBLIOGRAPHY

The material of a great many books and articles has been drawn on in writing *A Scientific View of Reality*. Most of the sources have contributed only to some small facet of the text. Those that are more widely relevant or would be useful for readers wishing to check facts or enquire further into the various topics covered are listed here. Also included are books cited in the text.

The book list is divided into sections by general topic. A **D** against a book indicates that it is demanding reading, and a **U** indicates it treats its subject at a university undergraduate level. The easier books, I hasten to add, may still require mental exertion.

SCIENTIFIC METHOD (Chapter 2 and Section 9.5)

Baggott, J.E. *Farewell to Reality.* Constable, 2013.
Chalmers, A.F. *What Is This Thing Called Science?* Open University Press, 1999.
French, S. *Science: Key Concepts in Philosophy.* Continuum International Publishing Group, 2007.
Popper, K. *Objective Knowledge: An Evolutionary Approach.* Oxford University Press, 1972. **D**
Popper, K. *The Logic of Scientific Discovery.* Routledge, 2002. **D**

PHYSICS (Chapters 1, 3 and 4, and Chapter 10, Sections 1-3)

Burbridge, G. and Narlikar, J.V. *Facts and Speculation in Cosmology.* Cambridge University Press, 2008.
Clark, S. *The Unknown Universe.* Head of Zeus, 2015.
Coughlan, G.D., Dodd, J.E. and Gripaios, B.M. *The Ideas of Particle Physics.* Cambridge University Press, 2006. **D**
Einstein, A. and Infeld, L. *The Evolution of Physics.* Simon & Schuster, 1988.
Ferguson, K. *Measuring the Universe.* Headline Book Publishing, 1999.
Gagnon, P. *Who Cares About Particle Physics?* Oxford University Press, 2016.

Garlick, M.A. *The Story of the Solar System*. Cambridge University Press, 2002.

Jones, M.H. et al (eds). *An Introduction to Galaxies and Cosmology*. Cambridge University Press/The Open University, 2015. **U**

Kittel, C., Knight, W.D. and Ruderman, M.A. *Mechanics*. McGraw-Hill, 1965. **U**

Krauss, L.M. *A Universe From Nothing*. Simon & Schuster, 2012.

Liddle, A. *An Introduction to Modern Cosmology*. John Wiley & Sons, 2015. **U**

Liddle, A. and Loveday, J. *The Oxford Companion To Cosmology*. Oxford University Press, 2008.

Martin, B.R. *Particle Physics: A Beginners' Guide*. Oneworld Publications, 2011.

Martin, B.R. and Shaw, G. *Particle Physics*. Wiley-Blackwell, 2008. **U**

Nelkon, M. and Parker, P. *Advanced Level Physics*. Heinemann, 1995. **D**

Oerter, R. *The Theory of Almost Everything*. Plume, 2006.

Quinn, R. and Nir, Y. *The Mystery of the Missing Antimatter*. Princeton University Press, 2008.

Rae, A.I.M. *Quantum Mechanics*. Taylor & Francis, 2007. **U**

Rae, A.I.M. *Quantum Physics: Illusion or Reality?* Cambridge University Press, 2004. **D**

Rowan-Robinson, M. *Cosmology*. Oxford University Press, 2003. **U**

Shipman, H.L. *The Restless Universe*. Houghton Miflin Company, 1978.

Stenger, V.J. *The Comprehensible Cosmos*. Prometheus Books, 2006.

Tabor, D. *Gases, Liquids and Solids*. Cambridge University Press, 2008. **U**

Tolansky, S. *Introduction to Atomic Physics*. Longmans, 1963. **U**

Weinberg, S. *The First Three Minutes*. Basic Books, 1993.

LIFE ON EARTH (Chapter 5)

Cairns-Smith, A.G. *Seven Clues to the Origin of Life*. Cambridge University Press, 1985.

Campbell, P.N., Smith, A.D. and Peters, T.J. *Biochemistry Illustrated*. Churchill Livingstone, 2005. **U**

Conn, E.E. and Stumpf, P.K. *Outlines of Biochemistry*. John Wiley & Sons, 1987. **U**

Dawkins, R. *The Selfish Gene*. Oxford University Press, 1976.

Jastrow, R. and Rampino, M. *Origins of Life in the Universe*. Cambridge University Press, 2008.

Miller, S.L. and Orgel, L.E. *The Origins of Life on Earth*. Prentice Hall, 1974. **U**

Whitfield, P. *Evolution*. Marshall Editions, 1998.
Yarus, M. *Life from an RNA World*. Harvard University Press, 2010.

NEUROBIOLOGY AND CONSCIOUSNESS (Chapters 6 and 7, and Sections 8.1 and 9.4)

Beaumont, J.G. *Introduction to Neuropsychology*. Guilford Press, 2008. **D**
Johns, P. *Clinical Neuroscience*. Churchill Livingstone, 2014. **U**
Kandel, E.R. et al. *Principles of Neural Science*. McGraw-Hill, 2012. **U**
Luria, A.R. *The Working Brain*. Basic Books, 1976.
Miller, R. *Meaning and Purpose in the Intact Brain*. Oxford University
    Press, 1982.
Pinel, J.P.J. *Biopsychology*. Pearson, 2010. **U**
Rose, S. *The Conscious Brain*. Weidenfeld and Nicolson, 1973.
Sommerhoff, G. *Logic of the Living Brain*. John Wiley & Sons, 1974. **D**
Squire, L. et al. *Fundamental Neuroscience*. Academic Press, 2008. **U**

RELIGION (Chapter 10, Sections 4 and 5)

Dawkins, R. *The God Delusion*. Bantam Press, 2006.
Smith, G. *Atheism: The Case Against God*. Prometheus Books, 1979.
Stenger, V.J. *God: The Failed Hypothesis*. Prometheus Books, 2007.

IDEOLOGY and MORALITY (Chapter 8, Sections 2-6, and Chapter 9, Sections 1-3)

I know of no books which treat ideas and morals the way I have treated them in these chapters. However, both of Dawkins' books listed above discuss ideas as gene-like replicators. Also a not dissimilar approach can be found in:

Blackmore, S. *The Meme Machine*. Oxford University Press, 1999.

Blackmore defines a meme as any piece of imitative behaviour. 'Imitative' implies copying, as does chapter 8's 'communication interaction'. Beyond that, the 'meme' concept is more all-embracing than chapter 8's 'idea'.

INDEX

neutron disintegration,
— by weak force, 100, 126-127,
  171-172, 173
— by X force, 174, 175
— in Big Bang, 186
neutrons, 82
Noether's Theorem, 20, 104
non-science, 70
noradrenalin (see adrenalin)
'now', subjective, 463, 465-466
nucleic acid bases, 213f
— and protein manufacture,
  226-230, 229f
— before life, 235
— complementary pairing, 212, 225
— sequence errors, 239-240, 241
nucleic acids, 210
— 210-212, 214, 224-225, 242, 393
nucleus, 81
— size, 83, 109

O
objective reality, 51
observation, primacy of, 61, 65,
  69-70, 294, 489, 493
Occam's razor, 61
— 69, 101, 108, 157
occipital lobes, 290
— 291f
olfactory bulbs, 286
oligodendrocytes, 275
organic molecules, 204
— 205, 207, 235
organisms, 205

P
pain and the reticular formation, 330
pain, role of
— as an emotion, 286, 318, 347
— in learning, 321-323
— in moral behaviour, 414, 425
— reflexes, 278
parallax, 141
— 141-142, 144, 145, 146, 147, 151
parietal lobes, 290

— 291f
parsec, 147
past, subjective, 463-466
Pauli Exclusion Principle, 91
periodic table, 80
— 82, 84, 86, 130/131t
— and chemical bonds, 197, 199,
  201
— and hydrogen bonds, 212
— and quanta, 91-93
permanent recallable memory, 356
permanent skill memory, 357
personality, 306, 478
phosphate/phosphoric acid, 201, 210,
  212, 214, 225
photons, 95
— 134t
photosynthesis, 246
physical level of complexity, 248
pions, 110
pituitary gland, 282f, 286, 344
Planck's constant, h, 90, 96, 176
— and energy of waves, 87
— $\hbar$, 'h bar', 104
planets of the solar system, 143t
plants, 245
— 214, 215, 218, 256
polymers, 207
— 210, 217
pons, 282
— 282f, 283, 284
potassium ions, K^+, 234, 264-265,
  268, 269, 270
potential (mass-)**energy**, 18
prayer, 480, 481, 482
pre-frontal cortex, 291f, 294
— function, 305-306
— role in consciousness, 325, 326,
  347
— role in speech, 302
pre-motor cortex, 291f, 294, 299,
  301-302, 306, 318, 321
— and volition, 339, 341
primary cortex, 291f, 293
— auditory, 293, 297, 302

508

www.ingramcontent.com/pod-product-compliance
Lightning Source LLC
Chambersburg PA
CBHW061701240326
41458CB00162B/6911/J